Mike Dietrich

Business Performance Measurement in the Metal Finishing Industry

AF827716

Controlling und Management

herausgegeben von / edited by

Prof. Dr. Ronald Gleich
Strascheg Institute for Innovation and Entrepreneurship
EBS Business School
EBS Universität für Wirtschaft und Recht

und / and

Prof. Dr. Andreas Wald
School of Business and Law
University of Agder

Band / volume 10

LIT

Mike Dietrich

Business Performance Measurement in the Metal Finishing Industry

An Approach Towards a Sustainable and Integrated Management System for Contract Processing Businesses

LIT

Hochschulsignatur: 1540.

Bibliographic information published by the Deutsche Nationalbibliothek
The Deutsche Nationalbibliothek lists this publication in the Deutsche Nationalbibliografie; detailed bibliographic data are available on the Internet at http://dnb.d-nb.de.

ISBN 978-3-643-90750-9
Zugl.: Oestrich-Winkel, European Business School, Diss., 2016

A catalogue record for this book is available from the British Library

© LIT VERLAG GmbH & Co. KG Wien,
Zweigniederlassung Zürich 2016
Klosbachstr. 107
CH-8032 Zürich
Tel. +41 (0) 44-251 75 05 Fax +41 (0) 44-251 75 06
E-Mail: zuerich@lit-verlag.ch http://www.lit-verlag.ch

Distribution:

In the UK: Global Book Marketing, e-mail: mo@centralbooks.com
In North America: International Specialized Book Services, e-mail: orders@isbs.com
In Germany: LIT Verlag Fresnostr. 2, D-48159 Münster
Tel. +49 (0) 2 51-620 32 22, Fax +49 (0) 2 51-922 60 99, e-mail: vertrieb@lit-verlag.de

In Austria: Medienlogistik Pichler-ÖBZ, e-mail: mlo@medien-logistik.at
e-books are available at www.litwebshop.de

For my girls Yvonne and Johanna.

EDITORIAL

With our series "Controlling and Management" we focus on work that addresses current topics in the fields of management accounting & control as well as general management. Purpose of this forum is the discussion and diffusion of new insights in the respective research fields.

Since the exchange between research and practice is of special concern to us, the main focus of the series will be on work with high practical relevance. Thus, different contributions should either result from a direct cooperation with business partners or originate from research findings that lead to new methods of management accounting & control as well as management concepts. In addition, the new series will offer a forum to present empirical findings obtained from different methodological approaches, either of quantitative or qualitative nature.

A lot of works in this series provide findings from outstanding dissertations and postdoctoral thesis dealing with managerial questions or with challenges concerning the field of management accounting & control. Furthermore, the series picks up topical areas of interest in edited volumes, including contributions from renowned researchers and practitioners.

We kindly invite you to send in manuscripts – either in German or in English. The editors will decide jointly on their publication.

We hope you will enjoy the reading. If you have any suggestions or questions please do not hesitate to contact us directly.

Prof. Dr. Ronald Gleich
EBS Business School
EBS Universität für Wirtschaft und Recht
ronald.gleich@ebs.edu

Prof. Dr. Andreas Wald
School of Business and Law
University of Agder
andreas.wald@uia.no

Foreword

The literature on performance measurement systems has suggested a variety of comprehensive, theory-based performance measurement models. Many models are characterized by a high degree of rigor and sophistication. Whilst these attributes are highly valued in academia, they can be a barrier to a wider diffusion in practice.

In fact, managers hardly pay attention to performance measurement research. Performance measurement systems in organizations are often the result of habit, the availability of (financial) performance measures and the imitation of competitors. This gap between theory and practice not only can be observed in management accounting research but in many other disciplines of management research. However, in management accounting research this fact is particularly deplorable as there is much to be gained by making generic performance measurement systems available to practitioners and by helping to adapt these systems to the specific requirements of individual industries and companies.

The volume at hand presents the results of the doctoral dissertation written by Mike Dietrich. The author develops a concept for facilitating the selection and adaptation of generic performance measurement models and applies this concept to a specific company case within the metal finishing industry.

The contribution to knowledge is twofold: For the practitioner it provides access to the self-contained world of academic research by offering a support and guideline for the selection and adaptation of generic models. For scholars it informs about important aspects concerning the further, more practice-oriented development of generic models.

Prof. Dr. Ronald Gleich

Acknowledgements

I warmly thank my doctoral supervisor Prof. Dr. Ronald Gleich for his expert guidance and his excellent ideas throughout the whole dissertation project that led to this book at hand – I really appreciate the open door mentality and the creative, stimulating and cooperative atmosphere within his SIIE institute. I wish also to thank my second doctoral supervisor Prof. Dr. Stefan Smolnik for his numerous and valuable suggestions, especially in structuring the whole dissertation project. Special thanks go out to Prof. Dr. Andreas Wald for being a great academic sparring partner.

I am grateful that my employer gave me the chance, time and support for the research project; special thanks go out to my colleagues and "bosses" Stéphane and Wilfried for their ongoing support during the whole research project.

I owe to the whole Management Team of MeFiCom for their time and efforts supporting me in my qualitative data collection with emphasis on the fruitful expert interviews and vivid group discussion.

For their support of my research, especially the quantitative data collection, I thank

- the VDMA (Verband Deutscher Maschinen- und Anlagenbau - German Engineering Federation, DE), especially Dr. Martin Riester,
- the EFDS (Europäische Forschungsgesellschaft Dünne Schichten – European Society of Thin Films, DE), especially Dr. Kristin Brzezinski,
- the SEA (Surface Engineering Association, UK), especially Mr. Dave Elliott and
- the IMF (Institute of Materials Finishing, UK), especially Mr. Clive Larson for his strong interest and support of the research project.

For the lectorate of my dissertation and their great suggestions I warmly thank my American family, namely Leeann & Franz Grossman.

Furthermore I owe so much to my parents Ursula and Manfred, as without them I would have never reached this point my life. Last but not least I thank my wife Yvonne, not only for the lectorate of the dissertation but for her support and encouragement in helping to bring this project to fruition; without her ongoing love, interest in and indulgence for my academic priorities over the past four years, this project would not have been possible.

Table of Contents

List of Figures xi
List of Tables xiv
List of Abbreviations xvii
Abstract xxi
1 Introduction 1
1.1 Problem Statement 1
1.2 Structure of this Dissertation 2
2 Theoretical Framework 5
2.1 Agency Theory 5
2.2 Perspectives Within the Strategic Theory of the Firm 5
2.2.1 Resource-Based View 6
2.2.2 Relational View 7
2.2.3 Process-Based View 8
2.3 Management Control Systems Framework 9
2.3.1 Management Control Systems – The Rationale Behind 9
2.3.2 Management Control Systems Research 10
2.3.3 The Levers of Control 11
2.4 Business Performance Measurement Framework 15
2.4.1 Business Performance Measurement – The Rationale Behind 15
2.4.2 Business Performance Measurement Research 16
2.5 Synthesis of the Research Framework 19
2.5.1 Research Framework – BPM in the MFI 19
2.5.2 MCS and BPM – Evolving Academic Concern 21
2.5.3 Theoretical Conceptualization of Feasibility 23
2.5.4 Research Questions related to the Metal Finishing Industry 26
3 Metal Finishing Industry 29
3.1 General Classification 29
3.2 Contract Processing and Outsourcing 32
3.3 Environmental Performance – US EPA Case 34

3.4 International Nomenclatures of Economic Activities 34
3.5 International Statistical Databases of Economics 38
3.5.1 European Union – SBS and SBR Databases 39
3.5.2 USA – ASM and CBP Databases 42
3.6 Characteristics Clusters 44
4 BPM in the MFI – Current State of Research 47
4.1 Literature Review 47
4.2 Research Gap 51
5 BPM in the MFI – Research Methodology 53
6 BPM in the MFI – Qualitative Research Process 55
6.1 Research Setting 55
6.1.1 Ethnographic Research within an Embedded Single Case 55
6.1.2 MeFiCom as Proxy for the MFI 57
6.2 Preliminary Studies 59
6.3 Semi-Structured Expert Interviews 62
6.3.1 Preparation 62
6.3.2 Data Acquisition 67
6.4 Group Discussion 68
6.5 Qualitative Data Analysis 71
6.5.1 Transcription 72
6.5.2 Analyzing and Coding Process Using CAQDAS 73
6.6 Results – MFI Cause and Effect Model as Base for BPM 80
6.6.1 CEM – Metal Finishing Industry 85
6.6.2 CEM – MeFiCom's Belief Systems 86
6.6.3 CEM – MeFiCom's Boundary Systems 87
6.6.4 CEM – MeFiCom's Diagnostic Control Systems 89
6.6.5 CEM – MeFiCom's Interactive Control Systems 95
6.7 Quality Criteria and Assurance – Qualitative Research 97
7 BPM in the MFI – Quantitative Research Process 101
7.1 MFI Survey 101

7.1.1 Survey Sample ... 101
7.1.2 Design and Operationalization ... 102
7.2 Quantitative Data Analysis and Results ... 115
7.2.1 MFI Survey Part I – Statistical Data ... 116
7.2.2 MFI Survey Part II – Overall Characteristics of the MFC ... 120
7.2.3 MFI Survey Part III – Business Performance Measurement ... 128
8 BPM in the MFI – Synthesis ... 137
8.1 Adaption of Generic BPM Frameworks and Models ... 137
8.1.1 Metal Finishing – A Contract Processing Industry ... 139
8.1.2 Metal Finishing – A Job Shop Industry ... 141
8.1.3 MFI's Industrial Service Nature ... 142
8.1.4 MFI's Environmental Impact ... 144
8.2 Modeling the MFI Levers of Measurement ... 145
8.2.1 LOM – Knowledge Controls Systems ... 148
8.2.2 LOM – Process Controls Systems ... 151
8.2.3 LOM – Relational Controls Systems ... 154
8.2.4 LOM – Success Controls Systems ... 158
8.2.5 MFI-BPM System – Integrating the LOM Information ... 160
8.3 Further MFI-BPM Recommendations ... 163
9 Conclusion ... 167
9.1 Common Managerial & Academic Contributions ... 167
9.2 Discussion & Limitations ... 168
9.3 Research Agenda for the Future ... 170
References – Literature ... 173
References – Internet ... 190
References – Applied Software ... 193
Appendices ... 194
A.1 Classification of the MFI within the G20 ... 195
A.2 Eurostat – Detailed MFI Enterprise Statistics in 2012 ... 199
A.3 US Census – Detailed MFI Enterprise Statistics in 2013 ... 201
A.4 Literature Review within Scientific Databases ... 202

A.5 Semi-Structured Expert Interviews – Interview Guide ... 205
A.6 Semi-Structured Expert Interviews – Invitation Letter ... 211
A.7 Semi-Structured Expert Interviews – Informed Consent ... 213
A.8 Group Discussion Presentation ... 214
A.9 CAQDAS – Document System ... 215
A.10 CAQDAS – Code System ... 216
A.11 MFI Survey – Invitation Email – English ... 222
A.12 MFI Survey – Invitation Email – German ... 223
A.13 MFI Survey – Invitation Letter – English ... 224
A.14 MFI Survey – Invitation Letter – German ... 225
A.15 MFI Survey – Questionnaire – English ... 226
A.16 MFI Survey – Questionnaire – German ... 240
A.17 MFI Survey – Analyses ... 252
A.18 MFI Survey – Scale Reliability ... 266

List of Figures

Figure 1: Structure of the Dissertation Research Project 2
Figure 2: Selected Perspectives within the Strategic Theory of the Firm 6
Figure 3: Process-Based View 8
Figure 4: Elements of a controlled process 9
Figure 5: The LOC-Framework 11
Figure 6: Balancing intended & emergent strategies 12
Figure 7: Research Framework – MCS & BPM in an Organizational Context 19
Figure 8: The Industrial Organization Paradigm 24
Figure 9: Top-Down vs. Bottom-Up approach in BPM system design 25
Figure 10: Chalice-shaped bottom-up approach to improve BPM feasibility 25
Figure 11: Example of a Metal Finished Drill Bit – PVD Process 30
Figure 12: Distribution of Surface Finishing Processes in Germany in 2005 30
Figure 13: The Surface Engineering Roadmap 32
Figure 14: Metal Finishing Core Process linked to AT and the Values Added Chain 33
Figure 15: Example of Google Scholar Alerts 50
Figure 16: Research Design BPM in the MFI 54
Figure 17: Interview Guide Design 64
Figure 18: Adopting the Content-Structuring Qualitative Content Analysis Approach 74
Figure 19: Two Areas of Interest in the MAXQDA Code System 75
Figure 20: Adoption of the four LOCs within the MAXQDA Code System 75
Figure 21: Example for the Belief Systems Coding Process in MAXQDA 76
Figure 22: Example for a Boundary Systems Code in MAXQDA 77
Figure 23: Example for the Hierarchical DCS Code Structure in MAXQDA 78
Figure 24: Example for an ICS Code in MAXQDA 79
Figure 25: Example for a Code Memo in MAXQDA 79
Figure 26: Number of Transcript Passages assigned to the MAXQDA Code System 80
Figure 27: MFI Core Process adopted in System Dynamics Notation 81
Figure 28: From the CAQDAS Code System to the Cause and Effect Model 82
Figure 29: Example for the CAQDAS Code System Transformation 83
Figure 30: Information Systems Design at MeFiCom 84

Figure 31: Overview on the MFI CEM ... 84
Figure 32: Industry CEM Container ... 85
Figure 33: Belief Systems CEM Container ... 86
Figure 34: Boundary Systems CEM Container ... 88
Figure 35: Diagnostic Control Systems – BPM – Measure Output CEM Container ... 89
Figure 36: CAQDAS Code System – Business Measures vs. KPIs ... 90
Figure 37: Diagnostic Control Systems – Information Supply CEM Container ... 92
Figure 38: Diagnostic Control Systems – Success Factors CEM Container ... 93
Figure 39: Boundary Systems CEM Container ... 96
Figure 40: Overview on Propositions Derived from the MFI CEM ... 103
Figure 41: Screenshot of the MFI Survey Introduction Site ... 104
Figure 42: Screenshot of the MFI Survey Part 2 Introduction Site ... 106
Figure 43: Screenshot of the MFI Survey Part 3 Introduction Site ... 109
Figure 44: Structure of the BPM System Questionnaire Part ... 111
Figure 45: Example for the Back-Translation Process of Scale Items ... 113
Figure 46: Distribution of the MFI Survey Participant Experience ... 117
Figure 47: Distribution of the MFI Survey Organization's Years of Operation ... 118
Figure 48: Regular Order Backlog at the MFI Survey Sample MFCs ... 128
Figure 49: Distribution of Fixed and Variable Costs ... 130
Figure 50: MFI Survey – KPIs and Business Measures ... 131
Figure 51: Generic BPM Framework Comparison ... 138
Figure 52: MFC as Supply Chain Element ... 139
Figure 53: SCOR-Model Adapted for the MFI ... 140
Figure 54: Classifying Quantitative CEP Measures ... 144
Figure 55: MFC Model following the Research Framework ... 145
Figure 56: The Levers of Measurement - BPM Model for the MFI ... 146
Figure 57: Relation between the Strategic Theories of the Firm and the LOM ... 147
Figure 58: LOM – Knowledge Controls Systems ... 148
Figure 59: LOM – Process Controls Systems ... 151
Figure 60: LOM – Relational Controls Systems ... 154
Figure 61: Service Failure Domino Effect ... 155
Figure 62: LOM – Success Controls Systems ... 158

Figure 63: Performance Measures Identification and Selection Process 163

All figures are my own illustration as far as the source is not stated different.

List of Tables

Table 1: BPM System Types ... 18
Table 2: MFI identified in the UN ISIC rev. 4 ... 35
Table 3: MFI identified in the EU NACE rev. 2 ... 36
Table 4: MFI identified in the North American NAICS 2012 ... 37
Table 5: MFI identified in the Australian / New Zealand ANZIC 2006 rev. 2 ... 37
Table 6: Share of Personnel Costs in European Service Sectors ... 40
Table 7: MFI Enterprise Size Classes 2011 in Germany ... 41
Table 8: MFI Enterprise Size Classes 2012 in USA ... 44
Table 9: Literature Review – Dictionary ... 47
Table 10: Literature Review – Terms for Database Search ... 48
Table 11: Literature Review – Results in Scientific Databases ... 48
Table 12: Literature Review – Selected Journals ... 49
Table 13: Applied German and English Search Terms in Journal Content Search ... 49
Table 14: Google Scholar Alert Terms ... 50
Table 15: Expert Interviews – Recording Length ... 68
Table 16: Group Discussion – Time Flow ... 69
Table 17: Group Discussion – Expert's Subjects ... 71
Table 18: Overview Resulting Transcripts Expert Interviews and Group Discussion ... 73
Table 19: MFI Survey – Sample Composition ... 102
Table 20: Response and Completion Rates in the MFI Survey ... 115
Table 21: Composition of the Survey Sample – Focus on Participants' Function ... 116
Table 22: Composition of the Survey Sample – Focus on Organizations ... 117
Table 23: Composition of the Survey Sample – Focus on Metal Finishing Processes ... 118
Table 24: Turnover and Employees per Country and Enterprise Class ... 119
Table 25: Turnover and Employees Compared to Overall MFI Population ... 119
Table 26: Descriptive Statistics – Customer Focus Scale ... 121
Table 27: Descriptive Statistics – Closer to Customers Scale ... 121
Table 28: Descriptive Statistics – Competitive Priorities Scale ... 122
Table 29: Descriptive Statistics – Competitive Priorities Click Ranking ... 123
Table 30: Descriptive Statistics – Environmental Uncertainty Scale ... 123

Table 31: Descriptive Statistics – Technology Uncertainty scale 124
Table 32: Descriptive Statistics – Demand Uncertainty scale 125
Table 33: Descriptive Statistics – Exploratory and Exploitative Innovation Scales 126
Table 34: Descriptive Statistics – Flexibility Dimensions 127
Table 35: Descriptive Statistics – Responsibility BPM Rules and Guidelines Setting 128
Table 36: Descriptive Statistics – Performance Scale 129
Table 37: Descriptive Statistics – SME Performance Measures 129
Table 38: Descriptive Statistics – Applied External BPM Indicators 132
Table 39: Descriptive Statistics – BPM Information Services 133
Table 40: Descriptive Statistics – BPM Functions 133
Table 41: Descriptive Statistics – Customized BPM System Characteristics 134
Table 42: Descriptive Statistics – Reasons for BPM System Absence 134
Table 43: Descriptive Statistics – Potential BPM System Characteristics 135
Table 44: The Results & Determinants Framework and BPM in the MFI 143
Table 45: MFI-LOM-KCS – Health & Safety Measures 149
Table 46: MFI-LOM-KCS – Compensation & Incentives Measures 149
Table 47: MFI-LOM-KCS – Human Performance Measures 150
Table 48: MFI-LOM-KCS – Human Sustainability & Learning Measures 150
Table 49: Examples for Hard and Soft Quality in the MFI 151
Table 50: MFI-LOM-PCS – Quality Measures 152
Table 51: MFI-LOM-PCS – Delivery Reliability Measures 152
Table 52: MFI-LOM-PCS – Utilization Dimensions Measures 153
Table 53: MFI-LOM-PCS – Environment Measures 153
Table 54: MFI-LOM-RCS – Customers Measures 156
Table 55: MFI-LOM-RCS – Competitors Measures 156
Table 56: MFI-LOM-RCS – Market Measures 157
Table 57: MFI-LOM-RCS – Share- & Stakeholders Measures 157
Table 58: MFI-LOM-SCS – Financials Measures 159
Table 59: MFI-LOM-SCS – Innovation Measures 160
Table 60: MFI-LOM-SCS – Other Organizational Performance Drivers 160
Table 61: CPV Engineering from LOM-PCS and -SCS 161
Table 62: CPV Engineering from LOM-KCS and -RCS 162

Table 63: CPV Engineering from LOM-RCS and -SCS 163

All tables are my own illustration as far as the source is not stated different.

List of Abbreviations

AMJ..................	Academy of Management Journal
ANZIC..............	Australian and New Zealand Standard Industrial Classification
ASIG.................	German Occupational Safety Act – Arbeitssicherheitsgesetz
ASM..................	Annual Survey of Manufactures – USCB Database
AT....................	Agency Theory
B......................	Billion (1,000,000,000)
BAM..................	British Academy of Management
BDSG................	German Federal Data Protection Act – Bundesdatenschutzgesetz
BEUR................	Billion Euro
BI	Business Intelligence
BJM..................	British Journal of Management
BOS..................	Business Operating Systems
BPM	Business Performance Measurement
BPMS................	Business Performance Measurement System
BSC..................	Balanced Scorecard
BUSD................	Billion US-Dollar
CAQDAS............	Computer-Aided Qualitative Data Analysis Software
CBI...................	Confederation of British Industry
CBP..................	County Business Patterns – USCB Database
CEM..................	Cause and Effect Model
CEO..................	Chief Executive Officer
CEP..................	Corporate Environmental Performance
CFO..................	Chief Finance Officer
CP	Contract Processing
CPV..................	Critical Performance Variable
CRM..................	Customer Relationship Management
CVD..................	Chemical Vapor Deposition
DCS..................	Diagnostic Control Systems
DE....................	Deutsch (German)
DESTATIS..........	German Federal Statistical Office
DOE..................	UK Department of the Environment

EBIT……………….	Earnings Before Interest and Taxes
EBITDA………….	Earnings Before Interest, Taxes, Depreciation and Amortization
EBS………………..	European Business School, Oestrich-Winkel, Germany
EC………………..	European Communities
EFDS……………..	Europäische Forschungsgesellschaft Dünne Schichten e. V. – European Society of Thin Films, DE
e.g. ………………	Exempli gratia
EN………………..	English
EPA ……………..	US Environmental Protection Agency
ERP ……………....	Enterprise Resource Planning
EU ……………….	European Union
EUR………………	Euro
EUROSTAT ……..	Statistical Office of the European Union
FPM……………...	Financial Performance Measurement
G20 ……………...	Group of Twenty
GDP………………	Gross Domestic Product
GENESIS………..	Gemeinsames neues statistisches Informationssystem – German Common new statistical information system
GSR……………...	Gross Sales Revenue
h, hh…………….	Hour, hours
HR……………….	Human Resources
HRPM……………	Human Resource Performance Measurement
HSE………………	Health, Safety, Environment
IC………………...	Industry Class
ICS……………….	Interactive Control Systems
i.e. ……………….	Id est
IMF ……………...	Institute of Materials Finishing, UK
IPO……………….	Initial Public Offering
ISIC ……………...	International Standard Industrial Classification of all Economic Activities of the United Nations
ISIN……………...	International Securities Identification Numbers
IT ………………..	Information Technology
KBV……………...	Knowledge-Based View
KCS……………...	Knowledge Controls Systems
KPI………………	Key Performance Indicator

LOC..................	Levers of Control
LOM..................	Levers of Measurement
LTA..................	Lost Time Accident
M.....................	Million (1,000,000)
M2K..................	Mode 2 Knowledge
MCS..................	Management Control System
MeFiCom	Metal Finishing Company (pseudonym for the single-case company studied in this dissertation)
MFC..................	Metal Finishing Company (in general)
MFI	Metal Finishing Industry
MFI-BPM...........	Metal Finishing Industry - Business Performance Measurement
min..................	Minute
MIS	Management Information System
mm..................	Minutes
NACE	Nomenclature générale des Activités économiques dans les Communautés Européennes
NAICS	North American Industry Classification System
NSIs	National Statistical Institutes
ODBC...............	Open Database Connectivity
p., pp.	Page, pages
PBV..................	Process-Based View
PCS..................	Process Controls Systems
PPM..................	Parts per Million
PRODCOM	European Statistics on the Production of Manufactured Goods
PVD..................	Physical Vapor Deposition
R&D..................	Research and Development
RBV..................	Resource-Based View
RCS..................	Relational Controls Systems
RoA, RoI, RoS......	Return on Assets, Return on Investment, Return on Sales
RV....................	Relational View
s.	See
SBR..................	Statistical Business Register – DESTATIS Database
SBS	Structural Business Statistics – EUROSTAT Database
SCOR................	Supply Chain Operations Reference

SCS....................	Success Controls Systems
SEA....................	Surface Engineering Association, UK
SIIE....................	Strascheg Institute for Innovation and Entrepreneurship of the European Business School, Oestrich-Winkel, Germany
SMART...............	Strategic Measurement And Reporting Technique
SME..................	Small and Medium Sized Enterprises
SME-BSC..............	Balanced Scorecard for Small and Medium Sized Enterprises
SPSS....................	IBM SPSS Statistics 20
ss......................	Seconds
SSP....................	Sector Strategies Program of the US EPA
STI....................	Short Term Incentive
T.......................	Thousand (1,000)
TQM...................	Total Quality Management
TUSD.................	Thousand US-Dollar
TVSS..................	Total Value of Shipments and Receipts for Services
UK....................	United Kingdom
UN	United Nations
US / USA	The United States of America
USCB.................	United States Census Bureau
USD..................	US Dollar
VDMA...............	Verband Deutscher Maschinen- und Anlagenbau e. V. – German Engineering Federation, DE
WEO	World Economic Outlook (International Monetary Fund)
ZVEI.................	Zentralverband Elektrotechnik- und Elektronikindustrie e. V.
ZVO..................	Zentralverband Oberflächentechnik e. V. – German Central Association of Surface Treatment Professionals

Abstract

Management Control System (MCS) models can be combined with an ample set of generic Business Performance Measurement (BPM) models. Criticism evolving in past years has indicated that inadequate academic observance has been spent on the adaption and implementation of theoretical BPM models to specific industry and company contexts leading to a suboptimal practical feasibility.

In reflecting these concerns, a bottom-up approach to improve BPM feasibility is developed based on an industry's specific characteristics. These characteristics are then linked to corresponding generic BPM models. In this vein, practitioners shall be assisted in designing and implementing a sustainable BPM system within their MCS and industry context.

Correspondingly and in the sense of a meta-theory, BPM becomes integrated within an organization's MCS, controlling the success of the organization's strategy implementation based on the organization's competitive advantages as well as on its organizational environment relations.

Applying an exploratory sequential mixed methods research design, this theoretical conceptualization is empirically proved for the Metal Finishing Industry (MFI). The research findings derived from a three-year ethnographic research in a major German MFI organization, following a Mode 2 knowledge production research policy, are triangulated with the results of a survey in the German and British MFI. These triangulated results as well as a selection of MFI-suitable generic BPM frameworks are synthesized into a BPM model reflecting the MFI's characteristics and BPM requirements.

This research project is grounded on Simons' *The Levers of Control* as MCS framework and extends this framework with *The Levers of Measurement*, representing the integrated MFI-BPM model. It allows the adaption for other industry branches by reflecting their industry characteristics. In addition, it satisfies the demand for a BPM model in the MFI coming from single MFI organizations and from MFI associations.

Keywords:
Management control systems, Business performance measurement, Meta-theory, Metal finishing industry, Ethnography, Mode 2 knowledge production, Levers of control, Levers of measurement, Germany, United Kingdom

1 Introduction

> "The question is: 'Is this what a manager should consider important?' […] If these questions are not being asked in designing 'controls', we will end up by making business essentially uncontrolled […]." (Drucker in Bonini et al., 1964, p. 289)

More than fifty years ago Peter F. Drucker made the above cited statement which guided me through the last three years of my dissertation research project. The "controls" mentioned by Drucker are a synonym for business performance measures or indicators, either financial or non-financial, on which Business Performance Measurement (BPM) systems are built up (s. chapter 2.4). These BPM systems again are elements of Management Control Systems (MCS) (s. chapter 2.3).

How can a statement - being older than 50 years - be a driver for a dissertation project? One ought to think that a related research gap in BPM and MCS research should have already been addressed by the academic community in the past five decades. Indeed, it is quite surprising that an evolving concern can be identified in recent BPM and MCS literature stating that a significant part of academic research is not aligned with managerial needs (s. chapter 2.5.2). Neely (2011, p. 76), a worldwide recognized and guiding BPM researcher, for example states:

> "In spite of all of the research that has been undertaken, we actually know relatively little about what information managers use and how they use it."

In the next chapters of this introduction I will describe the problem for which the results of this research project shall offer solutions as well as the structure of the research project itself.

1.1 Problem Statement

The adaption and implementation of theoretical BPM concepts in *real world* settings, namely companies within their industry class context, is a serious and often not sustainably successful process (Gleich, 2011, p. 330).

The Metal Finishing Industry (MFI – s. chapter 3) is an example for an industry class in which an industry class related BPM system is not existing (s. chapter 4). Processes like metal coating or plating are performed by MFI organizations on a fee or contract basis (UN, 2008, p. 125). Although not readily apparent, the demand for such an MFI-BPM model is evident; the demand comes on the one hand from single companies like the German company *MeFiCom*[1], which represents the single case to be studied within this research project, and on the other hand from MFI associations, like the *Institute of Materials Finishing* in the United Kingdom:

> "Very little research is carried out on how our industry, virtually unknown outside itself, can improve its operations, and it is good for the industry to have this performance model work done."
>
> (Larson, 2014, p. 177 – commentary on this EBS-SIIE research project)

The challenge centers in the fact that the MFI was not yet analyzed from a social science view point focusing on MCS or BPM. Finding appropriate theories (s. chapters 2.1 to 2.4)

[1] "MeFiCom" stands for **Me**tal **Fi**nishing **Com**pany, a pseudonym for the company being the single case within the qualitative research part of this dissertation.

which help to explain MFI characteristics (s. chapter 3) is one step in order to build an MFI-BPM model. Another step is to revisit the design and adaption process for an industry class BPM system (s. chapter 2.5.3).

The guiding research question therefore is defined as follows:

The research is focused on the 'How' metal finishing companies of larger scale are measuring their business performance, also in comparison to other manufacturing companies.

1.2 Structure of this Dissertation

This dissertation is structured sequentially according to a research scheme by Riempp (2004, p. 316):

The research gap is deduced from the formulation of a practical problem and related theories. The research questions are deduced from the research gap. By using a suitable research methodology, the research process concludes with results leading to practical and theoretical contributions of the research project.

Adopting this scheme led to the following structure which shall guide the reader through the dissertation at hand:

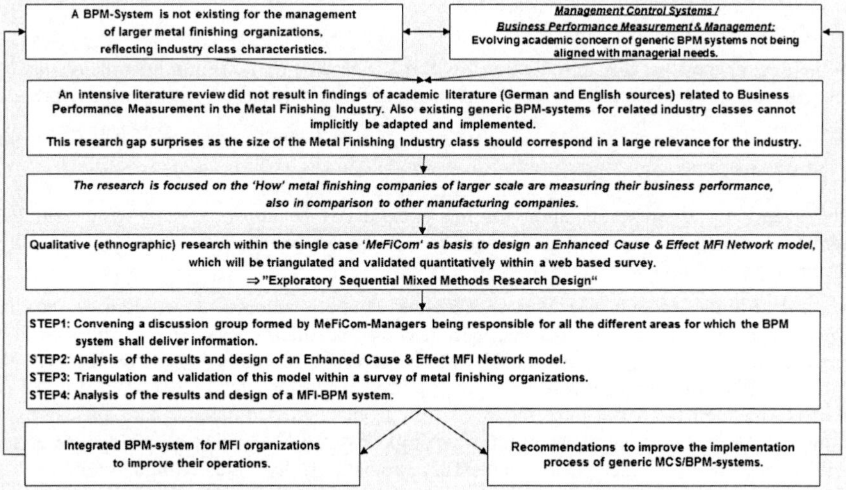

Figure 1: Structure of the Dissertation Research Project

Chapter 1.1 stated the practical problem on which the research centers.

Chapter 2 introduces theories on which the analysis of the practical problem is based.

Chapter 3 includes an analysis of the MFI representing the industry class of interest in which the practical problem is located.

Chapter 4 analyzes the current state of BPM research in the MFI and proves the existence of a research gap.

Chapter 5 describes the applied exploratory mixed methods approach combining sequentially qualitative and quantitative empirical research methods.

Chapter 6 includes the qualitative research part presenting the applied qualitative methods as well as the retrieved and analyzed data leading to an *Enhanced MFI Cause & Effect Network Model* as the result.

Chapter 7 includes the quantitative research part which shall triangulate and validate the qualitative research findings.

Chapter 8 combines the results of the qualitative and quantitative research to build the foundation for an MFI-BPM model.

Chapter 9 concludes with a reflection of the managerial and academic contributions as well as with a critical discussion of the research limitations and recommendations for a future research agenda.

2 Theoretical Framework

Chapters 2.1 and 2.2 present a selection of theories of the firm which shall help to explain several MFI characteristics with regard to BPM. Chapters 2.3 and 2.4 introduce the MCS and BPM rationale as well as related research on which this dissertation project is grounded.

Chapter 2.5 synthesizes the theories of the firm with the MCS and BPM framework in order to create a theoretical framework for this research project. Furthermore an evolving academic concern is presented stating that a significant part of academic research is not aligned with managerial needs. Chapter 2 concludes with a proposal for an improved BPM design approach which will be tested empirically within the MFI (s. chapters 5 to 7).

2.1 Agency Theory

Agency Theory (AT) is one stream within the New Institutional Economics (Coase, 1937, 1998; Williamson, 2000). Following Jensen & Meckling's (1976, p. 308) definition of an "*agency relationship*" a principal appoints an agent under a contract to perform a service on his behalf; the agent becomes the contractor of the principal. Under ordinary circumstances, both agent and principal are not in fact persons but organizations[2] (North, 1991, p. 109) in which different contracting relationships are performed among individuals (Jensen and Meckling, 1976, p. 310).

AT states that agency relationships between the principal and the agent create problems related to both parties trying to maximize their self-interests resulting in conflicting interests. Combined with an imbalanced flow of information, these agency problems result in agency / transaction costs (Coase, 1960, p. 15). It is the goal of AT to reduce those agency costs by identifying and eliminating the agency problems (Schwaiger and Meyer, 2009, p. 136).

AT is a widely accepted and applied but somewhat controversial economic theory (Eisenhardt, 1989a, p. 57; Porter Liebeskind, 1996), especially regarding its practical relevance and contribution for managers (Anthony, 1989, p. 16). However AT delivers important starting points for the usage of BPM; BPM can be applied as a tool in order to create transparency within the agency dilemma and can therefore be related to AT (Gleich, 2011, p. 33; Neely, 2011, p. 350; Franco-Santos et al., 2012, p. 97).

Nevertheless the cost-benefit ratio must be kept in mind when thinking of BPM solely for reducing agency costs: it would be a perverting situation if the cost for running a BPM system in order to eliminate potential agency costs would be higher than the agency costs themselves (Widener, 2007, p. 758).

2.2 Perspectives Within the Strategic Theory of the Firm

After having looked on agency relationships between entities within the AT, I will proceed by considering theoretical perspectives related to competitive advantages of the firm. The following figure shows the perspectives within the strategic theory of the firm that will be introduced to further explain characteristics of the MFI within the cause of this research project:

[2] Within my research project I built on AT's inter-firm relevance. Nevertheless AT is regularly applied in BPM research as an explanatory model for the relationship of managers to their subordinates and vice-versa (Franco-Santos et al., 2012, p. 97).

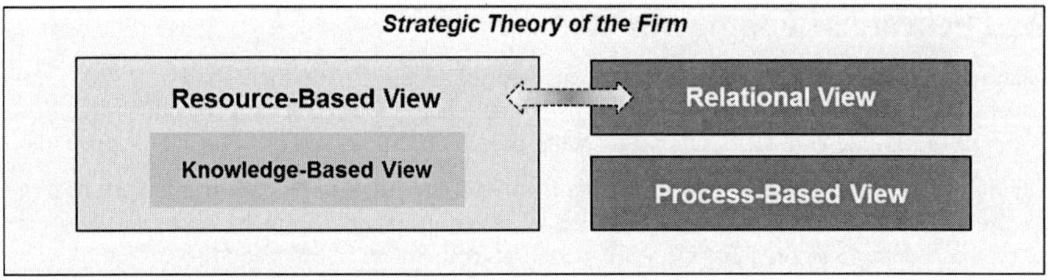

Figure 2: Selected Perspectives within the Strategic Theory of the Firm

Each view has a *special perspective*, i.e. resources, interorganizational relations and processes. Applying these different perspectives, as mentioned by Peteraf and Barney (2003, p. 321), is supposed to generate a holistic explanation of the MFI characteristics within the empirical research part (s. chapters 6 and 7).

Importance is laid on the Resource-Based View theory which is the starting point for several spin-off theories such as the Knowledge-Based View. The Relational View will then be introduced and linked to the Resource-Based View. The Process-Based View is founded within the supply chain management research field and therefore is intended to include important theoretical explanations for conditions within the MFI.

2.2.1 Resource-Based View

The Resource-Based View (RBV) theory is centered on the idea that a firm's competitive advantages arise from the resources this firm owns and applies (Penrose, 1955, p. 542). Resources of a firm are defined as "*all assets, capabilities, organizational processes, firm attributes, information, knowledge, etc. controlled by a firm*" (Barney, 1991, p. 101). Resources can also be linked to the previously introduced transaction costs (Grant, 1996; Foss and Foss, 2005; Kim and Mahoney, 2005). These linkages may help to reduce scholarly dissatisfaction with the transaction cost theories based on the missing relevance of the exploitation of resources therein (Porter Liebeskind, 1996, p. 93).

Following Barney's RBV core tenets (Barney, 1991, p. 105) resources have to be "*valuable, rare, inimitable, and non-substitutable*" in order to generate competitive advantages for the firm. A resource not necessarily needs to be tangible to create a competitive advantage; moreover it was empirically demonstrated in RBV studies that firms building their strategies on intangible and socially complex assets perform better than comparable firms focusing merely on their tangible assets (Barney, 2001, p. 648).

Armstrong and Shimizu (2007) have analyzed 125 empirical RBV studies published in major academic journals within the time frame 1991 to 2005. Based on their analyses they found several issues which they addressed by recommendations for future RBV research; these include adopting a qualitative research approach employing in-depth interviews where an industry to be studied is new and/or unexplored (Armstrong and Shimizu, 2007, p. 976). Furthermore they recommend controlling for confounding factors within empirical RBV research like *"parent effects"* which often arise when a firm is part of an industrial diversified conglomerate. The corporate parent may own resources which may be transferred industry-independently to its corporate children; in three empirical RBV studies the positive impact of these parent effects vary from 1% up to 15% of the firm's profitability (Armstrong and Shimizu, 2007, p. 974).

The Knowledge-Based View (KBV) is a RBV spin-off (Barney et al., 2011, p. 1301; Schwaiger and Meyer, 2009; p. 67). Grant (1996, p. 111) declares "knowledge" to be the most strategically important resource of a firm; therefore the firm's management is enjoined to integrate knowledge into the firm in order to sustainably generate competitive advantages (Grant, 1996, p. 120).

Grant's principles especially hold true for high-technology sectors which depend on such knowledge (Bingham and Davis, 2012, p. 634). In addition, Arend et al. (2014) have found that a firm's knowledge development activities correlate with higher performance regarding survival, RoA and Tobin's q.

Besides the development of the firm's internal knowledge sources, the usage of external knowledge sources positively influence a firm's ability to recognize and exploit new opportunities in order to gain competitive advantages (Foss et al., 2013, p. 1467). Important external knowledge sources are suppliers as well as customers (Laursen and Salter, 2006, p. 138), which implies the importance of a vital relationship of the firm to these stakeholders.

2.2.2 Relational View

Vital interorganizational relations are also the tenet within the Relational View (RV) theory (Dyer and Singh, 1998). Thus, Dyer and Singh (1998) focus on a firm's relations to and within its network environment rather than on the inward directed view of the firm within the RBV.

However the differentiation between RV and RBV oftentimes blurs (as is seen regarding the usage of external knowledge sources within the KBV): Lavie (2006) applies RV to extend RBV to a theory that includes network resources and the firm's behavior to use such resources, although they are not fully owned by the firm (Lavie, 2006, p. 638). Mesquita et al. (2008) compare RBV with RV. They found that in learning dyads (i.e. "*joint buyer-supplier knowledge acquisition*") a firm can improve its overall performance by having additional learning-dyad-exclusive "*relational performance*" (Mesquita et al., 2008, p. 935).

Lee et al. (2014) have found a positive influence on the overall supply chain performance by creating visibility in the supply chain in cases where information (equals knowledge) is shared via interorganizational information systems. This supply chain visibility is defined by Barratt and Oke (2007, p. 1218) as *"the extent to which actors within a supply chain have access to or share information which they consider as key or useful to their operations and which they consider will be of mutual benefit"*. This collaborative behavior regarding knowledge is contrasted by opportunistic behavior in cases where members of a supply chain maintain some level of information asymmetry in order to generate isolated firm advantages within a network environment (Lee et al., 2014, p. 286). The theoretical explanation for this *hidden information* or imbalanced flow of information is likely to be found in AT.

But sharing knowledge and other resources within interorganizational networks does not necessarily create additional performance boosts; it may also induce the hazards inherent in opportunism. Martinez-Noya et al. (2013) have analyzed the performance effects of R&D outsourcing. They argue that managers have to carefully evaluate which information they contract-out and to whom. A balance has to be found between a) the required amount of necessary information that the outsourcing firm has to deliver to the supplier in order to do the job and b) the right safeguard measures which should be applied in order to protect the transferred know-how. If safeguards are too few, the outsourcing company risks losing a competitive advantage. If safeguards are too many, the outsourcing company risks a reduced effectiveness of R&D outsourcing (Martinez-Noya et al., 2013, p. 85).

2.2.3 Process-Based View

In the previous two chapters, I introduced the RBV focusing mainly firm-inwards on resources and the RV focusing firm-outwards on interorganizational relations. In a further perspective Stölzle and Hofmann (2007) propose the Process-Based View (PBV) which focusses on processes related to strategy formulation and implementation within a firm. PBV describes competitive advantages arising from "*unique processes*" within a firm's internal (inwards directed) *as well as* external (outwards directed) context like an outstanding logistics process (Stölzle and Hofmann, 2007, p. 1064):

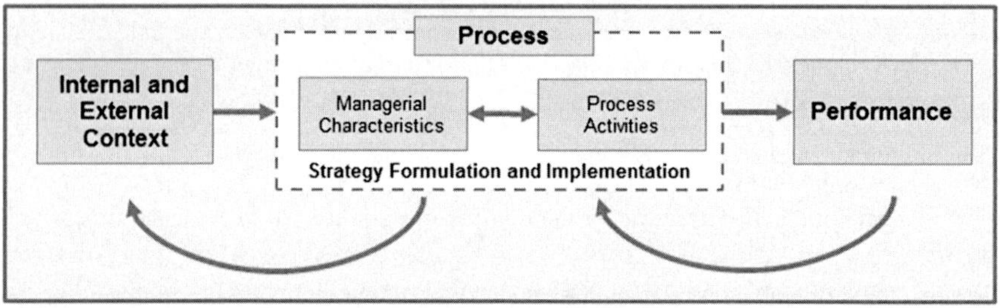

Figure 3: Process-Based View (adopted in modified form from Stölzle and Hofmann, 2007, p. 1062)

Supplementary to the "Processes follow Strategies and Structures" approach (Porter, 1981, p. 616; s. chapter 2.5.3) PBV delineates a "Strategies and Structures follow Processes" approach. A *process* within the PBV is based on a) managerial characteristics and b) process activities:

a) Stölzle and Hofmann, (2007, p. 1063) argue that competitive advantages are not identified rather than realized by the firm itself ("*a black box*"), but by the firm's managers. Therefore the characteristics of these managers ("*upper echelons perspective*"; Hambrick and Mason, 1984) have a significant relevance in creating processes within the firm's strategy context.
b) Following Teece et al. (1997, p. 518) organizational process activities have three roles:
   - *"coordination/integration (a static concept);*
   - *learning (a dynamic concept); and*
   - *reconfiguration (a transformational concept)".*

   These roles are vital in the strategic factor combination leading to new opportunities for the firm (Stölzle and Hofmann, 2007, p. 1064).

PBV is grounded within the Strategy Process Research (Pettigrew, 1992; Van de Ven, 1992). It is related to Supply Chain and Logistics Management (Hofmann, 2014, p. 50).

2.3 Management Control Systems Framework

"*Management control* is the process by which managers assure that resources are obtained and used effectively and efficiently in the accomplishment of the organization's objectives." (Anthony[3], 1965, p. 17)

Having introduced theories related to strategic aspects to the firm in chapters 2.1 and 2.2, I will continue by introducing Management Control Systems (MCS) for planning and controlling processes linked to these strategic aspects.

2.3.1 Management Control Systems – The Rationale Behind

Anthony et al. (2014) introduce MCS by presenting five basic elements for a cybernetic regulating loop:

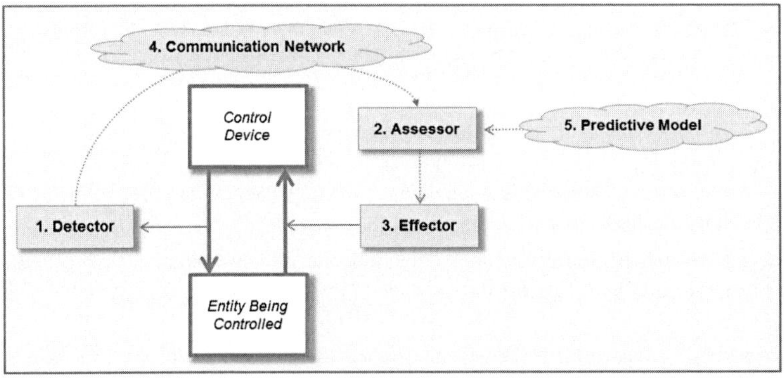

Figure 4: Elements of a controlled process (adopted in modified form from Anthony et al., 2014, p. 37)

A detector (1) is required in order to acquire data from the system to be controlled. These data are transmitted via a communication network (4) to the assessor (2). The assessor compares the received data with a standard in order to control for any deviations. Based on predictive models (5) the assessor decides the extent of behavioral changes necessary to be performed by the effector (3) in the event the standard is not met. Accordingly the assessor informs the effector via the communication network to react accordingly to the decisions made.

In an organization, the assessor is represented by the managers who use information in order to control the results of the decisions made to be consistent with the organization's strategic goals (Anthony et al., 2014, p. 39; Cugueró-Escofet and Rosanas, 2013, p. 34).

Reflecting solely these controlling and regulating aspects would mean a shortcoming of MCS, as MCS also includes strategy formulation (Simons, 1995, p. 18) as well as strategy implementation (Anthony et al., 2014, p. 4). Therefore a MCS cannot be seen as a single system, but a bundle or a "*package*" (Malmi and Brown, 2008) of different management, accounting, controlling, communication, reward etc. sub-systems or "*levers*" (Simons, 1995, p. 7; Widener, 2007, p. 757).

[3] Robert Newton Anthony (1916-2006) is widely seen as the founding father of the Management Control discipline and "*a pioneering thinker in Management Accounting*" (Birnberg, 2011).

2.3.2 Management Control Systems Research

MCS research is challenging as a generally accepted definition of MCS is nonexistent (Malmi and Brown, 2008, p. 289; Tessier and Otley, 2012, p. 171; Strauß and Zecher, 2013, p. 234). This lack of consistent MCS conceptualization is also illustrated by a variety of MCS frameworks[4] (to name a few: Simons, 1995; Otley, 1999; Malmi and Brown, 2008; Ferreira and Otley, 2009; Broadbent and Laughlin, 2009). Furthermore Ferreira and Otley (2009, p. 264) argue that the term MCS (related to Anthony's above cited Management Control definition) "*has become a more restrictive term than was the original intention*" and therefore preferred the term "*Performance Management System*" as alternative (see also Otley, 1999, p. 365).

However in this dissertation I will continue to use the term MCS due to two reasons:

1. It is necessary to clearly distinguish between "Performance *Management* Systems" and "Performance *Measurement* Systems" (Lebas, 1995, p. 35; Otley, 1999, p. 364; Watts and McNair-Connolly, 2012, p. 229) in order to avoid any confusion of the terms.
2. Risks have to be avoided, for instance in mistaking "Performance Management Systems" in the sense of MCS, with "Performance Management Systems" in other senses. In the sense of human resource management (Broadbent and Laughlin, 2009, p. 284), as one example.

Besides these inconsistencies the different system frameworks share common and process oriented grounds i.e. *"planning, **performance measurement**, rewards, and feedback and feed-forward information loops"* (Strauß and Zecher, 2013, p. 264).

Business Performance Measurement (BPM - s. chapter 2.4) is a crucial component within the different MCS frameworks[5]:

- Simons' "*Levers of Control*" (s. chapter 2.3.3) framework links BPM to the "*Diagnostic Control System*" in order to control for "*Critical Performance Variables*" (Simons, 1995, p. 59).
- Within Otley's "*Performance Management*" framework BPM is related to the application of Kaplan and Norton's (1992, 1996) "*Balanced Scorecard*" approach (Otley, 1999, p. 374).
- "*Cybernetic Controls*" form a sub-system of Malmi and Brown's "*Management Control Systems as a Package*" framework; the cybernetic controls include budgets and financial, non-financial as well as hybrid measurement systems (Malmi and Brown, 2008, p. 291).
- Ferreira and Otley's "*Extended Performance Management Systems*" framework is based on twelve questions of which question 5 "*Key Performance Measures*", question 6 "*Target Setting*" and question 7 "*Performance Evaluation*" are directed to BPM (Ferreira and Otley, 2009, p. 271).

[4] For an overview on influential MCS literature in the past 50 years please refer to Strauß and Zecher (2013) – especially to their "*Map of the MCS landscape*" (Strauß and Zecher, 2013, p. 265).
[5] The selection of MCS frameworks was chosen accordingly to Strauß and Zecher, 2013, p. 256.

- Broadbent and Laughlin's "Conceptual Performance Management Systems" is related to Ferreira and Otley's (2009) work and founded on the concept that an organization has to control for both the "*desired ends [...] and the chosen means to achieve these ends*"; means include "*Key Performance Indicators*" and ends include "*Performance Evaluation Systems*" (Broadbent and Laughlin, 2009, p. 292).

2.3.3 The Levers of Control

Simons (1995) introduced the MCS framework "Levers of Control" (LOC) which is centered on an organization's "Business Strategy":

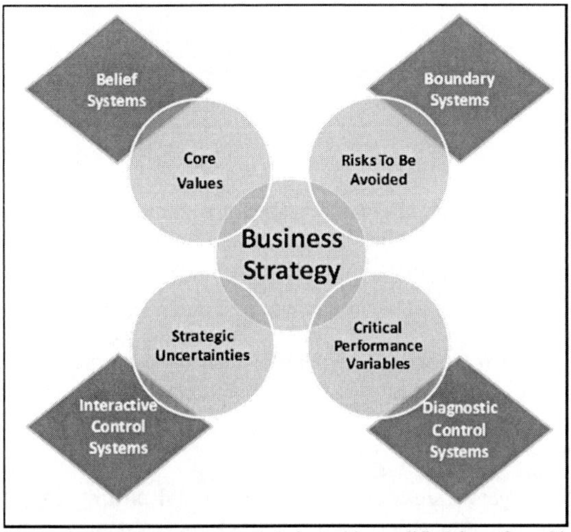

Figure 5: The LOC-Framework (adopted in modified form from Simons, 1995, p. 7)

Within an industry, the character and dimension of competition is generally influenced by five forces (Porter, 1979, 2008): *"Threat of Entry, Power of Suppliers, Power of Buyers, Threat of Substitutes and Rivalry Among Existing Competitors"*.

Based on their *Mission*, organizations apply *Strategies* to position their business within such a dynamic and competitive industrial environment. *Goals and Plans* are deduced from these strategies; *Performance Measures* are applied to analyze the extent and success to which the goals and plans are translated into *Action* (Simons, 2014, p. 42).

Strategies do not always follow this top-down hierarchy; they can also emerge inordinately:

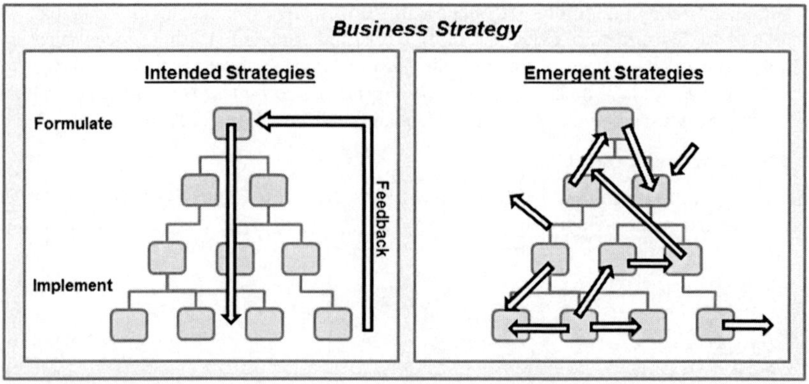

Figure 6: Balancing intended & emergent strategies (adopted in modified form from Simons, 1995, p. 20)

Emergent strategies arise unexpectedly and unanticipated – and are sometimes due to serendipity (Simons, 2014, p. 44). As shown in figure 6 intended and emergent strategies are not mutually exclusive – a business's strategy is built on both perspectives:

> "Likewise, there is no such thing as a purely deliberate strategy or a purely emergent one. No organization – not even the ones commanded by those ancient Greek generals – knows enough to work everything out in advance, to ignore learning on the route." (Mintzberg, 1987, p. 69)

As shown in figure 5, LOC consists of four control system areas, "levers" that form the overall MCS:

First, within the "*Belief Systems*" the organization defines its purpose and positive ideals including its core values, mission, credos and/or purpose statements. Belief systems are the transport vehicles for information regarding the organization's value creation process, the desired level of performance, as well as the desired behavior of the individuals within the organization (Simons, 1995, p. 34).

Second, managers implement "*Boundary Systems*" to set limits for their subordinates in order to guide them in those boundaries, within which they are supposed a) to create value for the organization, and b) to avoid risks. Therefore, and in contrast to the first lever, Boundary Systems have a more negative or proscriptive than positive peculiarity; an analogy can be found within the Ten Commandments from the Old Testament which are supposed to "*establish clear limits on behavior*" (Simons, 1995, p. 41). Besides the internal limits (like permitting doing business with a certain supplier e.g.), managers are also supposed to reflect external limits such as legislation within their business parameters (Simons, 1995, p. 42). From a practitioner's viewpoint it is important to understand what Simons (1995, p. 55) describes as "*Actions Speak Louder Than Words*": even the best formulated and thought beliefs and boundaries become worthless if they are only a nice sheet of framed paper hanging in some meeting rooms while being inconsistent with the actual actions performed by and within the organization[6].

[6] This holds true also for BPM: Melnyk et al. (2014) have found the paradox that in dynamic environments firms do change their strategy but do not necessarily adapt their existing BPM system to this new strategy.

Third, managers apply "*Diagnostic Control Systems*" (DCS) as formal information systems "*to monitor organizational outcomes and correct deviations from preset standards of performance*" (Simons, 1995, p. 59). As shown in figure 4, it requires three devices to enable a controlled process: a detector, an assessor, and an effector. These devices correspond to Simons' three DCS types. DCS and business strategies are linked with each other by "*Critical Performance Variables*" (Simons, 1995, p. 63) representing output variables which are related to performance dimensions of the given strategy. These variables include financial measures as well as non-financial indicators like customer satisfaction or employee satisfaction (Simons, 1995, p. 78). Following this concept, it becomes clear that the critical performance variables have to be changed or at least their significance has to be re-evaluated in case an organization plans to change its business strategies (Simons, 1995, p. 66). Management attention, motivation and incentives are additional supporting functions within DCS to enhance the overall success of the management control process (Simons, 1995, p. 72).

Fourth, managers employ "*Interactive Control Systems*" (ICS) in order to build "*internal pressure to break out of narrow search routines, stimulate opportunity-seeking, and encourage the emergence of new strategic initiatives*" (Simons, 1995, p. 93). It is especially the latter which is important: as shown in figure 6 emergent strategies may arise inordinately. In the event subordinates on lower levels have innovative ideas but are frustrated in making their way to the senior management, the chance for new strategies arising from such innovative ideas is rather low. Therefore ICS help to involve managers within these decision processes of their subordinates (Simons, 1995, p. 95) and to stimulate information sharing (Simons, 1995, p. 103). Whereas the information in DCS are directed to both past and present, in ICS it is directed to the present and future. This difference becomes vital in the sense of "*Strategic Uncertainties*", i.e. "*the uncertainties and contingencies that could threaten or invalidate the current strategy of the business*" (Simons, 1995, p. 94). Managers have to be aware of these contingency factors in order to a) safeguard the sustainable prosperity of their organization as well as, b) to find a balance between the flexibility needed to deal with these strategic uncertainties and the consistency needed to ensure an efficient standardization of operations (Malmi and Brown, 2008, p. 296).

The foundation for the reliability of and the resulting reliance on these four management control system categories are "*Internal Controls*" (Simons, 1995, p. 84): all of the organization's internal safeguards including "*structural safeguards*" (like the definition of authorization levels), "*staff safeguards*" (like expertise and training of finance employees), and "*system safeguards*" (like limited admittance to the organization's databases) are "*essential to ensure the integrity of the other systems that managers use to implement strategy*". Furthermore Internal Controls safeguard the organization's "*assets from theft or accidental loss*" (Simons, 1995, p. 181). Internal Controls can be seen as a crucial internal process-related support function which benefits the managers; however not the managers but the staff professionals like independent auditors are responsible for the installation and maintenance of Internal Controls (Simons, 1995, p. 85).

LOC unfolds its power as soon as all four levers are orchestrated together – interactively and diagnostically (Simons, 1995, p. 156; Widener, 2007, p. 782; Koufteros et al., 2014, p. 334):

On the one hand, "*Belief Systems*" and "*Interactive Control Systems*", as *positive* systems, are supposed to create intrinsic motivation inspiring employees to learn and share information. And likewise on the other, the *negative* control systems "*Boundary Systems*" and "*Diagnostic Control Systems*" are supposed to create extrinsic motivation by offering incentives for the achievement of objectives as well as "*delimiting the domain for opportunity-seeking*" (Simons, 1995, p. 158).

As a result of this management system orchestration, business strategy receives a dynamic character as the levers influence each other by having different viewpoints on the organization's strategy (Simons, 1995, p. 159):

- In Belief Systems, business strategy is perceived as the organization's "*Perspective*". Belief Systems aim for "*obtaining commitment to the grand purpose*".
- In Boundary Systems, business strategy is perceived as the organization's "*Position*" (in the market, in the industry, in the competitive environment, e.g.). Boundary Systems are "*staking out the territory*".
- In ICS, business strategy is perceived as "*Patterns in Action*". ICS help to position the organization "*for tomorrow*".
- In DCS, business strategy is perceived as the "*Plan*". DCS has an active role in "*getting the job done*".

LOC has been criticized as it is supposed to be designed for the senior management level and therefore would *only* apply to this group of recipients (Berry et al., 2009, p. 6). Berry et al. are right regarding Simons' usage of the term "*senior management*" (Simons, 1995, p. 214); however I doubt if Berry et al.'s viewpoint holds true: by relating LOC to "*the ongoing operations of* ***any*** *business*", Simons (1995, p. 10) directs his research not exclusively to large, hierarchically complex corporations but to *any* business, which include low-hierarchical small and medium sized enterprises as well. Furthermore Simons clearly distinguishes between a *"Corporate Strategy"* and *"Business Strategies"* within a 1:n-relationship (Simons, 2014, p. 27). As the LOC is centered on *business strategy*, the framework itself is directed to the management of *the business* rather than to its *corporation's senior management*.

Besides these and possibly other existing limitations (Tessier and Otley, 2012, p. 173), I have chosen the LOC framework as the leitmotif for my research for the following reasons:

- LOC it is an "*action-oriented theory of control"* grounded on ten-year research analyzing management practices of over 100 mainly innovative companies (Simons, 1994, p. 171; Simons, 1995, p. ix). Accordingly LOC has a more innovation than command-and-control orientation (Simons, 1995, p. 4). This holds especially true regarding emergent strategies, which are supposed to be an important factor for the research within the innovative MFI.
- LOC allows integrating informal as well as interactive controls and is balanced in the sense that it is not narrowly focused on accounting-figures (Anthony, 1965; Otley et al., 1995, p. 32).
- LOC depicts the organization holistically (Berry et al., 2009, p. 6) and is therefore predestined as a framework to observe all aspects of a company, including intangible aspects like strategic uncertainties in order to create integrated models of organizations.
- LOC is broad enough to allow the integration of further control systems that have not been regarded in the original mid-90's LOC. Gond et al. (2012) for example propose

the integration of *sustainability* within organizational strategy by linking "*Sustainability Control Systems*" to LOC.

- BPM can clearly be linked to the third LOC lever "Diagnostic Control Systems" including BPM information tools allowing managers to control and adjust the performance of their business.
- Finally LOC is an established and academically accepted MCS framework on which BPM research can be built (Franco-Santos et al., 2012, p. 81; Strauß and Zecher, 2013, p. 256).

2.4 Business Performance Measurement Framework

> "*Performance measurement* can be defined as the process of quantifying the efficiency and effectiveness of action." (Neely, 2005, p. 1229)

As shown in chapter 2.3.2 Business Performance Measurement (BPM) is a crucial component within all influential MCS frameworks. BPM also represents the research field and main orientation within my research project.

2.4.1 Business Performance Measurement – The Rationale Behind

In order to understand the difference between MCS and BPM it was important for me to understand the difference between "*Control*" and "*Controls*" as delineated by Drucker (1964, p. 286):

> "The synonyms for 'controls' are measurement and information. The synonym for 'control' is direction. [...] 'Controls' are analytical and operational, concerned with what was and is. 'Control' is normative, concerned with what ought to be, with significance rather than with meaning. [...] For, in the task of a manager, 'controls' are purely a means to an end; the end is 'control'."

As presented above MCS shall enable managers to *control* their business by using *controls* i.e. measures. These singular financial measures have evolved into a more complex *system* of hierarchical measures interrelating to one primary financial performance measure – a "*pyramid of financial ratios*" (Otley in Neely, 2011, p. 14). The origin of these *Financial Performance Measurement* (FPM) systems can be traced back to the DuPont company (Dale, 1957) in the early 1920s which applied a pyramid of financial ratios topped by the *Return of Investment* as pyramidion. This DuPont-Model (Van Voorhis, 1981) can be regarded as the blueprint for further FPM systems like the German "*ZVEI-Kennzahlensystem*" (Gleich, 2011, p. 10).

For decades these FPM systems were applied in order to analyze the overall performance of the firm (Neely and Bourne, 2000, p. 3) – a rather *imbalanced* process reflecting only financial, in time backwards oriented measures and neglecting non-financial indicators like employee satisfaction, response times, customer satisfaction (Kaplan and Norton, 1996, p. 44). Since the mid-1980s criticism was centered on this circumstance (Eccles, 1991); correspondingly Gleich (2011, p. 13) identified nine shortfalls related to FPM systems:

1. Purely financial systems are rather past than future oriented and therefore stimulate *re*acting rather than acting.
2. Focusing solely on internal aspects and neglecting external aspects (s. chapter 2.2.3) increases for instance the risk of missing important (market) developments.

3. Aggregated financial measures within pyramids of financial ratios do not reflect a finer degree of granularity and therefore miss information related to employees or processes.
4. Concentrating on reaching a global optimum (like the financial performance top measure in a pyramid of financial ratios) creates the risk to miss local optima and as a result to reduce the overall firm performance.
5. Highly aggregated FPM systems are failing to include "soft-facts" like the quality of processes or relevant information on competitors' behaviors.
6. Traditional control systems do not include early warning systems that allow spotting any uprising erroneous trend.
7. FPM systems lack the direct relation to the firm's strategy.
8. Incentives for Managers applying FPM systems are based on the reduction of gaps to planned or budgeted values rather than on continuous improvement processes.
9. Intangible values of the firm get neglected, especially the skills of the employees which could better be controlled via "human resource accounting".

In response to this criticism, BPM researchers have developed "*Contemporary Performance Measurement"* (Hall, 2008, p. 143) frameworks within in the past quarter century (Franco-Santos et al., 2012, p. 79) which benefited the transformation of FPM systems into BPM systems (Nanni et al., 1992).

2.4.2 Business Performance Measurement Research

Within the context of my dissertation I apply the term *Business Performance Measurement (BPM)* as a synonym for *Contemporary Performance Measurement* which includes improvements to the aforementioned FPM system shortfalls. The word *Business* is added to *Performance Measurement* throughout my dissertation in order to distinguish *Performance Measurement of the Firm* from the *Performance Measurement of Individuals* like the Human Resource Performance Measurement (HRPM) (Gawron et al., 1990; Feigh et al., 2011); the latter is a crucial part of the former in many BPM frameworks. It could be argued that BPM is always the measurement of a firm's outcomes performed by and related to its individuals (s. chapter 2.2.3); nevertheless HRPM is supposed to evaluate the performance of one individual in order to assess for example the individual degree of compliance or possible skill enhancements (Berman et al., 1999, p. 6) and not the firm as a whole.

Comparable with MCS (s. chapter 2.3.2) BPM lacks a generally accepted definition (Franco-Santos et al., 2007, p. 787; Lange, 2009, p. 13; Micheli and Mari, 2014, p. 147) as BPM research is influenced by a large number of scholars coming from a diversity of research fields, each reflecting a different perspective on the subject. As a result Neely (2005) argues that BPM is "*a widely distributed and relatively immature field of academic study, which has relatively little consensus about its core theoretical foundations*"; a statement which is generally held today (Koufteros et al., 2014, p. 313).

Nevertheless the following *BPM frameworks* have been influential in BPM research[7]:

- Tableau de Bord: The first formalization of this framework in the sense of a (management) control cockpit was made by the French *Centre de recherches et d'études des chefs d'entreprise* dating back to the mid-1950s (Pezet, 2009, p. 113).

[7] The selection of BPM frameworks was chosen accordingly to Watts and McNair-Connolly (2012, p. 228) what corresponds to Neely's (2011, p. 144) review on existing BPM frameworks.

Its use and functions get compared to Kaplan and Norton's (1992) Balanced Scorecard (Eppstein and Manzoni, 1997).

- The Performance Measurement Matrix: Keegan et al. (1989) have introduced a 2x2-matrix in which indicators are classified either being "cost" or "non-cost" and either being "internal" or "external" in order to reflect a balanced view on the firm.
- The Strategic Measurement And Reporting Technique (SMART) Pyramid: Lynch and Cross (1991, p. 64) developed this performance pyramid which shall deliver top management objectives top-down to subordinated organizational units using "*business operating systems*" (BOS – Lynch and Cross, 1991, p. 45). SMART includes financial as well as non-financial measures in order to control the internal effectiveness as well as external efficiency (Lynch and Cross, 1991, p. 79).
- The Results and Determinants Framework: Fitzgerald et al. (1991) have analyzed BPM in service companies. In their framework they classify six generic performance dimensions into the two groups "*results*" (i.e. competitiveness and financial performance) and "*determinants*" (i.e. quality of service, flexibility, resource utilization, and innovation) of these results (Fitzgerald et al., 1991, p. 8). Linking both groups leads to the consideration of causality in the firm and as a result, to the identification of performance drivers (Neely, 2011, p. 146).
- The Balanced Scorecard (BSC): Kaplan and Norton's (1992) BSC links measurement to the organization's vision. The firm's vision and strategy is observed from the four perspectives of finance, internal-business, innovation & learning and customer; for each perspective goals and measures are defined (Kaplan and Norton, 1992, p. 73). The BSC is a very popular framework and can be seen as the starting point for contemporary BPM research and applications (Bourne, 2008, p. 67; Franco-Santos et al., 2012, p. 82).
- The Input-Process-Output-Outcome Framework: Brown (1996) focuses on the firm's core *process* and divides this process into five sequential steps: inputs, processing system, outputs, outcomes, and goal. For each process step, measures are defined (Brown, 1996, p. 96). The strength of this framework lays in the differentiation of the steps *output* and *outcome* what was well received in the public sector (Neely, 2011, p. 147).
- The Strategic Balanced Scorecard: Kaplan and Norton (1996) have further developed their BSC framework by including targets and initiatives in each of the four BSC perspectives. Relating the BSC closer to the firm's strategy by incorporating cause-and-effect chains is supposed to extend the 1992 BSC version from "*a tactical or operational measurement system*" to "*a strategic management system*" (Kaplan and Norton, 1996, p. 10).
- The Performance Prism: Neely's BPM framework (Neely et al., 2001) includes five interrelated facets: stakeholder satisfaction, strategies, processes, capabilities and stakeholder contribution. For each perspective questions are defined asking for necessary measures. The strength of this framework lays in the stakeholder orientation requiring managers to identify their main stakeholders (internal as well as external) and to reflect their wants and needs in order to align their firm accordingly (Neely et al., 2001, p. 6).

Further *BPM models and concepts* have been developed either by a) firms, b) consultancies, c) associations/not-for-profit organizations or d) the public sector including[8]:

a) the Skandia Navigator, the Hewlett-Packard-Concept, the Caterpillar-Concept, the J. I. Case-Concept, the Festo Innovation Balanced Scorecard,
b) the Ernst & Young Performance Measurement-Concept, the Horváth & Partners Process Oriented Performance Measurement, the Quantum Performance Measurement-Concept (Arthur Andersen), the Business Management Window (consulting branch of IBM UK),
c) the EFQM Excellence Model (Wongrassamee et al., 2003), the SCOR-Model (Supply Chain Council; SCOR, 2010),
d) the DART-Concept (Dallas Area Rapid Transit) and the Performance Measurement Model (applied by the U. S. Army Materiel Command).

The BPM frameworks can be classified as *generic*, as they all share the characteristic "*that they contain a standardized core supplemented with a customizable part*" (Silsand and Ellingsen, 2014, p. 178). As a result firms as well as industry sectors have deduced, customized, adopted and implemented *BPM systems* from these *generic BPM frameworks* accordingly to their requirements (for the adoption and implementation of BSC for example see Speckbacher et al., 2003).

Franco-Santos et al. (2012) have analyzed 76 empirical studies related to BPM systems published in 15 high-quality academic journals in the time frame 1992 (the year of the publication of Kaplan and Norton's BSC framework) to 2011[9]. They claim that a BPM "*system exists if financial and non-financial performance measures are used to operationalize strategic objectives*" (Franco-Santos, et al., 2012, p. 80); following this definition the authors have categorized the analyzed studies according to four BPM system types presented in the following table:

	BPM-A	BPM-B	BPM-C	BPM-D
Components	◦ Financial and non-financial performance measures implicitly or explicitly linked to strategy	◦ Financial and non-financial performance measures explicitly linked to strategy ◦ With explicit cause-and-effect relationships among measures	◦ Financial and non-financial performance measures explicitly or implicitly linked to strategy	◦ Financial and non-financial performance measures explicitly or implicitly linked to strategy
Use / Purpose	◦ Inform decision-making ◦ Evaluate organizational performance	◦ Inform decision-making ◦ Evaluate organizational performance	◦ Inform decision-making ◦ Evaluate organizational and managerial performance (without links to monetary rewards)	◦ Inform decision-making ◦ Evaluate organizational and managerial performance ◦ Influence monetary rewards

Table 1: BPM System Types (adopted in modified form from Franco-Santos et al., 2012, p. 82)

Besides these characteristics, Hall (2008, p. 143) adds that BPM systems have to be dynamic in the sense that they have to respond on emergent circumstances across functional

[8] For further details of origin, history, benefits and drawbacks of these BPM models and concepts please refer to Gleich (2011, pp. 67-125).

[9] As a source for *International Business* Performance Measurement research please refer to Hult et al. (2008) who have examined the measurement of performance described within 96 articles published in high-quality academic journals within the time frame 1995-2005.

boundaries affecting the firm's strategy, especially in competitive environments (s. chapter 2.3.3).

Furthermore Franco-Santos et al. (2012, p. 97) argue in their findings that the relation between BPM systems and inter-firm performance has received little attention by the academic community, and that "*more research in this area is required especially given the importance of buyer–supplier relationships in our current business environment*". This finding is important within my research as the inter-firm performance is supposed to be a vital aspect within the MFI.

2.5 Synthesis of the Research Framework

Within this chapter I will synthesize the theories presented previously in order to model a theoretical framework on which my research project is grounded. Furthermore I found an evolving academic concern in the MCS and BPM literature indicating that a significant part of academic research is not aligned with managerial needs; in order to improve the situation related to this concern I have developed a concept to advance the feasibility of BPM system adaption and implementation.

2.5.1 Research Framework – BPM in the MFI

In order to synthesize the previous chapters I have modeled the following framework for my research project:

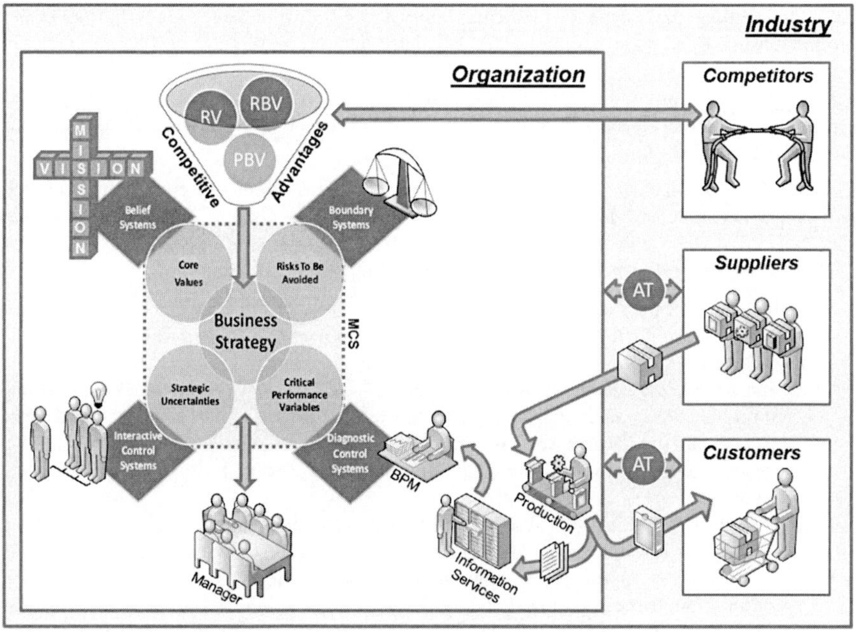

Figure 7: Research Framework – MCS & BPM in an Organizational Context

The organization is embedded within its industrial environment and therefore is influenced by the industry's characteristics (Porter, 1980, p. 30; Schmalensee, 1985, p. 349; McGahan and Porter, 1997, p. 29; Porter and Kramer, 2011, p. 67). It has one or more competitive advantages that can be explained using different theories of the firm like RBV, RV, or PBV (s. chapter 2.2). These competitive advantages are the foundation for the organization's strategy which embodies the heart of the MCS (represented by the LOC-MCS-framework being the leitmotiv of my research project; s. chapter 2.3.3). It is the managers' job to use their control systems in order to determine if the company is still on track fulfilling the business strategy. This means information receiving, information evaluation and feedback sending processes including:

- Controlling interactively for emergent ideas and innovation throughout the organization's hierarchy in order to respond to uncertainties.
- Controlling a suitable set of core values which are deduced from the organization's mission and vision.
- Controlling the verification that the organization operates within its boundaries.
- Controlling diagnostically if the performance of the organization is congruent with the targets and goals set.

BPM is applied in the DCS in order to evaluate the organization's performance. Exemplarily, the *production performance* is measured within figure 7; according to the organization's needs, this could include measures relating to quality (defect rates in PPM, e.g.), cycle-times (parts produced per time interval, e.g.), or monetary aspects (turnover per production head and time interval, e.g.). All these data are gathered via a communication network (s. figure 4) in information services (data-warehouses, ERP systems, e.g.). A person or a department (regularly the controlling department) or an automatic system processes these data in order to provide the results of the production BPM in the DCS for the managers' evaluation.

Regarding the organization's external environment within its industry context, the relation to its suppliers and customers becomes associated with AT (s. chapter 2.1). As both external stakeholders of the organization, suppliers as well as customers, are part of the production cycle, AT has a direct influence on the organization's BPM – two examples:

1. The customer (as the principal) knows, that the product prices of the organization (as the agent) are far too low compared to market prices due to an ex ante product cost miscalculation. In this case an information asymmetry exists benefitting the principal's self-interest. In case the agent performs ex post product cost calculations within his BPM system, he is supposed to find that the margin of the concerned order lies beneath the targets set. Accordingly the organization is supposed to claim a supplemental cost, raise prices for future orders, or end the relation with the customer.
2. The supplier (as the agent) knows that he cannot deliver the quality expected by the organization (as the principal). Nevertheless the agent delivers the cheap parts as input resources to the principal without telling him about the quality issues. The principal then processes the delivered input resources into his products causing the quality defect rate to rise and overrun a maximum target set within the BPM system. The organization is then supposed to claim warranties, get substitute resources in the demanded quality, or end the relationship with the supplier.

As a third external stakeholder party, the competitors have direct influence regarding the organization's competitive advantages. It is the managers' task to stimulate and to protect the organization's competitive advantages in order to sustain profitably on the market.

In order to reduce complexity within figure 7, further BPM processes like controlling the performance of the administrative department, the performance of individuals (HRPM) or the performance of the R&D department especially in innovative organizations are not presented although they certainly represent parts of BPM. For the same reason further industry classes are not presented, although an organization can have customers or suppliers from different industry branches (like automotive, aerospace, public sector, e.g.). In the same way further stakeholder groups like shareholders or higher-ranking corporate structures are not included, although they have undeniably important influences on the firm (for example *parent effects* – s. chapter 2.2.1). However the influence of such stakeholder groups is unlikely for the majority of MFI firms, which mainly consist of micro-, small- and medium-sized enterprises (SMEs) (s. chapter 3.5.1). Finally it is indisputable that a clear distinction between stakeholders groups is not always feasible as customers can simultaneously be suppliers; a circumstance that is typical for the MFI (s. chapter 3.2) but does not change the significance of AT associated to the organization's relation to its stakeholders.

My research framework follows Franco-Santos et al. (2012, p. 99) recommendation for the creation of a BPM meta-theory[10] including elements of different theories related to BPM. Based on this research framework I define BPM as follows:

Business Performance Measurement is an integral Management Control Systems element reflecting the success of an organization's ongoing strategy implementation.

2.5.2 MCS and BPM – Evolving Academic Concern

By reviewing MCS and BPM literature I could identify an evolving concern within the academic community:

- Anthony (1989, p.16) critically scrutinizes the use of theories, mentioning AT, in dissertations and in scholarly journal publications only for promoting academic careers rather than supporting "*those interested in improving practice*".
- Simons (1995, p. ix), regarding management literature, argues that *knowing* much about formulating winning strategies and building plans does not necessarily result in any practical contribution as these plans cannot be *implemented* in an organization.
- Bourne et al. (2003) call for further research regarding the success or failures of BPM implementation. They argue that although the *success stories* of BPM implementations are presented at business conferences and in related academic publications, not enough academic attention is paid to the BPM implementation *failure*. And the latter is of special interest for the practitioners.
- Nixon and Burns (2005, p. 261) have identified a *"perceived gap between practice and the management control literature"* based on fast developments within organizations that make aspects of traditional MCS literature antiquated.

[10] Following Franco-Santos et al. (2012, p. 97), this BPM meta-theory could include further theories not explicitly introduced within this dissertation like Contingency or Goal-Setting Theory (s. chapter 9.2).

- Merchant and Van der Stede (2006, p. 126) demand the creation of "*usable knowledge*": *"Practitioners are problem- and issue-focused, not paradigm- or discipline-focused."*
- Franco-Santos et al. (2007) argue that within their literature review of 300 documents they solely found 17 documents including definitions for BPM what makes it difficult for the reader of the other documents to understand what the researchers are investigating therein.
- Bourne (2008, p. 69) summarizes his concern by relating seven academic issues referring to the "*gap between academia and practice*":
    1. Academics tend to talk with each other or "*to themselves*" and therefore risk the practical relevance of their research.
    2. Being focused on details leads to refining existing frameworks within simplified models of the world which reduces the practical relevance of the research within the real-world.
    3. BPM and MCS involve cross-disciplinary actions in organizations; a circumstance that should be reflected by scholars in learning from colleagues from different academic disciplines.
    4. Research in medicine and engineering gives an answer on the question of "*how to do things*"; management research seems to lack this characteristic.
    5. Yet research has not sufficiently answered the question of the BPM benefits to organizations. Bourne relates this gap to quantitative research methods rather than to longitudinal research methods within a focused set of organizations.
    6. *"The incentives are wrong – Academic promotion is based on publishing in academic journals. We do not reward the link to practice in the same way."*
    7. Practice seldom emerges from academic research related to the *"relatively few academics wrestling with the problems of business compared to the millions of practitioners"*.
- Neely (2011, p. 76) argues that despite the existing BPM research the academic community knows *"relatively little about what information managers use and how they use it"*.
- Gleich (2011, p. 330) states that practitioners are only sporadically and at most generally supported by the academic community with advice for a successful and sustainable BPM system implementation.
- Nudurupati et al. (2011) contend that existing BPM systems are out of date and do not respond to the dynamic business world (an issue that can be related to the aforementioned Nixon and Burns criticism). Furthermore only a few BPM systems include Management Information Systems (s. *Information Services* in figure 7) that are supposed to enable efficient reporting processes.
- Micheli and Mari (2014, p. 147) argue that practitioners as well as academics *"suffer from an underdeveloped conceptualization of the notion of performance measurability, and of the derived measurement processes"*.

- Melnyk et al. (2014) find evidence for Bourne's third academic issue that BPM "*needs to be researched in different contexts*". They also argue that research has to be reinforced regarding strategy design processes and the linkage of strategy to BPM (s. chapter 2.3.3).

The tenor of these selected criticisms is that a significant part of academic research has no practical relevance as it is not aligned with managerial needs.

This finding is not new and has been addressed already 50 years ago (Drucker in Bonini et al., 1964, p. 289; s. chapter 1). However, recent research has not resolved the issue and linking theoretical concepts to practical applications is not a trivial issue. Grosswiele et al. (2013) for example introduce a decision framework for BPM systems. By triangulating their conceptual model in a "*real-world*" setting they faced several challenges like limited data access and suboptimal data quality with regard to different time scales as well as time lags of available measures. As one limitation and aspect for future research they recommend improving the practical application of their framework's input parameters by involving industry experts (Grosswiele et al., 2013, p. 1028).

The British Academy of Management (BAM) tried to bridge the gap between management science and practice by promoting the *Mode 2 Knowledge Production* (M2K) research policy (Gibbons et al., 1994) as an advance in the future of management research (Tranfield and Starkey, 1998). The rationale of M2K research policy is based on focusing on the practical research relevance and to connect researchers and practitioners in interdisciplinary joint research projects (Starkey and Madan, 2001). However, BAM's attempt to promote a M2K research policy in the UK took place in the end of the last millennium – but the above delineated concern in MCS and BPM literature, including the internationally renowned UK BPM research community with Neely and Bourne, mainly evolved thereafter. In consequence the question arises if adopting M2K research designs would be an appropriate approach to satisfy this evolving concern; Bartunek (2011, p. 557) has analyzed the status of M2K and states that M2K is stuck in a phase of discussion between relevance, rigor and research modes instead of being actively applied in management science:

> "It seems to me that a good deal of what started out as discussions of promising ways of bridge-building in the BJM special issue [referring to the 2001 special issue including the Starkey and Madan study] have become mired in deliberations about how to get started with the work, or if it is even feasible. [...], but I suggest that it may be worthwhile *to do some empirical data collection.*"

I will pick up Bartunek's recommendation for M2K empirical data collection within my research setting section (s. chapter 6.1.1).

2.5.3 Theoretical Conceptualization of Feasibility

In this chapter[11] of my theoretical research framework I focus on the earlier introduced missing practical relevance of generic BPM frameworks and models. The question remains as to how to overcome the gap between academia, represented by generic BPM frameworks within a MCS context, and practice, represented by the implementation and application of these frameworks and systems in a business context.

[11] This chapter is based on the British Academy of Management conference contribution "The Challenge of the Adaption and Implementation of Management Control Systems: A Bottom-Up Approach to Improve Feasibility" by Dietrich and Wald (2014).

This gap may be aggravated by the fact that generic frameworks and models have to be collapsed and spread to the needed criteria and characteristics of an industry – as it is the nature of generic frameworks and models to abstract and condense. For instance the Skandia-Navigator BPM model (Edvinsson, 1997) primarily reflects financial as well as intellectual capital dimensions (Gleich, 2011, p. 119) and is therefore not generally the first choice for manufacturing companies. But the Skandia-Navigator includes the "Human Focus" BPM element (Mouritsen et al., 2001, p. 405) that principally fits manufacturing businesses.

From a practitioner's perspective three questions remain: what are the generic BPM frameworks and models from which to choose? Which elements of those BPM frameworks and models shall then be chosen to model my individual BPM system? And how shall they be adapted to the context at hand? Choosing between generic BPM frameworks and models bears the risk of creating an information overload and possibly inhibit rather than enable a successful BPM system implementation (Bawden and Robinson, 2009, p. 182).

I propose a different perspective: an organization is embedded in and influenced by its industry (s. chapter 2.5.1); more specifically, the challenge is to fit generic BPM frameworks to the BPM requirements of firms within their industry class boundaries (Simons, 1995, p. 39). In accordance with Porter's "*Updated Version of the Industrial Organization Paradigm*" (1981, p. 616) the interrelation between a firm's strategy apropos its industry structure is vital for the firm's performance:

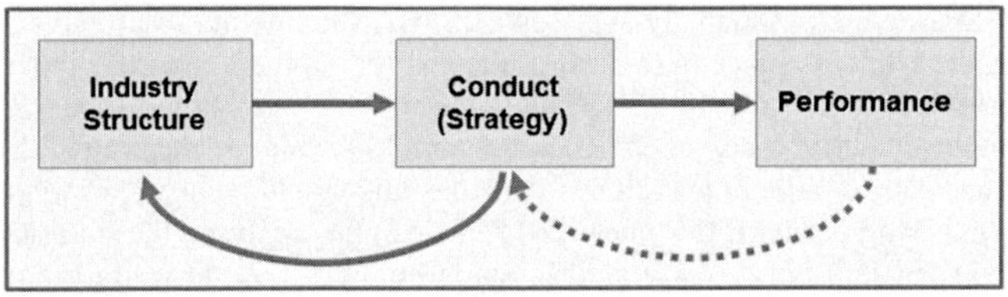

Figure 8: The Industrial Organization Paradigm (adopted in modified form from Porter, 1981, p. 616)

Interestingly generic MCS and BPM frameworks (s. chapters 2.3 and 2.4) are regularly designed focusing solely on the organization as unit of analysis; *top-down* approaches are applied starting at the management level of organizations in order to structure the control and measurement systems according to the organization's hierarchical structures (s. SMART Pyramid in chapter 2.4.2). Including external stakeholder groups is a novelty in BPM research (s. Performance Prism in chapter 2.4.2) but industry characteristics are regularly neglected and are therefore a missing dimension.

I assume that a MCS & BPM system that suits specific business requirements within an industry class will make a successful implementation (Neely and Bourne, 2000) and use more likely and therefore enhance the practical relevance:

> "The first step in structural analysis is an assessment of the competitive environment in which the company operates [...]. The success of a company's competitive strategy depends on how it relates to its environment. [...], the key aspect of the company's environment is the industry or industries in which it operates" (Porter, 1980, p. 30)

Therefore and in contrast to the design of generic MCS & BPM frameworks, I pursue a *bottom-up* approach in order to design an industry class BPM system:

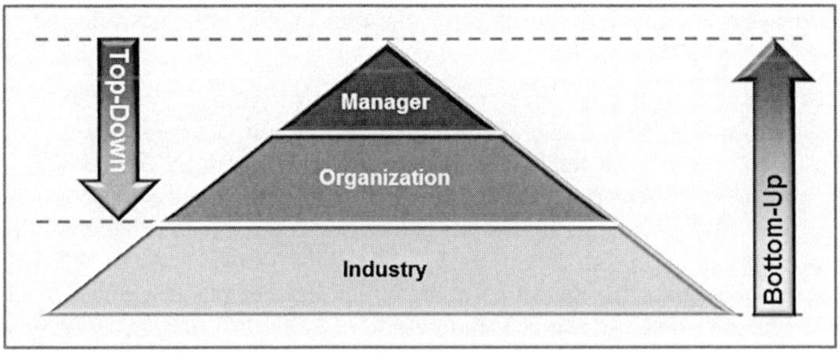

Figure 9: Top-Down vs. Bottom-Up approach in BPM system design

This bottom-up approach starts with the specific characteristics of a single industry. Generic BPM framework, model, or system elements reflecting those characteristics are then joined to an individual BPM system:

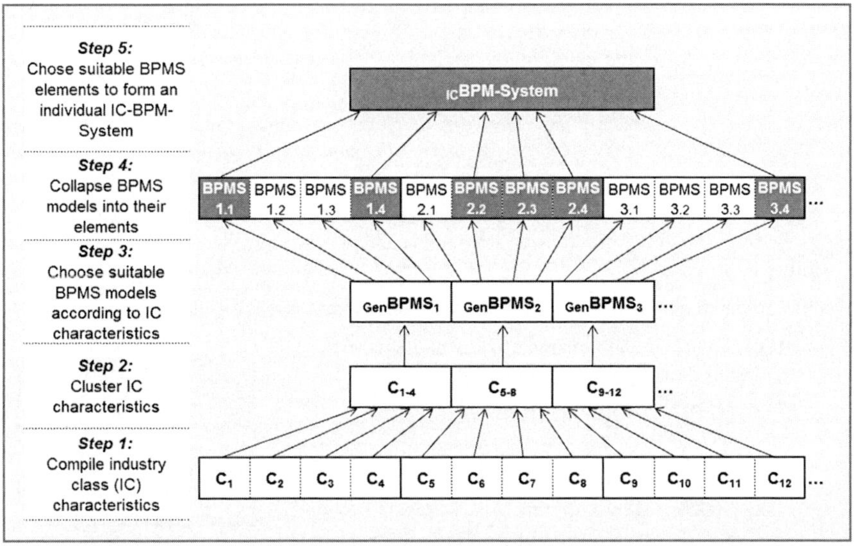

Figure 10: Chalice-shaped bottom-up approach to improve BPM feasibility

The first step in this chalice-shaped concept (s. figure 10) is the compilation of industry class characteristics in order to reflect the industry class as congruent as possible. In a second step those characteristics are bundled to clusters forming the overall character of the industry class. The adaption comes in the following steps from the selection of generic BPM frameworks, models, or systems being suitable to this bottom-up created characteristics clusters. Selecting appropriate elements from those chosen generic BPM frameworks, models, or systems will result in an applicable industry class BPM system. This approach may also result in a combination of different elements of several generic frameworks, models, or systems. The dots to the right in figure 10 symbolize that each step consists of an unlimited number of elements. Furthermore step 4 is open to include further BPM system elements in order to allow the adaption of industry class specialties not yet reflected in generic BPM systems (Gleich, 2011, p. 325).

Note that figure 10 symbolizes the *research process,* whereas figure 7 symbolizes the *research framework* for an MFI-BPM model. Within my thesis I will describe in detail the adaption of this research process, especially regarding the compilation and clustering of industry characteristics, as corresponding results are the foundation for the selection of appropriate generic BPM model elements for the MFI. This detailed approach shall result in a blueprint for the adaption of the chalice-shaped concept for further industry classes: besides the BPM recommendations for the MFI that are in the focus of this research project, I will contribute with this approach to the academic community by offering an alternative research process as one answer to the evolving academic MCS and BPM concerns. The first step in this detailed approach is the adaption of the BPM in the MFI centered research questions according to the research process following the chalice-shaped concept.

2.5.4 Research Questions related to the Metal Finishing Industry

As presented in chapter 1.1 the problem on which my research centers is related to a missing Business Performance Measurement System for the Metal Finishing Industry; accordingly I defined the guiding research question within my dissertation project as follows:

The research is focused on the 'How' metal finishing companies of larger scale are measuring their business performance, also in comparison to other manufacturing companies.

I deduced the following research questions related to the MFI from the guiding research question according to my previously introduced chalice-shaped bottom-up approach to BPM feasibility improvement.

The first research question cluster RQ.1 is related to steps 1 and 2 shown in figure 10:

- **RQ.1:** What are the characteristics and specifics of the Metal Finishing Industry (MFI)?
  - ***RQ.1a:*** How can the MFI as an industry class be identified?
  - ***RQ.1b:*** How can "metal finishing companies of larger scale" be defined?
  - ***RQ.1c:*** How can the MFI be described and differentiated from other industry classes?

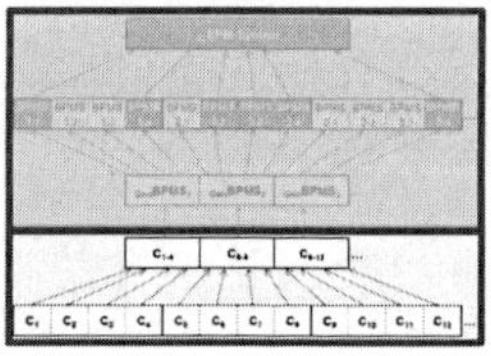

RQ.1a and *RQ.1b* will be processed within chapter 3 in which I analyze international nomenclatures of industry classifications, statistical databases as well as literature related to the MFI. Chapter 3.6 presents first insights related to *RQ.1c*; within my empiric research I will establish an MFI Cause & Effect Model (s. chapter 6.6) in order to give further, more detailed answers to *RQ.1c*.

The second research question cluster RQ.2 is related to steps 3, 4 and 5 shown in figure 10:

- **RQ.2:** Which generic MCS/BPM-models can be chosen and/or combined in order to create an integrated and sustainable Industry Class Business Performance Measurement System for the MFI?
  - ***RQ.2a:*** What are suitable generic BPM models based on the MFI characteristics?
  - ***RQ.2b:*** Which elements of generic BPM models can be combined in order to create an MFI-BPM model?

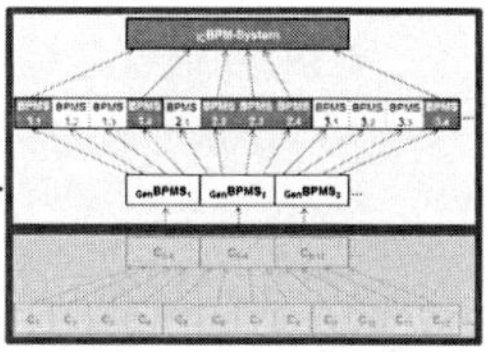

RQ.2a and RQ.2b will be processed in chapter 8 in order to give recommendations for the adaption and implementation of a Business Performance Measurement System for the Metal Finishing Industry.

3 Metal Finishing Industry

The Metal Finishing Industry (MFI)[12] is the target industry for which a BPM system shall be developed. The general MFI characteristics will be described in chapter 3 and in-depth analyzed within the subsequent empirical research part following the research process presented in figure 10.

Chapter 3.1 includes a general classification of the Metal Finishing Industry (MFI). Chapter 3.2 analyzes contract processing and outsourcing activities as major characteristics of the MFI. Chapter 3.3 illustrates the MFIs environmental impact on basis of the US EPA. Chapters 3.4 and 3.5 include an analysis of the MFI based on international statistical data. Chapter 3.6 concludes with the presentation of the main MFI industry characteristics clusters retrieved in this chapter (s. step 2 in figure 10).

3.1 General Classification

> "Broadly speaking, *metal finishing* is the process of coating an object with one or more layers of metal so as to improve its wear and corrosion resistance, control friction, impart new physical properties or dimensions, and/or alter its appearance. Applications range from jewelry, to common hardware items and automotive parts, to communications equipment and aerospace technologies." (EPA, 2004, p. 31)

According to the UK Department of the Environment (DOE, 1995, p. 1) the earliest form of metal plating can be traced back to the "*decorative gilding of ecclesiastical or domestic vessels during the Middle Ages*".

Today the term *metal finishing* is globally applied as a synonym for the treatment and coating of metals (Lee, 1977, p. 3; Industry Canada, 2000, p. 3; Larson, 2012a). Within the MFI, metal surface engineering and enhancement processes are performed by *autonomous job shops* on a contract or fee basis (UN, 2008, p. 125; EC, 2008, p. 161) or by *captive in-house* metal finishing departments of larger manufacturing companies (Larson, 2012a, p. 184), like the paint shop departments of automotive manufacturers.

The metal surface engineering and enhancement processes are mainly aiming for the improvement of wear resistance, corrosion protection, performance enhancements, or appearance improvements of the metal substrates on which the processes are applied (s. also Process-Based View in chapter 2.2.3); in addition metal coatings allow the combination of rather cheap base material like steel with the characteristics of an more expensive material like titanium without producing the whole component fully out of this expensive material (Matthews et al., 1992, p. 25).

Figure 11 shows an example for a steel drill bit which was coated with a titanium coat layer using a Physical Vapor Deposition (PVD – Matthews et al., 1992, p. 57) coating process in order to improve the wear resistance of this tool; main advantages are a) the cost avoidance compared to using solely titanium for the drill bit and b) the cost effectiveness as the drill bit has an improved endurance due to its enhanced wear resistance. Further advantage is an enriched appearance of the drill bit as the metal finishing process results in a gold colored coat layer. The golden look signalizes the drill bit customer a higher quality grade compared to regular grey steel drill bits.

[12] In German "Oberflächenveredelungsindustrie".

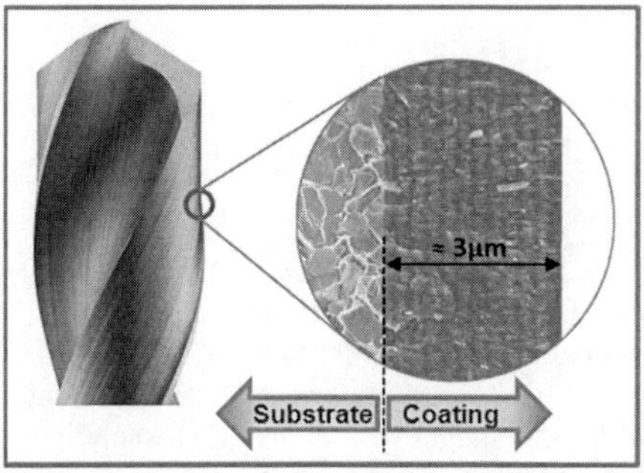

Figure 11: Example of a Metal Finished Drill Bit – PVD Process[13]

Main metal finishing processes[14] include (EPA, 1992, p. 5; s. chapter 3.3):

- Paint finishing (including wet and powder painting)
- *C*hemical / *P*hysical *V*apor *D*eposition (CVD / PVD)
- Electro plating (including chrome plating) and anodizing
- Galvanization (including hot-dip galvanization, phosphate conversion)
- Heat treatment (including nitriding)
- Thermal spraying (including plasma spraying)

The following figure shows the distribution of MFI processes applied in Germany in 2005 according to the German Central Association of Surface Treatment Professionals (ZVO):

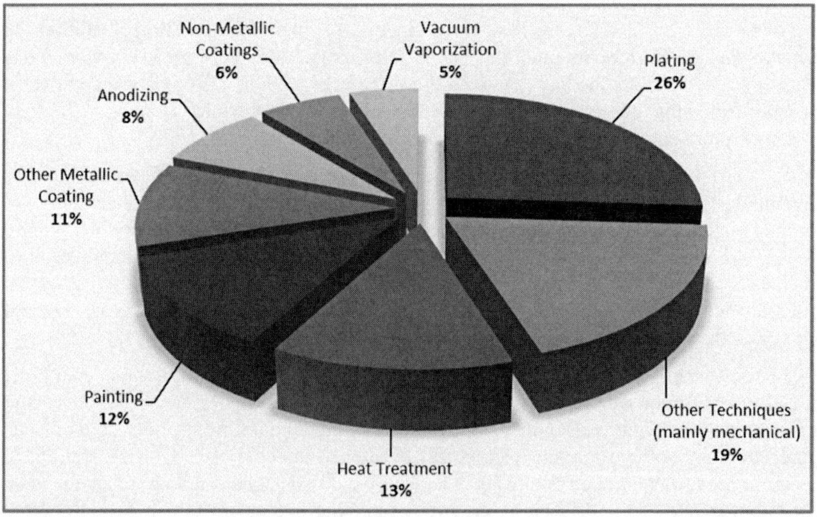

Figure 12: Distribution of Surface Finishing Processes in Germany in 2005 (adopted in modified form from König, 2010, p. 67)

---

[13] Photo material was provided courtesy of MeFiCom (s. chapter 6.1.2).

[14] Please refer for example to Matthews et al. (1992, p. 26-61) for detailed descriptions of metal finishing processes.

Figure 12 shows plating as the highest share of all MFI processes followed by heat treatment (hardening processes, e.g.) and paint processes. CVD and PVD processes are parts of the vacuum vaporization portion. As shown by Larson (2012b, p. 236), the distribution of MFI processes varies between the different European countries.

MFI pre-, post- and coating-processes regularly require a) high amounts of energy as many processes are executed under high temperatures, as well as b) hazardous chemicals (for example used in plating processes). These environmentally negative characteristics brought the MFI into the focus of the US Environmental Protection Agency (EPA) in the mid-1990s (s. chapter 3.3).

Customers of these metal finishing processes, within a business-to-business context, belong to different industry branches (Artley and Matthews, 2010, p. 27-86) such as:

- Automotive industry (substrates are automotive components like piston pins, plunger, as well as car design parts and whole car body parts, e. g)
- Tool manufacturing industry (substrates are drill bits, milling cutter, gun drills, saw blades, e.g.)
- Mechanical engineering industry (substrates are gear parts or bearing parts, e.g.)
- Aerospace industry (substrates are high-duty aircraft engine parts, e.g.)
- Medical industry (substrates are precision components like trepans or surgical instruments, e.g.)
- Military supply industry (substrates are firearm components, e.g.)
- Consumer retail industry (substrates are for example, decoration elements like special colored bathroom fittings or jewelry, e.g.)
- Construction industry (substrates are for example, special construction steels which have to be protected against corrosion)

Regarding the variety of customer industries, finishing processes, and substrates, it is apparent that the MFI is versatile; a main reason for the high share of micro-, small- and medium-sized enterprises (SMEs) in the MFI (Larson, 2012a, p. 183) usually specializing in one metal finishing process and being located closely to their main customers. This job shop characteristic of the MFI (Brummet and Arnett, 1967, p. 5; Flannery and May, 2000, p. 650) will be further analyzed in chapter 3.5.

Bell (2000) has composed a surface engineering road map in order to illustrate the above shown interrelations of different MFI processes, customer industries, substrates, and surface enhancement advantages:

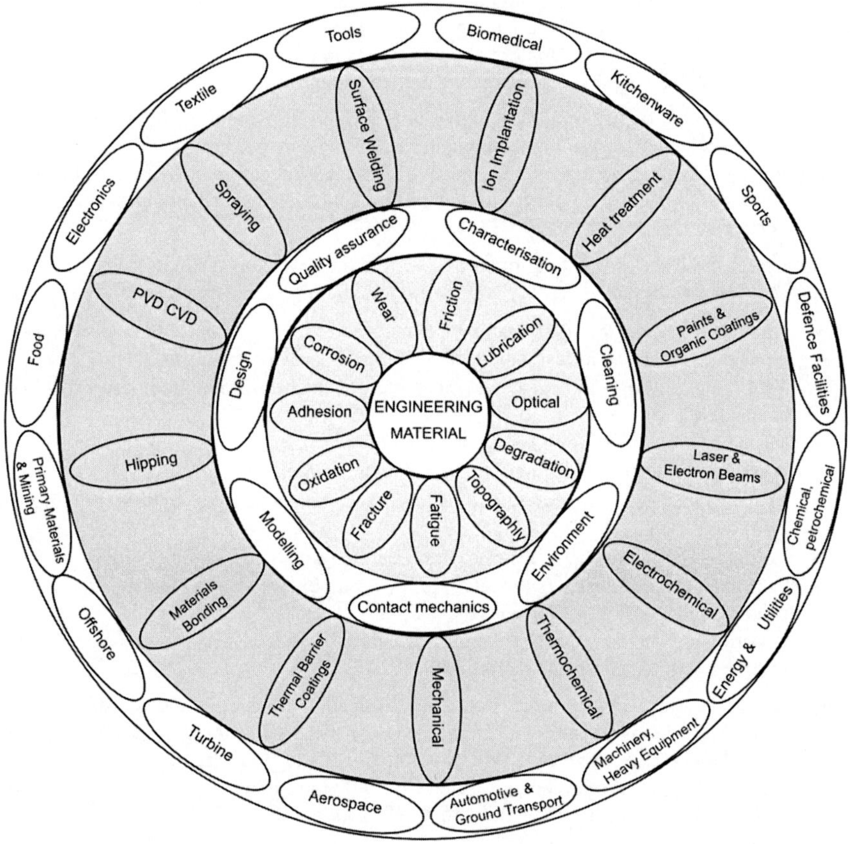

Figure 13: The Surface Engineering Roadmap (adopted in modified form from Bell, 2000, p. 89)

The road map centers on the engineering of materials which is surrounded by different motivations for which the MFI processes are applied. The suitable kind of MFI process has to be chosen according to the substrate properties and customer requirements shown next to the motivations. The next level shows several types of MFI processes and the outer ring a selection of different industry branches which benefit from the MFI.

3.2 Contract Processing and Outsourcing

The metal substrates (or raw products) on which the metal finishing processes are applied are typically provided to the Metal Finishing Company (MFC) by its customers. Therefore MFCs are more of an industrial service than a classical manufacturing nature (Brummet and Arnett, 1967, p. 5). Moreover the MFC and its customers are the two elements of a symbiotic relationship (Watkins, 1998, p. 99) as it is the combination of the surface enhancement *and* the substrate being a functionally graded system that allows a better cost effective performance compared to the performance of each element on its own (IOM3, 2000, p. 1).

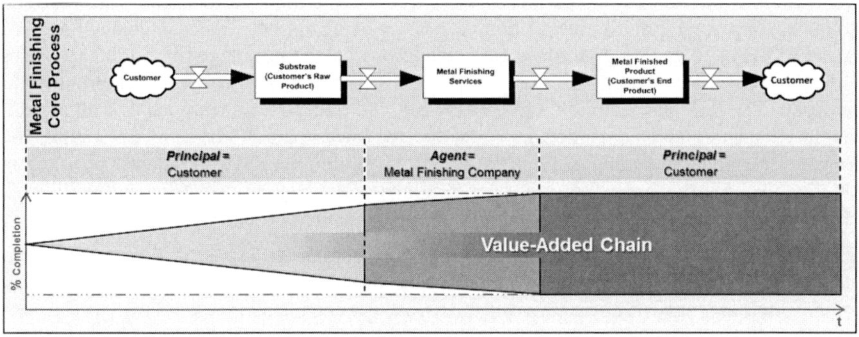

Figure 14: Metal Finishing Core Process linked to AT and the Values Added Chain

Figure 14 introduces the metal finishing core process[15] in which the customer as principal (s. chapter 2.1) delivers his raw product to the MFC as agent and subcontractor. The MFC applies then its metal finishing services in order to deliver the metal finished product back to the customer. The metal-finished product is regularly equivalent to the principal's end product as the metal finishing process is commonly the final production step within the value-added chain (Brummet and Arnett, 1967, p. 7). In other words, the largest share of the product's value was created *before* metal finishing processes take place.

Importance has to be laid therefore to the inherent property rights of the product which are regularly not transferred from the principal to the agent: both principal and agent should agree on terms and conditions in case the product gets damaged or destroyed during the metal finishing service including for example, warranties or allowed scrap rates.

As described in the previous chapter the metal finishing processes and services are predominantly performed on a contract or fee basis and can therefore be declared as contract processing and be related to outsourcing (Gray et al., 2009, p. 489; Hypko et al., 2010, p. 463; Selviaridis and Norrman, 2014, p. 155).

UN ISIC rev. 4 as well as EC NACE rev. 2 (s. chapter 3.4) relate the MFI to outsourcing processes (UN, 2008, p. 29; EC, 2008, p. 30):

NACE declares outsourcing as a *"contractual agreement according to which the principal requires the contractor to carry out specific tasks, such as parts of a production process or even the full production process, employment services or support functions"*. In this economic construct the principal is the *"unit that enters into a contractual relationship with another unit (contractor) for that other unit to carry out specific tasks, such as parts of a production process or even the full production process* [...]*"*. The contractor is the *"unit that carries out specific tasks, such as parts of a production process or even the full production process* [...] *on a contractual relationship with a principal"*.

[15] Please refer to chapter 6.6 for the system dynamics notation of the metal finishing core process.

3.3 Environmental Performance – US EPA Case

The MFI was one out of twelve sectors within the Sector Strategies Program (SSP) of the National Center for Environmental Innovation, a part of the US Environmental Protection Agency (EPA). It was the goal of the SSP to improve the environmental performance of US industry sectors acknowledging their economic contribution but also their environmental footprint. The SSP started in 2003 in cooperation with industry associations as well as with single companies and concluded in 2009; the MFI was part of the SSP until 2006 (R. Benson[16], personal communication, May 1st, 2013). Within the SSP the industry profile was created, the environmental footprint over a decade was analyzed and the opportunities for environmental improvements were identified for each industry sector (EPA, 2004, p. ii).

The MFI was already in the focus of the EPA even before the SSP started (s. for example EPA, 1982, 1984, and 1992) due to the fact that the majority of metal finishing processes require high amounts of energy as well as hazardous chemicals (s. chapter 3.1). The SSP identified three main opportunities for environmental improvements within the MFI (EPA, 2006, p. 55):

- managing and minimizing toxins and waste
- reducing air emissions
- conserving water

Improvements in these three areas would undoubtedly have a direct impact on the business performance of an MFC: in the US, all three areas are regulated by legislation (EPA, 1984). Breaching the limits set within those regulations could result in severe consequences for an MFC such as:

- the denial or the revocation of the MFC's operating license
- surcharges assessed
- reputational damage or
- other, criminal proceedings, related to the severity of the legislation violation.

All these consequences have severe implications for any business. Corresponding environmental indicators (Hassini et al., 2012, p. 77) should therefore be reflected in an environmental performance measurement section within an MFI-BPM model. However it is important to find an appropriate set of environmental indicators and management practices. Sarkis (2006) has analyzed the adoption of environmental and risk management practices in the MFI and found that increasingly investing in an ample set of such practices is not positively correlated with performance enhancements.

Finding such a set of indicators is also vital in order to supply the MFC managers with the relevant information to support their environmental decision making. MFC managers do not limit their environmental decisions on the above mentioned legislations or norms, but also on apparent personal characteristics of management (s. chapter 2.3.3) leading to ethical decisions as shown by Flannery and May (2000, p. 656).

3.4 International Nomenclatures of Economic Activities

This chapter presents the four most common worldwide nomenclatures and classification system types of economic activities (EC, 2008, p. 45) each starting with a general

[16] Mr. Robert Benson is the senior advisor for aquatic ecosystem strategies of the US EPA and was the director of the Sector Strategies Program.

introduction of the system followed by a presentation of the MFI specific classification within each. The chapter concludes with an overview of the MFI classification within the Group of Twenty countries.

As a result, this chapter shall answer research question RQ.1a (s. chapter 2.5.4): *How can the MFI as an industry class be identified?*

United Nations - ISIC

The International Standard Industrial Classification of All Economic Activities (ISIC) is founded on the UN resolution 149 A (VII) of August 27th, 1948 (UN, 2008, p. ix) which *"recommends that all Member Governments make use of the International Standard Industrial Classification of Economic Activities either by:*

> *(a) Adopting this system of classification as a national standard, or*
>
> *(b) Rearranging their statistical data in accordance with this system for purposes of international comparability".*

Therefore ISIC, today used in revision number four (UN, 2008), has a predominant importance within all international nomenclatures of economic activities.

The MFI is identified within ISIC rev. 4 as follows (UN, 2008, p. 49):

	Section	Division	Group	Class	Description
UN ISIC rev. 4	C				MANUFACTURING
		25			Manufacture of fabricated metal products, except machinery and equipment
			251		Manufacture of structural metal products, tanks, reservoirs and steam generators
			252	2520	Manufacture of weapons and ammunition
			259		Manufacture of other fabricated metal products; metalworking service activities
				2591	Forging, pressing, stamping and roll-forming of metal; powder metallurgy
				2592	Treatment and coating of metals; machining
				2593	Manufacture of cutlery, hand tools and general hardware
				2599	Manufacture of other fabricated metal products n.e.c.

Table 2: MFI identified in the UN ISIC rev. 4

According to ISIC rev. 4 class 2592 includes (UN, 2008, p. 126):

- plating, anodizing etc. of metals
- heat treatment of metals
- deburring, sandblasting, tumbling, cleaning of metals
- colouring and engraving of metals
- non-metallic coating of metals: plasticizing, enamelling, lacquering etc.
- hardening, buffing of metals
- boring, turning, milling, eroding, planing, lapping, broaching, levelling, sawing, grinding, sharpening, polishing, welding, splicing etc. of metalwork pieces
- cutting of and writing on metals by means of laser beams

ISIC rev. 4 class 2592 excludes:

- activities of farriers (class 0162)
- rolling precious metals onto base metals or other metals (class 2420)

European Communities - NACE

The General Industrial Classification of Economic Activities within the European Communities (NACE – Nomenclature générale des Activités économiques dans les Communautés Européennes) dates back to 1970 (EC, 2008, p. 13).

NACE, today used in revision number two (EC, 2008), is a *derived classification* of ISIC (UN, 2008, p. 37; EC, 2008, p. 39). The main difference of this *derived* compared to a *related classification* regarding NACE lies in a more detailed classification of economic activities, splitting the ISIC classes into more detailed NACE classes while sharing the same higher level data (EC, 2008, p. 14).

The MFI is identified within NACE rev. 2 as follows (EC, 2008, p. 67):

	Section	Division	Group	Class	Description
EU NACE rev. 2	C				MANUFACTURING
		25			Manufacture of fabricated metal products, except machinery and equipment
			25.1		Manufacture of structural metal products
			25.2		Manufacture of tanks, reservoirs and containers of metal
			25.3		Manufacture of steam generators, except central heating hot water boilers
			25.4		Manufacture of weapons and ammunition
			25.5		Forging, pressing, stampingand roll-forming of metal; powder metallurgy
			25.6		Treatment and coating of metals; machining
				25.61	Treatment and coating of metals
				25.62	Machining
			25.7		Manufacture of cutlery, tools and general hardware
			25.9		Manufacture of other fabricated metal products

Table 3: MFI identified in the EU NACE rev. 2

According to NACE rev. 2 group 25.6 refers to the following:

> "This group includes general activities for the treatment of metal, such as plating, coating, engraving, boring, polishing, welding etc., which are typically carried out on a fee or contract basis." (EC, 2008, p. 161)

NACE rev. 2 class 25.61 includes (EC, 2008, p. 161):

- plating, anodizing etc. of metals
- heat treatment of metals
- deburring, sandblasting, tumbling, cleaning of metals
- colouring, engraving of metals
- non-metallic coating of metals: plasticising, enamelling, lacquering etc.
- hardening, buffing of metals

NACE rev. 2 class 25.61 excludes:

- activities of farriers (class 01.62)
- printing onto metals (class 18.12)
- metal coating of plastics (class 22.29)
- rolling precious metals onto base metals or other metals (classes 24.41, 24.42, 24.43, 24.44)
- "while-you-wait" engraving services (class 95.29)

North America - NAICS

The North American Industry Classification System (NAICS) (Canada, 2012) is a *related classification* of ISIC[17] (UN, 2008, p. 37; EC, 2008, p. 41).

[17] The United States Census Bureau published a downloadable Microsoft Excel table in which the NAICS 2012 classes are translated into ISIC rev. 4 classes as NAICS has a completely different structure as ISIC or NACE (USCB, 2014a).

NAICS, today used in the 2012 edition, replaced the formerly used Standard Industrial Classification (SIC) system in 1997 and was jointly developed by Canada, USA and Mexico (Canada, 2012, p. 10).

The MFI is identified within NAICS 2012 as follows (USCB, 2014b, p. 156):

	NAICS Sector	NAICS Code	NAICS Title
North America NAICS 2012	31-33		MANUFACTURING
		332	Fabricated Metal Product Manufacturing
		3328	Coating, Engraving, Heat Treating, and Allied Activities
		33281	Coating, Engraving, Heat Treating, and Allied Activities
		332811	Metal Heat Treating
		332812	Metal Coating, Engraving (except Jewelry and Silverware), and Allied Services to Manufacturers
		332813	Electroplating, Plating, Polishing, Anodizing, and Coloring

Table 4: MFI identified in the North American NAICS 2012

According to NAICS 2012 code 33281 refers to the following:

> "This industry comprises establishments primarily engaged in one or more of the following: (1) heat treating metals and metal products; (2) enameling, lacquering, and varnishing metals and metal products; (3) hot dip galvanizing metals and metal products; (4) engraving, chasing, or etching metals and metal products (except jewelry; personal goods carried on or about the person, such as compacts and cigarette cases; precious metal products (except precious plated flatware and other plated ware); and printing plates); (5) powder coating metals and metal products; (6) electroplating, plating, anodizing, coloring, and finishing metals and metal products; and (7) providing other metal surfacing services for the trade. Establishments in this industry coat, engrave, and heat treat metals and metal formed products fabricated elsewhere." (USCB, 2014b, p. 156)

Australia and New Zealand - ANZSIC

The Australian and New Zealand Standard Industrial Classification (ANZSIC) (Australia, 2013) was developed in 1993 and is a *related classification* of ISIC (UN, 2008, p. 38; EC, 2008, p. 45).

Today ANZSIC is used in the 2006 Revision 2nd edition and was jointly developed by Australia and New Zealand (Australia, 2013, p. 7).

The MFI is identified within ANZSIC 2006 rev. 2 as follows (Australia, 2006, p. 173):

	Level 1	Level 2	Level 3	Level 4	Description
Australis / New Zealand ANZIC 2006 rev. 2	C				MANUFACTURING
		C22			Fabricated Metal Product Manufacturing
			C221		Iron and Steel Forging
			C222		Structural Metal Product Manufacturing
			C223		Metal Container Manufacturing
			C224		Other Sheet Metal Product Manufacturing
			C229		Other Fabricated Metal Product Manufacturing
				C2291	Spring and Wire Product Manufacturing
				C2292	Nut, Bolt, Screw and Rivet Manufacturing
				C2293	Metal Coating and Finishing
				C2299	Other Fabricated Metal Product Manufacturing n.e.c.

Table 5: MFI identified in the Australian / New Zealand ANZIC 2006 rev. 2

According to ANZIC 2006 level 4 code C2293 refers to the following:

> "This class consists of units mainly engaged in engraving, polishing, heat treating, plating, galvanising, anodising, colouring, plastic dipping, ceramic or other coating or *finishing of client supplied metals or metal products*. This class also includes units mainly engaged in metal coating of non-metal products not elsewhere classified." (Australia, 2006, p. 173)

MFI Classification within the Group of Twenty Countries

The above presented nomenclatures represent the most widely used systems of classifications: ISIC or NACE have been the blueprints for classifications of economic activities in more than 150 countries as of 2008 (EC, 2008, p. 45).

In addition I combined the MFI classifications within the Group of Twenty (G20) countries in an overview (s. appendix A.1), sorted according to the gross domestic product (GDP) of each country in order to illustrate the international existence of the MFI. Sources for this overview have been taken from:

- G20 list of countries (G20, 2013)
- G20 gross domestic product (GDP) according to the International Monetary Fund World Economic Outlook (IMF WEO) subject code NGDPD (IMF, 2013)
- G20 classification data according to each national statistical institution[18]

In conclusion and in order to answer research question *RQ.1a*, the MFI can clearly and globally be identified using international nomenclatures and classification systems of economic activities. This identification is necessary in order to establish a border between the MFI and other industry branches. As a result of this clear distinction it is possible to analyze this singular industry branch in more detail – for example using international statistical databases of economics.

3.5 International Statistical Databases of Economics

In this chapter I analyze MFI related datasets retrieved from statistical databases of economics for Europe and the USA representing the strongest systems regarding GDP worldwide (s. appendix A.1). Larson (2012c, p. 286) has analyzed the MFI on a global scale and approximated the global output of the MFI with 300 BUSD, with 150 BUSD accounting each for the autonomous job shops and for the captive in-house MFCs.

I decided to perform these structural analyses of the MFI (s. chapter 2.5.3) in order to find characteristics a) that allow a distinction between the MFI and other industry branches and b) that represent the foundation for further MCS and BPM (s. chapter 2.3 and 2.4) analyses. Furthermore the results shall help to answer research question RQ.1b (s. chapter 2.5.4): *How can "metal finishing companies of larger scale" be defined?*

The applied statistical datasets include data related to autonomous enterprises meaning that the MFI *job shops* are included in the data. The *captive in-house* metal finishing departments are not included, as they belong to manufacturing companies shown in other industry classes rather than the MFI. In general it is challenging to retrieve data related solely to captive MFI departments due to the missing autonomy as shown by Larson (2012a, p. 184).

[18] More information related to 774 (September 2013) worldwide nomenclature / classification systems can be found on a dedicated website of the United Nations Statistics Division (UNSD, 2013).

3.5.1 European Union – SBS and SBR Databases

The statistical office of the European Union (EUROSTAT) offers the Structural Business Statistics (SBS)[19] for the analyses of industrial branches within the EU which I have applied for an MFI analysis and overview:

> "Structural business statistics (SBS) and global business activities cover industry, construction, trade and services. Presented according to the NACE activity classification, they describe the structure, conduct and performance of businesses across the European Union (EU) [...]." (EUROSTAT, 2015a)

According to European Communities NACE nomenclature, SBS datasets have been retrieved using EUROSTAT's online data explorer (EUROSTAT, 2015d) for

- NACE rev. 2 class "C-25.61 - Treatment and coating of metals" and
- NACE rev. 2 section "C - Manufacturing"

from the SBS database "sbs_na_ind_r2: Annual detailed enterprise statistics for industry (NACE Rev. 2, B-E)" and

- NACE rev. 2 Service sectors

from the SBS database "sbs_na_1a_se_r2: Annual detailed enterprise statistics for services (NACE Rev. 2 H-N and S95)".

As of March 2015 complete SBS datasets have been available for the years 2008 through 2012; the following analyses are based on the latest 2012 datasets (s. appendix A.2 for the MFI dataset):

The retrieved data for 2012 show the total EU MFI turnover at 27.9 billion Euro with 250,891 persons employed in 22,755 enterprises. The MFI contributes 0.4% to the total turnover of the overall European manufacturing industry (NACE rev. 2 section "C") recorded with 7,437 billion Euro.

Germany, with an MFI turnover of 8.3 billion Euro is the largest MFI country in Europe employing 71,918 persons in 3,250 enterprises. Countries with an MFI turnover higher than 1 billion Euro are Italy (4.4 BEUR), France (3.3 BEUR), United Kingdom (2.0 BEUR), Spain (1.9 BEUR), Poland (1.7 BEUR) and Netherlands (1.4 BEUR). These seven countries have also the highest share of all MFI enterprises and employees in the EU MFI.

The Northern European countries Denmark, Sweden, Norway and Finland have the highest share of personnel costs in turnover averaging at 36.2%. Personnel costs are also the largest cost category in these four countries as well as in Germany and United Kingdom with each more than 50% of the total purchases of goods and services. The average share of personnel costs to the total purchases of goods and services in these six countries equates to 64.4% with a maximum in Denmark with 73.1% and a minimum in Germany with 55.6%. Personnel costs together with energy costs average at 71.3% of the total purchases in these six countries. Information related to other costs like consumables, rents, e.g. is not included in the SBS datasets.

[19] Please refer to the SBS reference metadata in Euro SDMX metadata structure (EUROSTAT, 2015b) as well as to the SBS national methodologies located on the Eurostat Reference And Management Of Nomenclatures server (EUROSTAT, 2015c) for further information regarding the SBS data structure or statistical dimensions.

The lowest share of personnel costs to the total purchases can be found in Eastern Europe with Latvia, Bulgaria, Poland and Romania averaging at 18.2%. The overall European MFI average share of personnel costs to the total purchases of goods and services is recorded as 41.2%. This average is significantly higher in comparison to the overall European manufacturing industry value in 2012 with 19.8%. One reason for this finding can be related to the MFI contract processing characteristic (s. chapter 3.2): the MFC is regularly provided with raw material (metal substrates) by its customers; raw material costs are therefore far less important compared to mechanical engineering companies.

Moreover MFI's high average share of personnel costs to the total purchases of goods and services is comparable to the rather high share of personnel costs in European service sectors in 2012:

NACE rev. 2 Service Sectors - Year: 2012	
European Union (28 countries)	
NACE rev. 2 Code - Service Sector	**Share of personnel costs in total purchases of goods and services [%]**
H - Transportation and storage	34.7
I - Accommodation and food service activities	51.8
J - Information and communication	41.0
L - Real estate activities	25.4
M - Professional, scientific and technical activities	57.2
N - Administrative and support service activities	62.1
S95 - Repair of computers and personal and household goods	44.2

Table 6: Share of Personnel Costs in European Service Sectors (SBS data - own illustration)

This circumstance illustrates the twilight zone in which the MFI is located:
Although the MFI is listed in the manufacturing sections of international nomenclatures (s. appendix A.1) as it produces tangible products, i.e. layers of coated metal, it also shows service industry characteristics, i.e. significant high personnel costs.

The number of persons employed per enterprise in the EU MFI averages at 11, between a minimum of 3 (Slovakia) and a maximum of 22 (Germany). In recent literature, Larson (2015, p. 64) has compared the size by firms and employees in the EU before and after the economic recession in 2008-2009; he found that although the number of MFCs recovered well till 2012 above the pre-recession number in 2005-2006, the number of employees in the same region barely exceeded the recession's lowest point. As one reason for this finding Larson refers to the companies having rethought their operations and invested in technical automation processes which require a lower number of employees.

A further in-depth analysis of the employment levels in classes of enterprises within the four-digit NACE class C-25.61 was challenging as related data within the European SBS databases are only created on a three-digit NACE class level (EUROSTAT, 2015b, chapter 3.1). However the three-digit NACE class C-25.6 includes the non-MFI sub-class C-25.62 "Machining" representing enterprises performing mechanical processes like *"boring, turning, milling, eroding, planing, lapping, broaching, levelling, sawing, grinding, sharpening, polishing, welding, splicing etc. of metalwork pieces"* as well as *"cutting of and writing on metals by means of laser beams"* (EC, 2008, p. 161). Therefore the EU SBS employment size class data are not applicable for analyses in the MFI.

To retrieve applicable employment size class data, at least for the largest European MFI country, I applied the GENESIS online database of the German Federal Statistical Office (DESTATIS). The corresponding data table in GENESIS is table 42271-0005 (DESTATIS, 2015a) for the 2,3- and 4-digit industry class code leading to annual data related to the employment size classes of the industry class of interest.

Surprisingly the retrieved data showed a total number of German MFI enterprises in 2011 of 835 employing a total number of 50,413 persons; the corresponding European SBS data for Germany show a total number of German MFI enterprises in 2011 of 3,295 employing 68,893 persons. To understand this significant gap between the data of both sources I contacted DESTATIS who explained that the data within the GENESIS database table 42271-0005 includes solely establishments with 20 or more employees and that more detailed data are not available online using GENESIS[20].

In order to get the full picture of the German MFI employment size classes and in order to have an EU comparable dataset I placed a data request (DESTATIS data request ID "E102/35211120-AUSW" as of Sept. 25th, 2013). The received data were generated by DESTATIS based on the German Statistical Business Register (SBR) and are presented in the following table:

Germany - MFI - Enterprise Size Classes - Year: 2011			
WZ-2008 class 25.61 - "Oberflächenveredlung und Wärmebehandlung"			
Employment Size	**Number of Enterprises**	**Number of Employees**	**Turnover** [MEUR]
Total	3,885	64,607	8,846
0-9	2,507	6,011	1,038
10-49	1,059	23,571	2,758
50-249	301	27,487	3,367
>=250	18	7,538	1,682

Table 7: MFI Enterprise Size Classes 2011 in Germany (SBR data - own illustration)

It is apparent that the number of enterprises and employees in table 7 is higher than the related European SBS data for Germany in 2011. The reason for these gaps can be found in the German SBR methodologies paper, referring to single person enterprises without employees subject to social insurance contributions that are included in the German SBR but not in specialized statistics like the SBS (DESTATIS, 2015b).

Nevertheless the figures in table 7 show the distinct structure of micro-, small- and medium-sized enterprises (SME) within the MFI following the Commission of the European Communities' definition of SMEs (EC, 2003, p. L124/39):

[20] Personal communication with Mr. Ralf Müller, B3 - Information Service of the German Federal Statistical Office, as of Nov. 13th, 2013.

- a microenterprise employs fewer than 10 persons and has an annual turnover and/or annual balance sheet total of 2 MEUR or less
- a small enterprise employs fewer than 50 persons and has an annual turnover and/or annual balance sheet total of 10 MEUR or less
- a medium enterprise employs fewer than 250 persons and has an annual turnover of 50 MEUR or less and/or an annual balance sheet total of 43 MEUR or less.

64.5% of all MFI enterprises fulfill the microenterprise characteristic of employing less than 10 persons. Although representing the largest share of all MFI enterprises, these MFI microenterprises employ only 9.3% of all persons being responsible for 11.7% of the overall German MFI turnover. This finding supports MFI's SME job shop characteristic (s. chapter 3.1).

91.8% of all German MFI enterprises employ less than 50 employees and account for 45.8% of all MFI employees and 42.9% of the entire MFI turnover. By contrast 8.2% of all German MFI enterprises employ 50 or more employees and account for 54.2% of all MFI employees and 57.1% of all MFI turnover; the share of enterprises with 250 and more employees (0.5% of all German MFI enterprises) account for 11.7% of all MFI employees and 19.0% of all MFI turnover.

Referring to research question RQ.1b (s. chapter 2.5.4), I have chosen the share of all medium and larger sized MFI enterprises employing 50 or more employees as *metal finishing companies of larger scale*. With more than 50 employees it is rather obvious that planning as well as performance measurement processes are performed more intensively in larger scale MFCs compared to smaller MFCs having less manpower to perform such supporting services. Furthermore the larger scale MFCs contribute more than half of the turnover to the MFI, Germany in 2011 being one example.

3.5.2 USA – ASM and CBP Databases

The USA, possessing the highest GDP of all single G20 countries in 2012 (s. appendix A.1), is statistically analyzed regarding the MFI in order to check the robustness of the aforementioned European MFI characteristics.

The United States Census Bureau (USCB) offers the Annual Survey of Manufactures (ASM)[21] database for the analyses of US industrial branches which I have applied for an MFI analysis and overview:

> "The Annual Survey of Manufactures (ASM) provides sample estimates of statistics for all manufacturing establishments with one or more paid employees. [...] The ASM provides the best current measure of current U.S. manufacturing industry outputs, inputs, and operating status, [...]." (USCB, 2015a)

As a second database I have applied the County Business Patterns (CBP)[22] database for the analysis of US MFI employment size classes:

> "County Business Patterns (CBP) is an annual series that provides subnational economic data by industry. This series includes the number of establishments,

[21] Please refer to the ASM Methodology (USBC, 2015b) for further information regarding the ASM data structure or statistical dimensions.

[22] Please refer to the CBP Methodology (USBC, 2015d) for further information regarding the CBP data structure or statistical dimensions.

employment during the week of March 12, first quarter payroll, and annual payroll." (USCB, 2015c)

According to the North American NAICS nomenclature the latest datasets have been retrieved using USBC's online data explorer "Fact Finder" (USCB, 2015e) for

- NAICS 2012 class "33281 - Coating, Engraving, Heat Treating, and Allied Activities" and
- the corresponding superordinate NAICS 2012 class "31-33 - Manufacturing"

each from the ASM data table "AM1331GS101" and from the CBP data table "CB1200A13"; as of March 2015 the latest ASM dataset includes data for the year 2013 and the latest CBP dataset includes data for the year 2012.

Corresponding to the retrieved ASM data (s. appendix A.3), the MFI contributes 0.5% to the total value of shipments and receipts for services (TVSS) of the overall US manufacturing industry.

The share of the annual MFI payroll equates to 20.1% of the TVSS. That is nearly twice as much as the personnel costs share for the whole US manufacturing industry and corresponds to the high MFI personal costs finding within the EU SBS dataset. In addition the share of total cost of materials in TVSS is recorded with 59.1% for the whole US manufacturing industry but only with 39.7% for the MFI. This corresponds with the MFI contract processing characteristic (s. chapter 3.2).

The ASM data compared to the European SBS dataset is more detailed concerning industry cost structures. I placed emphasis on those MFI cost elements showing significant deviations in shares compared to the corresponding overall US manufacturing cost elements:

The share of electricity cost in total cost of materials is with 4.5% three times higher than the share of the whole US manufacturing industry with 1.5%. In addition the share of refuse removal services cost (including hazardous waste) in total other operating expenses is with 6.1% significantly higher compared with the share of the whole US manufacturing industry with 3.7%. Both findings are indicators for the environmental negative characteristics of the MFI (s. chapters 3.1 and 3.3).

The share of temporary and leased employee expenditures in total other operating expenses is with 14.1% also significantly higher relative to the whole US manufacturing industry at 8.6%. This is an additional indicator for a personnel-intensive MFI and also an indicator for the MFI being versatile and flexible in order to react to customers' wishes (for example to rapidly react to higher capacity requirements due to higher quantities of substrates released by the customers).

The share of advertising and promotional services in total other operating expenses is with 1.1% rather low compared to the overall US manufacturing industry with 3.5%. A reason therefore can be found in the MFI contract processing characteristic; the MFC as subcontractor is regularly located close to its main customer(s) in order to avoid logistical costs and to reduce delivery times. In a sound symbiotic MFI-customer relationship (s. chapter 3.2) MFI's demand for new customer acquisition is rather low compared to other manufacturing industries.

In the following table the retrieved CBP data related to the US MFI employment size classes are shown, aggregated according to the EC SME classes (EC, 2003, p. L124/39) focusing on the number of enterprises:

USA - MFI - Enterprise Size Classes - Year: 2012			
NAICS-2012 class 33281 - "Coating, Engraving, Heat Treating, and Allied Activities"			
Employment Size	**Number of Enterprises**	**Number of Paid Employees**	**Annual Payroll [MUSD]**
Total	5,759	123,655	5,432
1-9	2,853	11,431	442
10-49	2,284	51,637	2,214
50-249	599	52,124	2,340
>=250	23	8,463	436

Table 8: MFI Enterprise Size Classes 2012 in USA (CBP database - own illustration)

In comparison to the German SBR data, the US CBP dataset includes annual payroll related data instead of turnover related data. The average US annual MFI salary equates to 43.9 TUSD whereat annually

- the microenterprises pay at average 38.7 TUSD
- the small enterprises pay at average 42.9 TUSD
- the medium enterprises pay at average 44.9 TUSD and
- the large enterprises pay at average 51.5 TUSD.

This positive relationship of wages to firm size can be explained when considering the labor union effects or valuable top-class employees working for larger enterprises (Idson and Oi, 1999, p.104). However a downward trend of this wage premium of larger US firms over the last decades was found in recent literature (Even and Macpherson, 2012).

The overall structure of US MFI employment size class distribution is similar to the German distribution, showing a tendency for larger MFI enterprises in the USA compared to Germany: 49.5% of all US MFI enterprises fulfill the microenterprise characteristic of employing less than 10 persons (Germany: 64.5%). The share of MFI enterprises employing less than 50 employees (including microenterprises) equates to 89.2 % (Germany: 91.8%) accounting for 51% of all US MFI employees. The share of *larger scale metal finishing companies* employing more than 50 employees is recorded as 10.8% (Germany: 8.2%). This finding further strengthens MFI's SME and job shop characteristics (s. chapter 3.1).

3.6 Characteristics Clusters

In the previous chapters I have identified several MFI characteristics which I have summarized into four main clusters.

These clusters serve as an initial and preliminary industry structure on which I build my further research for a BPM system (s. figure 10); the industry characteristics are related to research question RQ.1c[23] and will be analyzed and triangulated in more detail within the empiric research part of this dissertation (s. chapters 6 and 7).

1. Metal Finishing – A Contract Processing Industry

It is rather the MFI's size (s. chapter 3.5) that explains its economic importance, but the interdependence with other industry branches relying on MFI products and services (Brummet and Arnett, 1967, p. 2; Larson, 2012a, p. 179). Therefore Matthews et al. (1998, p. 56) describe metal finishing processes "*as an enabling technology in many industrial sectors*".

A limited number of main customers per MFC, the proximity of the MFC to these customers and the MFC's flexibility to react to their customer demands are characteristics of the contract processing character of the MFI. Therefore related agency and inter-firm relationships (s. chapter 2.1) have to be reflected in an MFI-BPM model.

2. Metal Finishing – A Job Shop Industry

The MFI statistical data retrieved from international databases (s. chapter 3.5) show a clear emphasis on SME, or more precisely on microenterprises. However the share of larger scale MFCs contributes significantly to the overall MFI output with regard to turnover as well as to employment rates. BPM frameworks and elements dedicated to SMEs are therefore in the focus of further MFI-BPM research.

3. MFI's Industrial Service Nature

The MFI is part of the overall industry class of *manufacturing* in all industry nomenclatures of the G20 countries (s. appendix A.1). Comparing MFI cost elements, especially expenditures for personnel and material, with those from its superordinate industry class *manufacturing* showed significant deviations; these cost elements are comparable to *service* industry classes:

> "[...], the proportion of operating expenditure accounted for by personnel costs was [...] the highest share among the activities covered in Subchapters 9.1 to 9.7 [metals and metal products], reflecting the *industrial service nature of the sector.*" (EUROSTAT, 2009, p. 235)

MFI's dichotomous characteristic of neither being 100% a manufacturing industry nor being 100% a service industry contributes to an important facet for a future MFI-BPM-system: BPM frameworks and systems created either solely for manufacturing or for service companies may not deliver sufficient BPM solutions for the MFI.

4. MFI's Environmental Impact

The MFI consumes high amounts of energy and requires the usage of hazardous chemicals due to its processes (s. chapter 3.1). The MFI has therefore a rather negative impact on the environment. Reducing and avoiding negative influences on the environment can generate additional costs – but it might also lead to business performance improvements (s. chapter 3.3). BPM elements related to the environmental performance of an enterprise are therefore further features of an MFI-BPM model.

[23] How can the MFI be described and differentiated from other industry classes? (s. chapter 2.5.4)

4 BPM in the MFI – Current State of Research

After having established a theoretical research framework including Management Control as well as Business Performance Measurement Systems in chapter 2 and after having analyzed the Metal Finishing Industry in chapter 3, I will continue with a review of pertinent literature in order to prove the existence of the perceived research gap related to a missing BPM systems for the MFI (s. chapter 1.1).

4.1 Literature Review

> "You are right to say contacting other researchers in economics of the metal finishing industry is challenging, probably because there are very few of us."
>
> (C. Larson, personal communication, July 19th, 2013)

The citation of Mr. Larson[24] refers to the challenge of finding appropriate MFI economics literature: in mid-2013 I have conducted searches in different English- and German-speaking electronic scientific databases as well as in selected controlling journals in order to maximize the likelihood of finding relevant studies of BPM in the MFI. I realized soon that economic literature in general is scarce with regard to the MFI[25] and literally nonexistent to BPM. After my review I decided to get in contact with a researcher in the economics of the MFI, namely Mr. Larson, in order to discuss my findings regarding MFI economics literature. He affirmed that the vast majority of MFI publications refer primarily to technological or legislative issues (C. Larson, personal communication, July 19th, 2013).

Although I did not find any relevant BPM publications for the MFI I will describe the literature review process itself within this chapter to illustrate the applied review steps.

As a first step within my literature review I composed a dictionary that included terms related to the MFI. I have collected the terms in German as it is my mother tongue. The German terms were translated into English using the *Dictionary of Retailing and Supply Chain* (Davies-Labeck, 2005). The following table shows the resulting translations:

German MFI Terms	English MFI Terms
Lohnveredlung	Contract Processing
Lohnunternehmen	Contract Processor
Oberflächenveredelung	Surface Finishing
Veredelung	Finishing
Veredelungsindustrie	Finishing Industry
Veredelungsverfahren	Finishing Process
Endfertigung	Finishing
Endfertigungsprozess	Finishing Process

Table 9: Literature Review – Dictionary

[24] Mr. Clive Larson is the executive editor of the international journal for surface engineering and coatings *Transactions of the IMF* and Fellow of the *Institute of Materials Finishing*, UK.

[25] I have incorporated the few economics sources that I found within chapter 3, especially the work of Brummet and Arnett (1967), Matthews et al. (1992, 1998), Bell (2000), Artley and Matthews (2010) and Larson (2012a-c).

Based on this dictionary, I created MFI search terms linked to BPM in the notation [term] as presented in table 10:

Terms Applied in Database Search	
["contract processing"]	
["surface finishing" "business performance"]	
["metal finishing" "performance measurement"]	
["metal finishing" "performance management"]	
["metal finishing" "business performance"]	
["finishing industry" "performance measurement"]	
["finishing industry" "performance management"]	
["finishing industry" "business performance"]	
["finishing" "Leistungsmessung"]	German terms applied in WISO and Google Scholar.
["Veredelung" "performance management"]	
["Veredelung" "performance measurement"]	
["Veredelung" "business performance"]	
["Veredlung" "business performance"]	
["Veredlung" "performance management"]	
["Veredlung" "performance measurement"]	
["Veredlung" "Leistungsmessung"]	
["Veredelung" "Leistungsmessung"]	

Table 10: Literature Review – Terms for Database Search

I have applied these terms in the following electronic scientific databases[26] (Jesson et al., 2011, p. 39; Booth et al., 2012, p. 73):

Scientific Database	Number of hits	Thereof relevant for BPM in the MFI
EBSCO Business Source Complete	10	0
Google Scholar	693	0
Elsevier - SciVerse ScienceDirect	105	0
Emerald Management Xtra	35	0
JSTOR Business I, II, III	19	0
SAGE Publications	12	0
WAO - World Affairs online	0	0
WISO (German)	47	0

Table 11: Literature Review – Results in Scientific Databases

The free text searches were performed in all available database contents. The searches resulted in different numbers of hits for each search term which are shown summarized per database within the second column of table 11. Each hit was analyzed (Jesson et al., 2011, p. 49) regarding its relevance for BPM in the MFI by evaluating the available content – mainly limited to titles and abstract excerpts. Following this approach I did not find any relevant publication related to BPM in the MFI (please refer to appendix A.4 for detailed results).

[26] The databases were chosen according to their availability within the electronic library of the EBS Universität für Wirtschaft und Recht, Oestrich-Winkel.

As a second step and in addition to the database searches I have analyzed the tables of contents of different management accounting journals. The reason for this additional search process was to eliminate the potential risk of missing relevant studies not listed within the electronic databases.

I downloaded the table of contents of the different journal issues and searched for terms related to the MFI therein. The available data for the content searches were titles and authors as well as partially keywords and abstracts of the publications. The selected journals have been chosen in accordance with controlling experts from my research institute.

The tables of contents of the following journals have been analyzed:

ISSN	Journal	Language	Timeframe of Search				Available Data for Content Search
			From		Till		
			Year	Vol. (Issue)	Till: Year	Vol. (Issue)	
1741-5039	**International Journal of Business Performance Management**	EN	1998/99	01 (01-04)	2013	15 (01-02)	Titles and authors
1741-0401	**International Journal of Productivity and Performance Management**	EN	2004	53 (01-08)	2013	62 (01-06)	Titles, keywords and authors
1463-5771	**Benchmarking: An International Journal** Formerly: *Benchmarking for Quality Management & Technology*	EN	1994	01 (01-03)	2013	20 (01-04)	Titles, keywords and authors
1092-8057	**Cost Management** Formerly: *Journal of Cost Management*	EN	1987	1987 (01)	2013	2013 (2)	Titles and authors
0935-0381	**CONTROLLING – Zeitschrift für erfolgsorientierte Unternehmenssteuerung**	DE	2002	14 (01-12)	2013	25 (01-06)	Ttitles and authors, 2013 also abstracts
1616-0495	**Controller Magazin** Entries prior 2011 were checked using the German WISO database.	DE	2011	2011 (04-06)	2013	2013 (01-04)	Titles and authors

Table 12: Literature Review – Selected Journals

Entries for the following search terms[27] (Davies-Labeck, 2005) in the notation [term] have been searched within the table of contents of the above cited journals:

Terms Applied in German Journal Content Search	Terms Applied in English Journal Content Search
[Veredelung]	[contract processing]
[Veredlung]	[contract]
[Veredeler]	[finishing]
[Veredler]	[surface]
[Lohnfertigung]	
[Lohnfertiger]	
[Finishing]	
[Oberfläche]	

Table 13: Applied German and English Search Terms in Journal Content Search

[27] It was necessary to apply more search terms within the German journal content search than in the English content search as the German translation for *finishing* can either be *Veredelung* or be *Veredlung*.

The searches resulted in only a very few hits; each hit was analyzed regarding its BPM-MFI relevance. Comparable to the database searches I did not find any further relevant publication related to BPM in the MFI.

As a third and final step, I have created several Google Scholar Alerts (Google, 2015) following the literature review in order to be informed about any forthcoming BPM-MFI developments to be included into my research project. Google Scholar Alerts were created for the following terms in the notation [term]:

Google Scholar Alert Terms
["contract processing"]
["performance measurement" "contract processing"]
["performance measurement" "lohnfertigung"]
[lohnveredelung]
["metal finishing industry"]
["finishing industry"]
[Veredelung]
[Veredlung]

Table 14: Google Scholar Alert Terms

These alerts resulted in approximately four automated emails coming from Google Scholar every other day (exemplarily presented in the left part of figure 15). Each email contained between one and five new Google entries related to the corresponding search terms. Each consists of the publication title, author(s) as well as an excerpt from the publication abstract if available (exemplarily presented in the right part of figure 15):

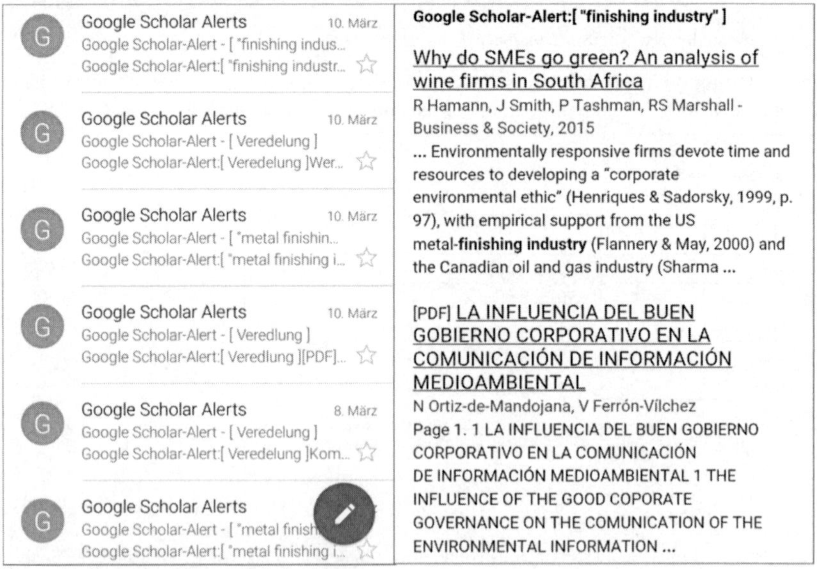

Figure 15: Example of Google Scholar Alerts

I checked those alerts in the German and English languages relating to my research field and tried to find the full publication or at least the full abstract in case an entry seemed relevant; but this happened only in a very few cases, for example with the study of Zhu and Zolkiewski (2015; s. chapter 8.2.3). All other studies were already considered within my research as several entries referred to the Academy of Management Journal (AMJ) publication by Flannery and May (2000; s. chapter 3.3).

4.2 Research Gap

As described in the previous chapters, the literature review led to no relevant BPM studies for the MFI; therefore the research is assumed to be unbiased with regard to the existing literature.

One could argue that the absence of MFI-BPM literature is an indication of the insignificance of this research field coupled with a lack of interest of the corresponding academic as well as managerial communities. I cannot reject this opinion as far as the past is concerned, but the interest regarding my research project illustrates the current significance of managerial as well as academic demand for a BPM system in the MFI.

Specifically interest in my work has come from

- several multinational MFI associations and research institutions (s. chapters 1.1 and 7.1.1)
- a major German MFC (s. chapter 6.1.2)
- personnel communications with UK MFI insiders like Mr. Larson (s. chapter 4.1) or Dr. Brace[28]
- the Strascheg Institute for Innovation and Entrepreneurship of the EBS Universität für Wirtschaft und Recht promoting and hosting my dissertation project

Existing generic BPM systems as well as BPM models and concepts (s. chapter 2.4) can also not be applied without any reservation due to the MFI characteristics (s. chapter 3.6). For instance, Lange (2009) developed a BPM model for industrial service companies, which he divided in the following three classes (2009, p. 31):

- product supporting services (mainly free of charge, for example technical customer services for consumer electronics),
- process supporting services (for example preventive maintenance of industrial machines), and
- customer supporting services (for example outsourcing the facility management to a corresponding service provider).

The MFI can be classified as offering customer supporting services due to its contract processing characteristic (s. chapter 3.2). However, Lange's BPM concept (2009, p. 81) is directed to all three industrial service classes, each showing different characteristics. As shown in chapter 2.5.3, my research project follows a bottom-up approach having the MFI as a distinct industry class to design a corresponding BPM model tailored to the MFI's characteristics.

[28] Dr. Arthur Brace is an internationally renowned MFI expert and known as the "Father of Anodizing". It was an honor as Dr. Brace contacted me after the announcement of my research project within the *Transactions of the IMF* (Larson, 2014, p. 177) in order to support my research by supplying me with his dissertation thesis (Brace, 1995) and further information related to the MFI (personnel communication, August 27th, 2014).

Due to the lack of appropriate literature and in order to close this research gap by finding answers to my research questions (s. chapter 2.5.4), I decided on an in-depth probe of the MFI utilizing an exploratory research methodology, cognizant of the evolving academic concern within the management accounting research community (s. chapter 2.5.2).

5 BPM in the MFI – Research Methodology

Within the previous chapters of my dissertation I have elucidated several MFI characteristics (s. chapter 3.6) differentiating the MFI from other industry classes. This differentiation is necessary in order to establish a BPM model tailored to the managerial demands within the MFI, following the research process described in chapter 2.5.3 and relating to research question RQ.1c (s. chapter 2.5.4): *How can the MFI be described and differentiated from other industry classes?* In order to triangulate and validate the hitherto MFI characteristics I have chosen to further analyze the MFI empirically; in this chapter I will present the research methodology which I have chosen for this empiric research in the MFI as well as the reasons for my decisions.

Finding the appropriate research methodology for such a project is not trivial: on one hand the necessary amount of data must be generated and analyzed using academically accepted and appropriate research field methods. And on the other hand the legendary "*death by data asphyxiation*" (Pettigrew, 1990, p. 281) has to be avoided, caused by a plethora of data and information which causes the research subject to lose its focus.

I decided that an appropriate research methodology for my project should a) be academically accepted for empiric research within the BPM field, and b) be practitioner-oriented as *"management accounting is a practical art"* (Anthony, 1989, p. 16). This would follow the evolving academic concern within the management accounting research community as presented in chapter 2.5.2.

For finding an academically accepted empiric BPM research design I referred to Franco-Santos et al. (2012, p. 84) who have analyzed the methodology of 76 empirical BPM studies published in high-quality academic journals in the time frame 1992-2011. They classified each study according to the applied research method:

- Archival – 5 studies
- Survey – 27 studies
- Experiment or quasi-experiment – 15 studies
- Case/field study – 20 studies
- Mixed methods – 9 studies

Franco-Santos et al. (2012) show that both qualitative and quantitative research methods (or a mix of both) are applied in BPM research.

Focusing on qualitative research methods, Merchant and Van der Stede (2006) have analyzed the utilization of field research in empirical studies published in major English-language accounting research journals in the time frame 1981 through 2004. Under the term "*field research*", the authors classified all research methods that "*involve the in-depth study of real-world phenomena through direct contact with the organizational participants*"; they found an increasing trend of applying qualitative research methods related to management accounting research.

Therefore it appears that the application of qualitative research methods is not only acceptable but moreover advisable for management accounting (including MCS and BPM) research.

After having analyzed the current state of BPM research in the MFI (s. chapter 4) it became clear that I would need a rather exploratory research design:

As shown in my research framework (s. chapter 2.5.1), BPM is influenced by different theories of the firm like RBV, RV, or PBV. Related to empirical RBV studies Armstrong and Shimizu (2007, p. 976) have recommended to *"adopt a qualitative approach when industry or a focal resource is new and/or unexplored"*. As presented in the previous chapters the MFI can be classified as *unexplored* concerning management accounting research and therefore qualitative research is a plausible research approach. Qualitative methods are in general the first choice in the event exploratory research should be performed (Patton, 2002, p. 193).

It is the nature of qualitative research to have rather small samples of cases to study (Yin, 2009, p. 17) or experts to interview (Gläser and Laudel, 2010, p. 36). But trying to establish a BPM system for a whole industry class means generalizing the qualitative findings coming from a small group of individuals to a larger sample. And this is exactly the purpose of an *exploratory sequential mixed methods design* according to Creswell and Plano Clark (2011, p. 86).

The following figure illustrates the resulting research design for my BPM research in the MFI:

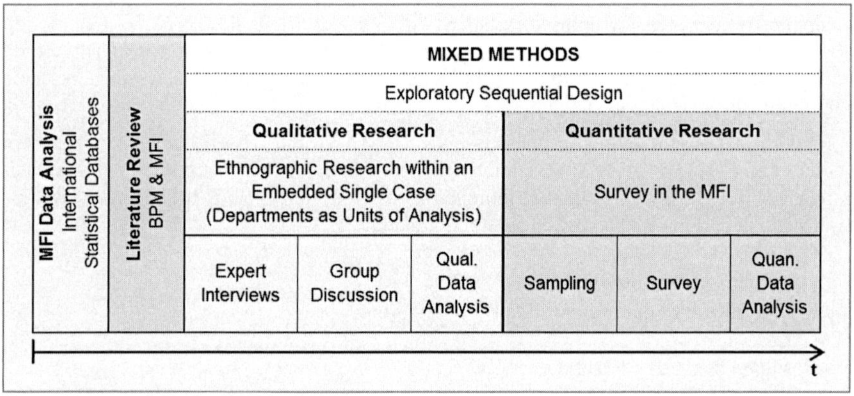

Figure 16: Research Design BPM in the MFI

The results of the MFI data analyses as well as the literature review were already presented in the previous chapters. The first and main phase of my exploratory sequential mixed methods research design is organizational ethnographic research (Hammersley, 1998; Ybema et al., 2009; Denzin and Lincoln, 2011, p.715) in a qualitative embedded single case study (Yin, 2009, p. 50) within a major German MFC. Based on the qualitative results, I will design an MFI enhanced cause and effect model (s. chapter 6.6) representing the foundation for BPM in the MFI.

The second phase is a quantitative survey to triangulate (Eisenhardt, 1989b, p. 537) the qualitative results a) as a proof of validity (Greene, 2007, p. 100) and b) to gain additional insights (Flick, 2011, p. 12) on BPM requirements within the MFI. Within this survey, experts in other MFCs will be asked for their assessment concerning the qualitatively established MFI model.

6 BPM in the MFI – Qualitative Research Process

As illustrated in the previous chapter, I have chosen an exploratory sequential mixed methods design starting within a qualitative research part. For my qualitative research of BPM in the MFI, I had the opportunity to research a major German MFC ethnographically by becoming a senior financial analyst in this company at the same time that my research project commenced. I will describe this research setting in more detail within chapter 6.1.

Besides the emic observations, an etic element (Patton, 2002, p. 267; Creswell, 2013, p. 96) is added to the qualitative research by having semi-structured expert interviews (Gläser and Laudel, 2010; s. chapter 6.3) followed by a group discussion (Bohnsack et al., 2010; s. chapter 6.4) to reach a qualitatively thick description (Geertz et al., 1987; Lapan et al., 2012, p. 201) or as Riemer (in Lapan et al., 2012, p. 165) noted: *"Interpretation of culture in its thick description requires both an insider's, or emic, perspective and an outsider's, or etic, perspective."*

Based on qualitative data analysis (s. chapter 6.5) I modeled an enhanced cause and effect model for the MFI as foundation for BPM (s. chapter 6.6).

6.1 Research Setting

> "[…], one should begin research with minimal assumptions so as to maximize one's capacity for learning. It is for this reason that ethnographers rarely begin their research with specific hypotheses. Rather, they have a general interest in some type of social phenomena and/or in some theoretical issue or practical problem." (Hammersley, 1998, p. 9)

6.1.1 Ethnographic Research within an Embedded Single Case

In November 2011, the *Strascheg Institute for Innovation and Entrepreneurship (SIIE)* of the *EBS Business School - EBS Universität für Wirtschaft und Recht, Oestrich Winkel* (EBS) in cooperation with the major German metal finishing company *MeFiCom*[29] published a nationwide job advertisement for a researcher and controlling expert in the BPM field combined with the possibility of starting a dissertation project within this field. At that time I was holding a degree in industrial engineering, focusing on controlling and management, from the University of Applied Sciences Trier. In addition I had been working in a controlling and management position within a medium sized enterprise (EC, 2003, p. L124/39) in the German industrial gearbox industry for seven years. I was convinced that being 32 years of age was the right moment in my life to combine my academic education with my managerial experience and to apply for this BPM research project within the MFI.

In April 2012 I started to work in MeFiCom's finance department in the position of a senior financial analyst; before that date I had no professional or academic relationship with the MFI. This date was also the starting point of my research project. The position was organized in a way that I worked at the German MeFiCom headquarters for four days a week, regularly Monday through Thursday. The remaining time I was released from work in order to pursue my research project.

---

[29] *MeFiCom* stands for ***Me**tal **Fi**nishing **Com**pany* and is a pseudonym for the company representing the single case within the qualitative research part of this dissertation.

In my position at MeFiCom I was mainly responsible for the composition and validation of investment proposals as well as for the controlling of capitalized expenditures. Additionally, during my three year research project, I got involved in several interdisciplinary projects as well as in regular finance tasks like the creation of business reviews (nine times), forecasts (35 times) or annual budgets. I was granted full access to any financial data and organizational information which I required in order to fulfill these tasks.

Being responsible for the investment proposals was especially beneficial, as it allowed me to get in contact and to work with colleagues of all MeFiCom departments including management, finance, production, sales, purchase, human resources (HR) and information technology (IT). Considering each of MeFiCom's departments was crucial, as these departments represent the embedded units of analysis (Yin, 2009, p. 46) within my information-rich (Patton, 2002, p. 46) single case (s. chapter 6.1.1).

Furthermore these personnel contacts allowed me to quickly become part of MeFiCom and thereby increase my credibility and confirm my integrity (Pratt, 2000, p. 460). This certainly enhanced the quality of the information retrieved in the interviews and group discussions; Moeran (in Ybema et al., 2009, p. 148) describes this situation with the switch from participant observation to observant participation in ethnographic field research: "[...], *observant participation leads to the kind of involved detachment that characterizes the very best of social scientific analysis*".

Being *one of them* reduced the necessity of gate keepers in their role of allowing externals the access to internal employees and data (Creswell, 2013, p. 94) during the research project. For sure, it did not reduce the importance of the management commitment to actively support and get involved into my research. The latter was a crucial element in order to retrieve reliable and rich data from management members within the data acquisition steps (s. chapters 6.3 and 6.4).

As of day one of my work, those for my research project relevant colleagues at MeFiCom were informed about my dichotomous position of being a colleague as well as a researcher. I decided to choose this overt ethnographic research approach (Hammersley, 1998, p. 35) after considering ethical concerns related to a possible contravention of the human rights of autonomy and dignity within covert ethnographic research approaches (Hammersley and Atkinson, 1995, p. 263; Lugosi, 2006).

This very sensitive personnel and researcher focused background information is relevant for understanding the setting of my research project in order to allow the academic community to assess my research (Hammersley, 1998, p. 136) and a fortiori, its reliability and validity (Miles et al., 2014, p. 312; s. also chapter 6.7) – or as Riemer (in Lapan et al., 2012, p. 182) noted: "[...] *validity-rich ethnography is transparent*".

Being actively involved in MeFiCom's everyday work allowed me to learn the daily routines, behaviors, processes and (organizational) language spoken in the MFI what represents the key idea of ethnographic research (Creswell, 2013, p. 90). Gläser and Laudel (2010, p. 148) noted that speaking the same organizational language like the experts in the research area is important for the realization of information rich expert interviews (s. chapter 6.3). The expert's usage of the term *benchmarking* serves as an example for a peculiarity in MeFiCom's language; the term is regularly applied in referring to a MeFiCom internal Microsoft Excel BPM *tool* in which cost groups are split into their respective details and not denoting the classical economic *process* of comparing one's own performance with that of the market leader (Gleich et al., 2008, p. 245). Hearing the term within a MeFiCom expert interview could therefore lead to misunderstandings in the event this internal denotation of the term is unknown.

I have chosen this qualitative research setting also in response to the evolving academic concern regarding management accounting research (s. chapter 2.5.2): being involved in the everyday life of MFI individuals over a three years' time span allowed me to create a qualitatively thick description (Geertz et al., 1987) of the single case which I studied. The resulting findings, grounded in the practitioners' world, will enhance the practical relevance of the research within the real-world. Furthermore the ethnographic research approach follows a longitudinal (Pettigrew, 1990, p. 271; Lapan et al., 2012, p. 247) orientation as the same organization and set of individuals was observed over a three year time period; longitudinal research is a further recommendation of the management accounting research community (Bourne, 2008, p. 69) to improve the practical relevance of corresponding research. Finally, asking MFI executives directly *"about what information managers use and how they use it"* in evaluating the business performance of their company falls within Neely's (2011, p. 76) concern.

My research setting follows the M2K research policy (s. chapter 2.5.2) in combining in an interdisciplinary manner an EBS-SIIE researcher with MeFiCom executives in a joint research project in order to solve a practical relevant problem i.e. a missing BPM model for the MFI.

However being actively involved in an overt ethnographic research setting bears the risk of researcher bias (Hammersley, 1998, p. 87). In order to reduce any such bias to a minimum and in addition to the overall exploratory sequential mixed methods research design (s. chapter 5), I have purposefully chosen further qualitative methods that a) are designed to reduce the researcher's influence during the data acquisition process (s. group discussion according to Bohnsack in chapter 6.4) and b) include a validation process of the qualitative findings together with the participants (s. member checking in chapter 6.7). Furthermore I followed Fetterman's (2010, p. 23) recommendation for a *"nonjudgmental orientation"* which demands the ethnographer to put himself in a neutral position during the research process without being led by his personal perspectives.

During the research process I logged any ideas, unique features or other important observations related to BPM in brief memoranda (Fetterman, 2010, p. 119) which I included within the modeling process of the enhanced cause and effect model of the MFI (s. chapter 6.6).

6.1.2 MeFiCom as Proxy for the MFI

In its ISO 9001:2008 quality management certification MeFiCom[30] describes itself as a *"provider of advanced tribological surface solutions through various global, strategically placed capabilities that include equipment design, sales and service, application of thin film, wear resistant coatings as well as heat-treating and surface preparation for a comprehensive range of industrial tooling and precision components."*

MeFiCom started its operations in Germany in the 1980s. Today MeFiCom employs more than 500 persons within more than ten production sites all over Germany specializing in

[30] It was the wish of MeFiCom's management to use a pseudonym instead of the real company name within my dissertation thesis and to present any company related figures only as approximations.

PVD and CVD coating processes of metal substrates (s. chapter 3.1). The company's 2014 turnover was above 50 MEUR.

Several characteristics of MeFiCom can be linked to the already identified MFI characteristic clusters (s. chapter 3.6):

1. ***MeFiCom – A Contract Processing Company***

   Each of MeFiCom's production sites is located near the top five customers in each region who directly supply the production site with the substrates to be coated. Although the sales department is centralized, many customers are in direct contact with the decentralized production sites, especially for logistics planning purposes. With a catalogue of more than 50 different coating types and coating process combinations, MeFiCom is able to respond to the different types of customer substrates and corresponding customer requirements. Substrates range from tiny automotive serial components like piston pins with a diameter of 10 millimeters up to heavy gear parts or tools like hobs, weighing more than 20 kilograms per part.

2. ***MeFiCom – A Job Shop Company***

   Although MeFiCom does not fall under the Commission of the European Communities' definition of SMEs (EC, 2003, p. L124/39) it shows the MFI's job shop capabilities by having more than ten decentralized production sites. In 2014, 80% of the productions sites on average employed around thirty persons and produced a turnover of 6 MEUR which corresponds with small enterprises within the EC SME definition. Although possessing a centralized management, MeFiCom is actually a conglomerate of decentralized and rather autonomous production sites representing MeFiCom's job shop character nested in the decentralization.

3. ***MeFiCom's Industrial Service Nature***

   In 2014 MeFiCom's share of personnel costs in turnover averaged at 30%; personnel costs are also the largest cost category within MeFiCom's cost of goods sold, accounting at 50%. These figures correspond with the statistical MFI data for Northern Europe regarding personnel costs analyzed in chapter 3.5.1. The rather high personnel costs reflect a) personnel intensive processes like the manual charging of metal substrates into coating fixtures and b) minimal material costs, as the metal substrates are not purchased by MeFiCom but are provided by MeFiCom's customers. A further example for MeFiCom's industrial service nature is their pickup service for substrates: for many customers MeFiCom organizes all logistics processes from the delivery of the metal substrates to the return delivery of the finished customer parts.

4. ***MeFiCom's Environmental Impact***

   MeFiCom needs to be particularly cautious in its use of hazardous chemicals, especially in cleaning and stripping processes of the metal substrates before the actual coating process takes place. The appropriate disposal and recycling of resulting wastes result in corresponding environmental costs. MeFiCom's PVD and CVD processes are executed in high vacuum and corresponding turbomolecular pumps require high amounts of electric energy. Environmental and energy costs account for about 10% of all MeFiCom cost of goods sold in 2014.

All these characteristics make MeFiCom a proxy model for larger scale MFCs (s. chapters 2.5.4 and 3.5.1) in the MFI.

MeFiCom is one of Germany's MFI vacuum vaporization market leaders. At the same time it exhibits typical SME and contract processing structures (s. chapter 3.2). Consequently MeFiCom lies between being an extreme (Lapan et al., 2012, p. 84) and a typical (Yin, 2009, p. 48) case and therefore can be interpreted as an intense case (Patton, 2002, p. 234): *"Using the logic of intensity sampling, one seeks excellent or rich examples of the phenomenon of interest, but not highly unusual cases."*

MeFiCom is part of an international corporation operating more than 100 production sites in more than 30 countries worldwide. All national companies share the same production processes, norms and quality requirements. Therefore MeFiCom can be regarded not only as a proxy model for the German but also for the international MFI.

6.2 Preliminary Studies

The first year within the ethnographic research at MeFiCom helped me to get familiarized with the daily routines and processes within the company. My position in the finance department offered me the opportunity to analyze MeFiCom's financial data and to build relationships with my colleagues. Simultaneously, I have analyzed MCS and BPM literature in order to build my research framework (s. chapter 2.5).

In order to obtain pilot data related to the processes of other MeFiCom departments I performed three open-ended and in-depth interviews (Lapan et al., 2012, p. 94) in January 2013. These interviews, in contrast to the semi-structured expert interviews later in the research process (s. chapter 6.3), were of a rather informal character: the data retrieved from these interviews was not used for any coding or deeper analysis processes but for understanding existing internal control (Simons, 1995, p. 84; s. chapter 2.3.3) structures and BPM enabling elements like information services (s. chapter 2.5.1). All three interviewees were not members of MeFiCom's top-management, but qualified experts in their fields being employed at MeFiCom each for more than ten years.

The interviews took on average 60 minutes and were of an informal conversation style (Patton, 2002, p. 342). Although, singular, non-dichotomous and open-ended interview questions (Patton, 2002, p. 353) were prepared and sent to the interviewees one day prior to the sessions. This was to allow the interviewees to get an overview on the topics to be discussed. I took field notes (Patton. 2002, p. 383) during all interviews and created interview protocols soon after each interview to log the results.

The following information was gathered by me in the three departments which were the focus of my preliminary studies:

1. ***MeFiCom's Internal Sales***
   Date of Interview: January 17th, 2013.

   The sales department operates a customer relationship management (CRM) system in order to administrate the contacts of prospective and existing customers as well as to store sales documents such as offers, purchase orders or invoices. Furthermore, the CRM is used for MeFiCom's complaint management; all sales representatives are responsible for logging in any complaints in the CRM in order to respond to the customers' wishes and requirements. Generating sales performance measures based on this complaint management process would allow evaluating the overall customer satisfaction with MeFiCom, as all information coming from the decentralized production sites is gathered in the CRM. The CRM system is also the source of existing sales indicators like the number of calls of sales representatives within a certain time frame. The head of internal sales is responsible for the completeness and quality of the data entered into the CRM system. Management reporting exists independently from the CRM system in which MeFiCom's management is informed about important sales information including aggregated turnover figures, trends in customer complaints, market volumes, e.g.

Especially pertinent is the sales perspective within the research process, as the MFI contract processing characteristic (s. chapter 3.2) emphasizes the importance of a vital relationship (s. chapter 2.2.2) between the MFC and its customers.

At the end of this interview I had an interesting ethnographic experience: I audiotaped this first of three interviews; as soon as I switched off the recorder, the interviewee said: *"Mike, off the records, please let me tell you some further details concerning our CRM system..."* Privately, and off the record, the interviewee told me some improvement potentials of the current CRM system, including a direct interface of the CRM system for MeFiCom's management. This additional information was important and interesting in the realm of BPM, but moreover this situation showed the switch from being in the ethnographic front stage to the back stage (Ybema et al., 2009, p. 148) in the sense of observant participation (s. chapter 6.1.1). In consequence I decided not to audiotape the remaining two interviews in order to avoid any resistance and to be in the ethnographic backstage from the beginning of each interview.

2. ***MeFiCom's Quality and Environment Audit Department***
   Date of Interview: January 17th, 2013.

MeFiCom is certified having adopted the following international norms related to quality and environment:

- General quality management system according to ISO 9001:2008
- Automotive related quality management system according to ISO/TS 16949:2009
- Environmental management system according to ISO 14001:2004

Measureable quality objectives are a requirement within ISO 9001:2008 (ISO, 2015a, section 5.4.1) including indicators subject to BPM. Accordingly, MeFiCom's quality performance measurement contains the following objectives (o) and related performance indicators (▪):

- Short and reliable delivery times
  - throughput time
  - number of customer complaints
  - delivery reliability
- Innovativeness
  - turnover with new products and services
  - number of patent applications
- Market leadership

    - market share growth
- Process enhancement and simplification
    - turnover per working hour
    - coating system utilization
- Employee satisfaction
    - employee satisfaction measured in employee opinion surveys
    - labor turnover rates
    - training hours per employee and month
    - absence rates
- Management involvement in enhancement processes
    - activities in continuous improvement processes
- Achievement of financial targets
    - actual costs of goods sold vs. budget
    - turnover trends per production site
    - actual EBITDA vs. budget

Furthermore MeFiCom's environmental performance measurement includes the following objectives (o) and related performance indicators (▪):

- Reduction of energy consumption
    - kilowatt hours per coating system
- Reduction of waste disposal cost
    - controlling for waste disposal cost
- Reduction of wastes being hazardous to water
    - controlling for wastes being listed in German water hazard class 1-3 (VwVwS, 2005)
- Environmental improvement of the supply chain
    - number of ISO 14001 certified suppliers
- Sustained environmental improvements
    - annual HSE investments
    - number of environmental improvement projects
    - results of internal environmental audits

In conclusion, the objectives and related indicators conform to the already identified MFI characteristics, especially with regard to reliable delivery time and environmental objectives. Furthermore it opened new BPM perspectives related to the role of employees in the MFI (represented by the employee satisfaction objective) and the demand for innovativeness.

3. ***MeFiCom's Information Technology and Information Services***

Date of Interview: January 21st, 2013.

Information technology (IT) and information services, including management information systems (MIS) and enterprise resource planning (ERP) software, are enabling factors in BPM (s. chapter 2.5.1).

MeFiCom applies a customized and integrated ERP system in which all company departments and processes are connected with each other. This includes the customer order management, order calculations, work schedules, shop floor orders, purchase order management, accounts receivable and payable systems, e.g.

Process related data can be retrieved from this ERP system in order to create production reports, for example relating to the reliability and relative process time of MeFiCom's coating systems. The ERP system is also the data source for the MeFiCom MIS, based in SAP Business Warehouse (SAP, 2015). The MIS is used by the finance department in order to create financial forecasts, variance analyses, e.g. to be included in management correspondence.

6.3 Semi-Structured Expert Interviews

In the second year of my ethnographic research at MeFiCom I have further analyzed MeFiCom's structures and processes. Toward the end of this second year I felt myself understanding how MeFiCom operates and therefore was in a position to develop and add in-depth information to my research project by applying semi-structured expert interviews (Gläser and Laudel, 2010; Lapan et al., 2012, p. 94) in order to reach a thick description (Geertz et al., 1987) of BPM in the MFI. Having in-depth interviews in the mid or end stage of a study is recommended by Fetterman (2010, p. 40) as the researcher is able to comprehend *"the fundamentals of a community from the 'insider's' perspective"*; Fetterman adds, that the resulting information of those interviews are closer to the reality of those observed than to the reality of the observer.

Interviews, in general, are one of the most important qualitative data sources within case studies (Yin, 2009, p. 106) as well as in ethnographic research (Fetterman, 2010, p. 40).

6.3.1 Preparation

The first step in the expert interview preparation process was the convening of a board of experts[31]. The board is formed by six MeFiCom management team members representing the main MeFiCom departments as units of analysis:

- Sales
- Production
- HR
- Accounting
- Finance & Controlling (including Purchase and IT)
- Management represented by the General Manager

I have chosen this management sample because it included managers of all main departments of MeFiCom and therefore allows a holistic and integrated interpretation of MeFiCom's BPM. Furthermore and in concordance with Gläser and Laudel (2010, p. 117) I assumed that the chosen experts:

- possess the relevant information for my research project,

[31] The term "informant" is regularly applied in social science literature for those individuals being the source of rich information, especially in qualitative case studies or ethnographic research (Hammersley, 1998, p. 87; Patton, 2002, p. 321; Lugosi, 2006, p. 541; Yin, 2009, p. 107; Ybema et al., 2009, p. 62; Lapan et al., 2012, p. 166). As this term has its roots within the anthropological research conducted in colonial settings (Fetterman, 2010, p. 67), some researchers prefer to apply equivalent terms as synonyms like "key actor" (Fetterman, 2010, p. 49) or "focus group participant" (Krueger and Casey, 2009, p. 2). I have chosen the term *expert* as it refers to the applied methodology of *expert interviews* within my research project.

- give precise information,
- are willing to give information, and
- are available for the interviews.

A semi-structured interview is built on open-ended questions; a major difference compared to fully structured interviews which are composed on questions that are combined with predefined answers (Gläser and Laudel, 2010, p. 41). Furthermore a semi-structured interview requires the development of an interview guide (Patton, 2002, p. 343) which helps to structure the interview, but which at the same time allows the development of a dynamic interview flow including a) switching the order of the questions to be asked and b) reflecting topics which evolve during the interview that have not been included in the interview guide (Gläser and Laudel, 2010, p. 42). I have purposefully chosen the semi-structured interview method due to the latter reason: I wanted the interviewed experts to answer the questions in any detail they liked and not to push them into predefined answers. However I wanted to have a basic structure for all six interviews that limited the BPM topics to the ones to be explored in-depth.

The resulting interview guide (appendix A.5[32]) is designed according to the next figure:

[32] In German as it is the original language of the interview guide. Any references to MeFiCom's real company name are blacked out.

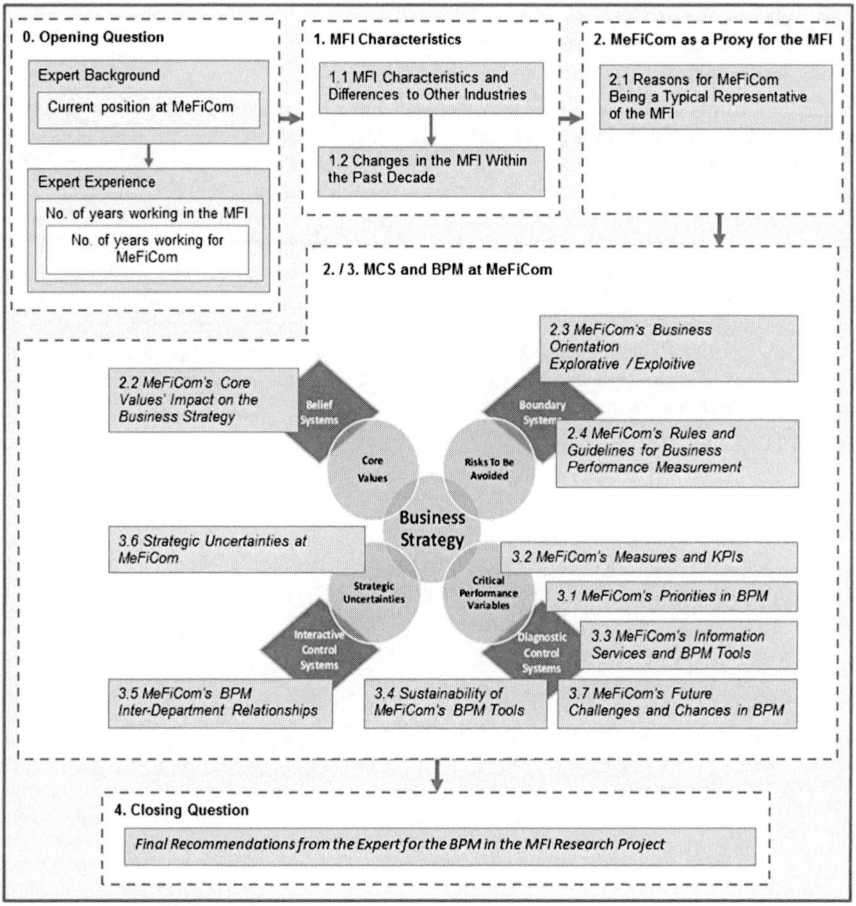

Figure 17: Interview Guide Design

The interview guide starts with an opening question (Gläser and Laudel, 2010, p. 147) related to the background of each expert asking for their individual function at MeFiCom, the number of years they have worked in the MFI and the number of years therefore worked for MeFiCom. This background and experience segment (Patton, 2002, p. 348) allows the expert to accommodate himself to the interview situation by talking about aspects he knows best (Patton, 2002, p. 352).

The first research content questions are related to the MFI characteristics and their differences to other industry branches; answers to these questions shall empirically triangulate the aforementioned MFI characteristics presented in the previous chapters. The question related to changes in the MFI within the past decade is included to analyze the characteristic of the MFI being either a more dynamic or static industry; the time span in the question was reduced in the event an expert had a less than ten years' experience in the

MFI. The sequence of asking a past-related question after a present-related question was chosen according to Patton (2002, p. 353) as interviewees are assumed to better answer past-related questions when the present situation was introduced previously as a perspective baseline.

In order to establish ties between the broader MFI and the more narrow MeFiCom perspectives I continued using a transition question (Gläser and Laudel, 2010, p. 127) by asking the experts for the reasons that make MeFiCom a typical representative of the MFI; answers to this question shall also triangulate the findings presented in chapter 6.1.2. I dedicated about 20% of the interview time to these first questions.

As described in chapter 2.3.3 I have chosen the LOC-framework (Simons, 1995) as the leitmotif within my research. In consequence the LOC-framework served as a guideline for structuring the main section of the interview according to the four LOC following a theory-based interview guide design approach (Gläser and Laudel, 2010, p. 115) – each open-ended question is deduced from the different LOC orientations as presented in figure 17:

- LOC Belief Systems:
  Question 2.2 refers to the influence and impact of MeFiCom's core values on MeFiCom's business orientation and strategy: are the core values influencing MeFiCom's workaday life actively or are they only printed on sheets of paper hanging in the meeting rooms (s. Simons "*Actions Speak Louder Than Words*" in chapter 2.3.3).
- LOC Boundary Systems:
  In order to check if MeFiCom's orientation is more explorative, exploitive or balanced in the sense of organizational ambidexterity (March, 1991; Simsek et al., 2009) I asked for the expert's corresponding evaluation in question 2.3. To emphasize the boundaries in MeFiCom's BPM, I asked for the BPM rules and standards for those setting these standards and for the sustainability of these standards in question 2.4.
- LOC Diagnostic Control Systems:
  I assigned a major share of interview time and questions to this lever as it is directly related to BPM. The experts were asked which groups or processes have the highest BPM priority in their responsible department in question 3.1.

  The following question 3.2 was related to measures and key performance indicators (KPI) which the experts control within their daily work related to the "Critical Performance Variables" (Simons, 1995, p. 63). Sub-questions referred a) to such KPI from which future developments of MeFiCom's business can be derived and b) to external indicators as BPM information source.

  In question 3.3 I asked the experts for the information services (s. chapter 2.5.1) and tools they have for the measurement of MeFiCom's performance (Simons, 1995, p. 190). Question 3.4 targets in detail the information services and asks for the sustainability[33] of MeFiCom's BPM tools within the past three years.

[33] Sustainability meant in the sense of introducing BPM tools (like specialized business queries) and their continued use within the company (referring to administration, update intervals, managerial relevance after introduction).

Question 3.7 is the last question before the closing question and is dedicated to the expert's perception of future challenges and chances for MeFiCom's BPM a) for the whole company and b) for the expert's department; following Patton, I have put this future-oriented question at the end of the interview as answers to this type of question tend to include speculations and therefore *"are typically less reliable than questions about the present or past"* (Patton, 2002, p. 353).

- LOC Interactive Control Systems:
  Question 3.5 depicts the interrelation and interaction of the expert's department with other MeFiCom departments; it asks for the importance of the expert's department BPM outcome for other MeFiCom departments and vice versa.
  Question 3.6 refers to MeFiCom's most important strategic uncertainties (Simons, 1995, p. 93) and the interaction between a) the expert and his supervisor and b) the expert and his staff members (Simons, 1995, p. 121).

The interview guide ends with a closing question (Patton, 2002, p. 379) giving each expert the chance to articulate his view related to the BPM-MFI research project and to receive additional information that have not been addressed within the interview guide or during the interview course (Gläser and Laudel, 2010, p. 149).

I have pre-tested the interview guide as a quality assurance step (Gläser and Laudel, 2010, p. 107) by sending it to three professors of my university research institute; the test resulted in wording corrections in order to clarify the context of the questions.

The interview guide helped me to give each interview a common structure as described above, but at the same time to stay flexible and true to the question sequence; I intentionally did not send the interview guide to the participating experts before the interview took place: I did not want them to focus and prepare themselves exclusively on the questions within the guide. In fact I asked them not to prepare anything at all, but just to attend the interview and to respond freely to the questions asked.

Besides the interview setting area and pre-testing portion of the interview guide, I was determined to have a useful and respectful informed consent form (Hammersley and Atkinson, 1995, p. 263; Patton, 2002, p. 407; Creswell and Plano Clark, 2011, p. 176; Lapan et al., 2012, p. 169) for the six experts taking part in the interviews. Also, following my overt ethnographic research setting (s. chapter 6.1.1), I created an information and invitation letter (s. appendix A.6) in which I described the

- extent of and responsibilities within the research project (Gläser and Laudel, 2010, p. 159),
- the interview process including audiotaping (Fetterman, 2010, p. 79; Lapan et al., 2012, p. 97),
- measures taken to guarantee confidentiality of the data (Hammersley, 1998, p. 139), including related information regarding the transcription and data analysis process,
- and protection of the experts related to their statements, including deletion of the audiotapes after transcription[34], anonymization of the expert's names and other

[34] To clarify the point in time when personal data, like interview audiotapes, are deleted is a requirement to be included in an informed consent (Helfferich, 2011, p. 190) following the German Data Protection Act section 4 in connection with section 20 (BDSG, 2015).

details that may relate statements directly to the experts (Helfferich, 2011, p. 191) as well as the expert's right to reject answers (Yin, 2009, p. 73).

This information and invitation letter was signed by my doctoral supervisor Prof. Dr. Ronald Gleich and me to emphasize the etic perspective within the interviews. The letter was sent to each expert one week before the first interviews took place.

6.3.2 Data Acquisition

All six interviews took place between February 5th and 13th, 2014 on MeFiCom's premises. I planned the interview schedule in a way that I had no more than two interviews per day and that they were separated by a break of several hours (Gläser and Laudel, 2010, p. 166). One expert and I took part in each interview. In the beginning of each interview I gave an introduction (Gläser and Laudel, 2010, p. 144) explaining the reasons for the interview and my role as a researcher. I emphasized that I personally would transcribe the interviews which guarantees the experts the confidentiality of their statements as no one other than me had access to the original audio files and non-anonymized raw data.

Although I knew the experts personally due to my position at MeFiCom, I asked them to answer all questions even though they understood me to anticipate the answer. I have chosen this approach in order to a) emphasize the etic perspective, i.e. my role as a researcher as against my role as a colleague, and b) include all necessary information within the subsequent interview transcripts.

Before the actual interview and audiotaping ensued, I asked each expert to sign an informed consent (s. appendix A.7) in order to document their approval in the interview process. This document included a reference to the information and invitation letter (s. previous chapter) and their approval to quote single interview sentences in publications (Hammersley, 1998, p. 139). In addition it included the approval to audiotape the interview, to transcribe the audio-files and to analyze the retrieved data. All six experts willingly signed the informed consent document.

The interviews were digitally audiotaped using a recorder of the type Sony IC Recorder ICD-PX312. Audiotaping bears the risk of increasing the possibility for socially desired answers (Kuckartz, 2014, p. 134), but having interviews without audiotaping assumes the higher risk of missing important information in the subsequent data analysis, which outweighs potential social desirability bias (Gläser and Laudel, 2010, p. 157). The resulting audio files have been saved in the mp3-file format for the subsequent transcription (s. chapter 6.5.1).

Besides audiotaping I also took field notes during the interviews (Patton, 2002, p. 383) and created field protocols shortly after the interviews in order to log the interview sequence, interview atmosphere and any additional noticeable circumstances like interruptions (Gläser and Laudel, 2010, p. 192; Helfferich, 2011, p. 193).

Following Gläser and Laudel's recommendations regarding active listening (2010, p. 173), I

- did not interrupt the experts during their answers,
- reacted flexibly to the experts' responses, including changing the question sequence according to the interview flow in the event the expert picked up a topic earlier than planned in the interview guide,
- clarified unclear answers that seemed uncertain,

- and requested details and further information in case their answers did not correspond to the actual questions.

The interviews took between 40min and 1h 20min as presented in table 15[35]. A total of nearly six hours interview recordings have been established with an average interview length of nearly one hour:

Expert	Interview Recording Length [hh:mm:ss]
B1	**01:19:58**
B2	**01:00:03**
B3	**01:24:58**
B4	**00:39:47**
B5	**00:47:05**
B6	**00:41:54**
Total Length	**05:53:45**
Average Length	**00:58:58**

Table 15: Expert Interviews – Recording Length

The atmosphere in all interviews can be described as friendly, relaxed, open and cooperative. The fact that the interview guide was not sent to the experts prior to the interview helped to create this positive atmosphere during which I could actively switch the sequence of the questions according to the nature of their responses. Assuming a potential expert's anticipation of a line of inquiry, I determined that switching the question sequence would have been more difficult if the experts were informed ex ante about the questions and their planned sequence.

6.4 Group Discussion

After having gathered information separately per each unit of analysis within the six expert interviews, I proceeded to establish a holistic view on BPM at MeFiCom by applying a group discussion method according to Bohnsack et al. (2010). In contrast to a focus group (Krueger and Casey, 2009, p. 85), the researcher within a group discussion plays a reflexive role (Naderer and Balzer, 2011, p. 293): a group discussion begins with an open exchange which is not influenced by the researcher but rather aims for a free discourse between the discussion participants (Bohnsack et al., 2006, p. 76). This approach should reduce any possible researchers' bias (s. chapter 6.1.1) as the discussion participants chose the subjects and their sequence according to the relevance they – and not the researcher – saw in the topics. A detailed questioning route as applied in focus groups (Krueger and Casey, 2009, p. 35) does not exist in group discussions; in fact questions posed by the researcher in this first part of the group discussion are only appropriate in the event the self-maintaining momentum of the discussion is at risk to get interrupted. The open discussion part is followed by a series of immanent questions posed by the researcher leading to the third and

[35] The experts' names have been replaced with a *B* for the German word *Befragter* (which means *interviewee* when translated into English) and a consecutive number.

last part of the group discussion, which consists of exmanent questions posed by the researcher to include all issues that have not been covered in the previous discussion parts (Bohnsack and Przyborski in Buber and Holzmüller, 2009, p. 499).

The six experts invited to the group discussion were the same participants as in the semi-structured expert interviews. They were all members of MeFiCom's top management team and therefore share a common collective profile regarding the management of MeFiCom. This similitude represents a prerequisite within the composition of the group discussion participant sample, following the method according to Bohnsack (Bohnsack et al., 2006, p. 76).

The group discussion at MeFiCom took place on February 27th, 2014 – two weeks after the last semi-structured expert interview (s. previous chapter). Having the discussion timed close to the interviews had the advantage of the experts already having made up their minds on BPM and therefore were cognizant of the research topic. An introduction to the research topic was therefore not necessary as it left the major share of the discussion time for the actual discourse between the experts. Two of the six experts on short notice had to cancel their participation within the group discussion[36]. The group discussion was audiotaped similar to the semi-structured interviews.

The following table shows the timeline of the different group discussion parts as taken from the corresponding transcripts:

Discussion Part	Timestamp in Transcript Start [hh:mm:ss]	End	Transcript Paragraph Number
0. Welcome	**00:00:00**	**00:02:19**	**1 - 6**
1. Presentation of Discussion Rules	**00:02:20**	**00:04:16**	**7 - 18**
2. Presentation of Initial Stimulus	**00:04:17**	**00:06:25**	**19 - 43**
3. Open, Unguided Discussion - a	**00:06:26**	**00:19:18**	**43 - 109**
4. Open, Unguided Discussion - b	**00:19:19**	**00:39:10**	**110 - 271**
5. Immanent Question Part	**00:39:11**	**01:00:17**	**272 - 482**
6. Exmanent Question Part	**01:00:18**	**01:47:00**	**483 - 882**
7. Wrap-Up	**01:47:00**	**01:55:41**	**883 - 960**

Table 16: Group Discussion – Time Flow

I opened the group discussion session with a short Microsoft PowerPoint presentation (s. appendix A.8) by introducing a set of discussion rules which were intended to be applied during the group discussion:

1. All experts have equal rights in the discussion.
2. The time allotted for speaking would not be assigned to the experts.
3. It shall be discussed consecutively and not in parallel.
4. Perspectives and positions stated within the expert interviews can be repeated within the discussion in order to share with the group.

[36] These two experts took part in the subsequent member checking session (s. chapter 6.7) and therefore got the chance to include any missing information related to their area of responsibility in order to create a holistic perspective on BPM at MeFiCom.

The goal in applying the first and second rules was to have an open discussion between all participating members as it embodies the basic format of this group discussion method. The third rule was necessary in the context of the subsequent transcription process: having a discussion of two or more experts in parallel bears the risk of missing information during the transcription process or incorrectly assigning statements to the experts. The fourth and last rule made clear that the group discussion was not limited to only aspects that each expert had not as yet stated, but which could in fact be repeated in the group for mutual discussion with the other experts.

I started the group discussion (open, unguided discussion part "a") with the following initial stimulus (Bohnsack et al., 2010, p. 157) in order to elicit the discussion:

> *What are the reasons for the absence of Business Performance Measurement at MeFiCom which*
>
> - *generates reliable data*
> - *as a base for future activities*
> - *for all functional areas?*

This initial question included the assumption that MeFiCom does not apply an integrated and holistic BPM system. This rather provocative assumption prompted the participants to quickly engage themselves into the discussion. One expert replied that MeFiCom actually has *some kind of BPM*, which was disputed by another expert who made specific reference to the characteristic of being holistic.

After 20 minutes of the open, unguided discussion the flow of ideas began to wane. In order to speed up the discussion again, I faded in the following second stimulus which was hidden in the Microsoft PowerPoint presentation before:

> *Which elements form such an integrated, holistic Business Performance Measurement?*

This second stimulus enhanced the open and unguided discussion (open, unguided discussion part "b") by an additional twenty minutes. The participants have discussed on the subjects and in the sequence as illustrated in the following table:

Discussion Subject (Open Discussion)	Timestamp in Transcript Start [hh:mm:ss]	End	Transcript Paragraph Number
Open, Unguided Discussion - a	**00:06:26**	**00:19:18**	**43 - 109**
Status Quo Diagnostic Control System at MeFiCom	00:06:26	00:10:51	43 - 62
Reliability of BPM Data	00:10:52	00:14:35	63 - 92
Challenge of Predicting the Future in BPM	00:14:36	00:17:38	92 - 103
Reflection of MeFiCom' s Business Model in BPM	00:17:39	00:19:18	103 - 109
Open, Unguided Discussion - b	**00:19:19**	**00:39:10**	**110 - 271**
Integrating Sales CRM Information in Other BPM Tools	00:19:19	00:24:06	110 - 150
Reflecting Required Production Capacity by Simulation	00:24:07	00:28:57	151 - 179
Reflecting BPM in Information Services like ERP	00:28:58	00:31:18	180 - 205
Diversity of BPM Systems and Resulting Redundancies	00:31:19	00:33:18	206 - 224
Worldwide Comparability of BPM in Information Systems	00:33:19	00:37:53	225 - 260
Market Situation (Emergent vs. Mature) Reflected in BPM	00:37:54	00:39:10	261 - 271

Table 17: Group Discussion – Expert's Subjects

The experts have chosen the reliability of BPM data and their benefits in predicting future trends as the major topic within the first open discussion segment. Information services (s. chapter 2.5.1) and related issues like a diversity of different BPM tools at MeFiCom were reflected in the second open discussion part.

I took field notes during the whole group discussion; I especially noted ambiguities in the initial open discussion parts which were clarified in the subsequent immanent question portion of the discussion; for example, the term *global* was partly used for MeFiCom's worldwide business activities but also for the German MeFiCom organization as a whole.

Within the subsequent exmanent question part I asked for BPM aspects at MeFiCom that were not tackled in the previous discussion. Referring to the leitmotif of my research project I asked the participants for the boundaries (s. LOC Boundary System in chapter 2.3.3) in which they manage MeFiCom; in particular I asked for external boundaries like environmental related legislation (s. chapter 3.3). A further question referred to the corporate strategy related to organizational ambidexterity (March, 1991; Simsek et al., 2009) and the existence of exploitation and exploration at MeFiCom.

In the evening after the group discussion I created a field protocol in which I noted the course of the group discussion session. The group discussion generated one hour and 56 minutes of audiotape as an additional part of the qualitative data material which I analyzed together with the semi-structured expert interviews.

6.5 Qualitative Data Analysis

> "We see analysis as three concurrent flows of activity: (1) data condensation, (2) data display, and (3) conclusion drawing/verification." (Miles et al., 2014, p. 12)

After having collected qualitative data at MeFiCom within the semi-structured expert interviews and within the group-discussion, I continued to condense this data by transcribing the generated audio files (s. chapter 6.5.1) and by coding them using a *computer-aided qualitative data analysis software* (CAQDAS – s. chapter 6.5.2). The resulting data will be displayed in a coding tree including the categories established within a qualitative content analysis. The conclusions drawn from this qualitative data analysis will be presented in chapter 6.6.

6.5.1 Transcription

Transcription is the process of transforming verbal data stored on audiotapes or videos into written data for further data analysis (Kuckartz, 2014, p. 133; Kowal and O'Connell, in Flick, 2014, p. 64).

During March 2014 I personally transcribed the seven digital audio files retrieved from the six expert interviews and the group discussion. I rejected the possibility of making use of external transcription services for this process as detailed MeFiCom business and company information have been collected during the interviews and group discussion that were not supposed to be processed by a third party. Furthermore I assured the participating experts to personally transcribe the audiotaped sessions to safeguard confidentiality of their personal statements (s. chapter 6.3.2). Although this approach resulted in several weeks of transcription work, it offered me the possibility of familiarizing myself with the data (Kuckartz, 2014, p. 139) and to identify parts within the transcripts[37] that appeared especially interesting for the subsequent coding process. For the transcription process I followed the seven-steps approach by Kuckartz (2014, p. 133):

1. *Transcription Notation:*
   I decided to apply the less complex transcription notion system[38] by Kuckartz, Dresing & Pehl (Kuckartz, 2014, p. 133) which focusses on the content of the verbal information (Dresing and Pehl, 2013, p. 19).
   Applying a less complex transcription notation system is applicable when dealing with expert interviews, as described by Helfferich (2011, p. 162).
2. *Transcription Process:*
   For the actual transcription process of the audio files I applied the transcription support software f4 (F4, 2015). The digital audio files, which were stored in the mp3 file format, were directly imported into f4. Although f4 is not a speech recognition software[39], it is designed to enhance the transcription process by offering several mechanisms that speed up the transcription like the possibility of reducing or increasing the audio playback speed in order to adapt to the individual transcription speed. It can also be configured to automatically include timestamps, a paragraph number and a changing prefix for interviewer / interviewee each time the return button is pressed. Key combinations can be assigned to recurring transcription elements like "(…)" for lulls in conversations which also reduces typing efforts.
   I transcribed one hundred percent of all verbal audio file data including questions I asked as well the experts' answers, following the recommendations of Schreier (in Flick, 2014, p. 174).

[37] I marked such interesting sequences by adding my initials in double brackets "((MD))" and the corresponding time stamp within the recording as a tag into the transcription text. Saldaña (2010, p. 16) refers to this process as "pre-coding".

[38] Please refer to Kowal and O'Connell (in Flick, 2014, p. 74) for an overview of other transcription notation systems like GAT2, DT or HIAT.

[39] Speech recognition software is at present not able to transcribe verbal data automatically without extensive manual rework (Dresing and Pehl, 2013, p. 30).

3. *Correction of the Transcripts:*
   I have read all transcripts as a quality assurance step in f4's synchro-mode: in that mode the audio file is replayed and the corresponding transcript text passage gets highlighted simultaneously. I corrected all remaining transcription errors resulting in the final, but not yet anonymized transcript version.

   The transcription process led to a total of 228 German DinA4 format pages (font type: Arial, font size: 12, single line spacing) as shown in the following table:

Expert	Interview Recording Length [hh:mm:ss]	Transcript	
		Digital Size [Bytes]	Number of Pages
B1	**01:19:58**	173.629	**43**
B2	**01:00:03**	76.669	**19**
B3	**01:24:58**	210.073	**51**
B4	**00:39:47**	58.215	**14**
B5	**00:47:05**	90.193	**22**
B6	**00:41:54**	60.803	**15**
Group Discussion	**01:55:41**	264.037	**64**
Total Length	**07:49:26**	**933.619**	**228**

Table 18: Overview Resulting Transcripts Expert Interviews and Group Discussion

   The richest information, correlated with the largest Microsoft Word files in bytes and the highest number of Microsoft Word transcription pages, were retrieved from those MeFiCom experts with the longest experience within the company.

4. *Anonymization of the Transcripts:*
   The transcripts have been anonymized following the process stated in the interview invitation letter to the participating experts (s. chapter 6.3.1). A translation file was created in Microsoft Excel in which the changes were logged during the anonymization process (Kuckartz, 2014, p. 140).

5. *Formatting of Transcripts to be Processible in CAQDAS:*
   f4 automatically includes formatting elements into the transcripts which are designed to be read by the CAQDAS MAXQDA (Kuckartz, 2014, p. 138) that I applied to code the transcripts.

6. *Store the Transcripts in RTF or DOC/X format:*
   f4 is preset to store transcripts in RTF file format.

7. *Import Transcripts into CAQDAS:*
   As a final step within the transcription process I deleted the digital audio files as a measure to protect the experts related to their statements (as described in chapter 6.3.1) and imported the transcripts in the CAQDAS MAXQDA.

6.5.2 Analyzing and Coding Process Using CAQDAS

The coding process embodies the heart of the qualitative data analysis or as Miles et al. (2014, p. 72) noted: *"[...] coding is deep reflection about and, thus, deep analysis and interpretation of the data's meaning"*. In order to condense the transcripts accordingly, I applied a *content-structuring qualitative content analysis* (Mayring, 2010, p. 98; Kuckartz, 2014, p. 77; Schreier, 2014a) of the transcripts during April and May 2014.

In contrast to the open coding approach in grounded theory (Thornberg and Charmaz, in Flick, 2014, p. 153) this method allows one to combine deductive (concept-driven) and inductive (data-driven) coding processes in one coding system (Schreier, in Flick, 2014, p. 170). Besides combining induction and deduction, Langley (1999, p. 708) adds a third dimension, i.e. the researcher's inspiration in theorizing from organizational process data:

> "In theorizing from process data, we should not have to be shy about mobilizing both inductive (data-driven) approaches and deductive (theory-driven) approaches iteratively or simultaneously as inspiration guides us."

Figure 18 illustrates the adoption of the content-structuring qualitative content analysis principle into my research process:

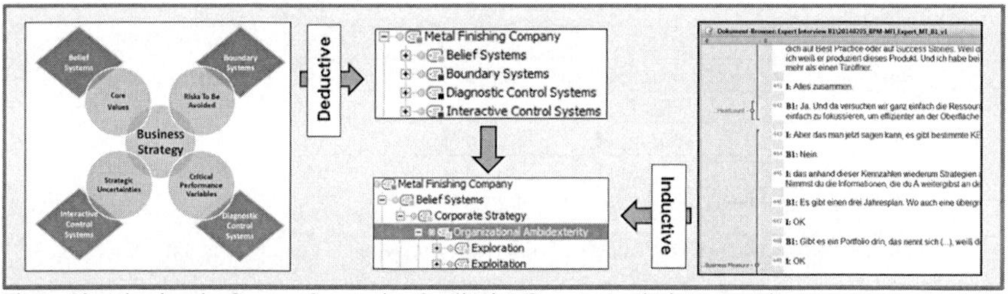

Figure 18: Adopting the Content-Structuring Qualitative Content Analysis Approach

The content-structuring approach allowed me to include the LOC-framework (s. chapter 2.3.3) as my MCS leitmotif into the qualitative data analysis: the four LOC form the four main categories of the coding tree related to MeFiCom (s. the left und upper mid part in figure 18). Further sub-categories are data-driven created during the qualitative data analysis process (s. the right and lower mid part in figure 18).

For this qualitative data analysis process I applied the computer aided qualitative data analysis software (CAQDAS) MAXQDA. The application of CAQDAS within refereed social science journal papers using qualitative methods has taken off in the year 2000 and significantly increased ever since as shown by Gibbs (in Flick, 2014, p. 279).

For setting up the analysis process (Gibbs, in Flick, 2014, p. 281) I created seven document folders within the MAXQDA document system, one folder for each of the six expert interviews and the group discussion; the transcripts, a photo of the interview / group discussion setting (in order *"to account for the contexts in which interview data are generated"* – Roulston, in Flick, 2014, p. 309), the field notes and the field protocols which have been imported in these folders (s. appendix A.9).

Following the expert interview guide (s. chapter 6.3.1), I divided the code system (Saldaña, 2010, p. 21) in the following two areas of interest as shown in the following MAXQDA screenshot:

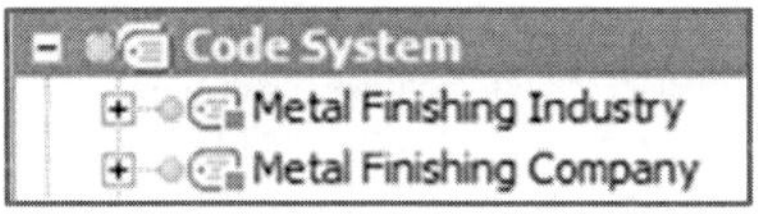

Figure 19: Two Areas of Interest in the MAXQDA Code System

The MFI area includes codes mainly related to the expert interview guide questions 1.1, 1.2 and 2.1 referring to the industry; further codes are included as soon as the experts referred to the MFI in their answers to other interview questions or within the group discussion. The MeFiCom area includes the majority of all codes referring in detail to the company. As outlined above, this second area is split into four deductively created main categories corresponding to the four LOCs:

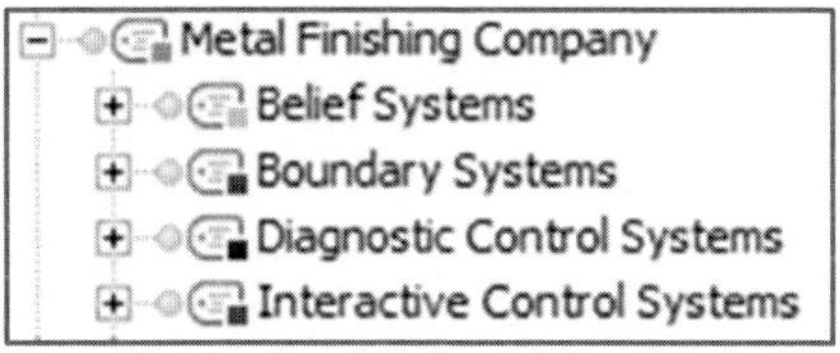

Figure 20: Adoption of the four LOCs within the MAXQDA Code System

I assigned a certain color to each main category in MAXQDA which was passed on to the corresponding transcript passage codes (Saldaña, 2010, p. 23). This procedure helped ease the analysis process as the codes and subcodes could at a glance for example, be identified as either being diagnostic (blue) or interactive (purple).

The data-driven sub-categories were allocated to the four theory-driven main categories during the first coding cycle (Saldaña, 2010, p. 45) by reflecting the sense of the codified transcript passage and by relating it to one of the LOCs. Miles et al. (2014, p. 74; s. also Maxwell and Chmiel, in Flick, 2014, p. 25) suggest *"descriptive codings"* as a coding method for being *"especially helpful for ethnographies"*. Following Miles et al., I applied nouns as labels to summarize *"the basic topic of a passage of qualitative data"*. With Saldaña (2010, p. 72) it is possible to assign subcodes ("children") to higher ranked general descriptive codes ("parents") which I adopted for the hierarchical coding process of the qualitative data. Therefore I partly treated sub-categories also as codes and assigned transcript passages to these sub-categories (s. figure 23).

I started the coding process with the B3 transcript being the largest of all expert interview transcripts (s. previous chapter) expecting that this would include the richest information. I continued with the other expert interviews and ended by coding the group discussion transcript. Although the transcripts were in German, I coded in English as the latter is the language I have chosen for my dissertation thesis; this is also the reason why I have chosen descriptive and not in-vivo codings (Lapan, 2012, p. 44; Saldaña, 2010, p. 74) for the coding process.

As an example of the coding process in the deductive *Belief Systems* main category, expert B2 was asked for his assessment on MeFiCom's orientation focusing on exploration and exploitation (s. chapter 6.3.1 – interview guide question 2.3). He answered as follows:

> B2: In my opinion MeFiCom uses a quite balanced way. This means on the one hand the sustainment of existing customers by satisfying them with our product portfolio

> by creating and offering new services, and on the other hand, a market explosion and expansion. (…) Here it gets very well monitored that both factors are balanced.
> (Translated from German transcript document "Expert Interview B2", paragraph 112)

I coded B2's answer as *Organizational Ambidexterity*, being a sub-category (or child code) of the category *Corporate Strategy* (or parent code) within the deductive main category *Belief Systems* as shown in the following MAXQDA screenshot:

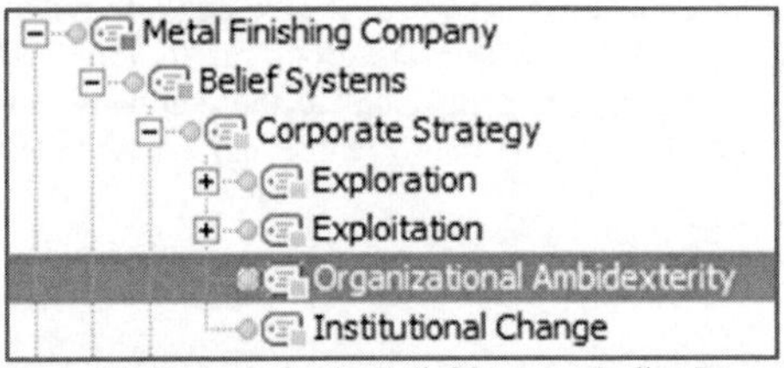

Figure 21: Example for the Belief Systems Coding Process in MAXQDA

In addition I coded B2's answer as an *illustrative example*[40] (Saldaña, 2010, p. 45) as reflecting exploration and exploitation simultaneously in a successful organization is a fundamental of the organizational ambidexterity concept (Simsek, 2009, p. 865). A total of 158 transcript passages have been marked as illustrative examples, many of which have been already marked with a ((MD))-tag originating from the transcription process (s. previous chapter).

Within the deductive main category *Boundary Systems*, I assigned the following group discussion transcript passage[41] to *Environment* referring to the *Legislation* category:

> I: How do you reflect the influence of EXTERNAL BOUNDARIES on our business performance measurement? (…) Is it obstructive? Are there guidelines which can be implemented without any problems? Are these subjects that can create uncertainties in the future and on which we have to react again?
> B1: Referring to environmental legislation, it is mainly influenced by one's personal perception, which one personally has. There are people who are very ENVIRONMENTAL ORIENTATED, which say: sure, that has to be. And you have to integrate this somewhere into business performance measurement.
> B3: Mhm (affirmative).
> B1: Others say: I don't care. This is part of it, yes. And then we have to look on what will be the result. (...) I think this is difficult to answer. From my point of view, I say, this has to take place. And this will be and should not be hindered by business performance measurement. Because it is part of it, as there are enough examples which one could read in the press in the past days. There are countries that cannot breathe anymore, because they do not measure such things, because they are only measuring the output. But (laughing) what happens in the environment – no one measures.
> B3: Mhm (affirmative).
> B1: Yes.
> B4: Ok, but the new laws bring new measures/ performance measures. For example we have now included LTA in our management report.
> B3: Mhm (affirmative).

40 In German: Ankerbeispiel (Mayring, 2010, p. 106; Kuckartz, 2014, p. 83).

41 "I:" in the transcript passage stands for *Interviewer*.

B4: HSE.
B3: Mhm (affirmative).
(Translated from German transcript document "Group Discussion", paragraph 485-494)

This passage shows a direct relation of the legislation being adapted into MeFiCom's management reporting as outlined by B4. The following screenshot shows this *Boundary Systems* code within MAXQDA:

Figure 22: Example for a Boundary Systems Code in MAXQDA

As an example of a *Diagnostic Control System* category, I coded the following transcript passage as *KPI-Engineering*[42]:

I: Ok. B4, a final question: if you reflect the topics and subjects which we have addressed: what do you think are further aspects and subjects, which we should examine within our research project "Business Performance Measurement in the Metal Finishing Industry"? What would you say: "Mike, investigate this. This is my wish; this is something I would like to have."

B4: Mhm (affirmative). (...) I believe in (...) complex KPI. I think that simple KPIs are established today. But we have too many ((MD))
KPIs. That means that we have to increase the complexity of KPIs and to reduce the number of KPIs. That is a challenge, because they have to be transparent and comprehensible. And the more complex they get, the more difficult they get. That means that one should not do work which ends in telling nothing. Therefore we have/ for example we currently try to define market indicators which give trends even if the people do not necessarily know which components form those complex indicators. But it needs a kind of KPI-Engineering ((MD))
which does not exist today."

(Translated from German transcript document "Expert Interview B4", paragraph 132-135)

This code became part of the sub-category *Challenges,* in the *Internal Controls* sub-category as part of the *Business Performance Measurement - Measure Output* category in the *Diagnostic Control Systems* deductive main category. *KPI Evolution and Maturing* is the sub-category of *KPI-Engineering* including *Sustainability* and *Interdependency* as further sub-categories as shown in the following screenshot:

[42] One may notice that in this case the descriptive code *KPI-Engineering* equals an in-vivo code as it is used by B4 in the interview. However this commonality happened rarely due to the different languages applied in the interview and in the coding process; related in-vivo code types are marked as ▨ in the MAXQDA code system (s. appendix A.10).

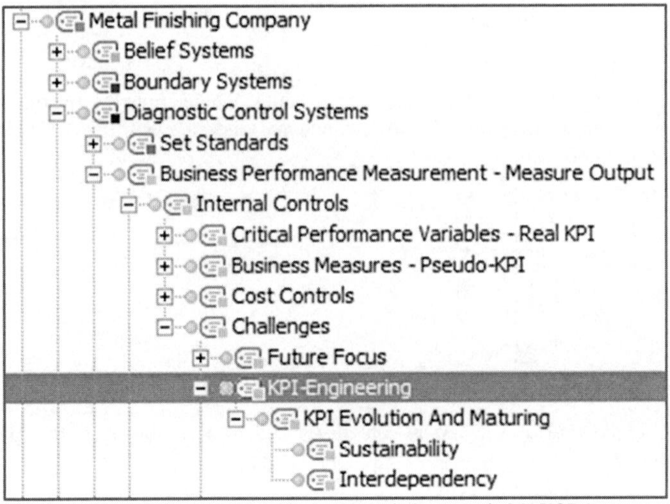

Figure 23: Example for the Hierarchical DCS Code Structure in MAXQDA

In the deductive *Interactive Control Systems* main category, I coded the following group discussion transcript passage as *Business Performance Measurement Results*, being the child code within the category *Interactive Intergroup Communication*:

B4: And that the people, too, for example the heads of production, know for which reason we make the reports. Why we do the benchmarking. It seems that this was missing. That the people understand why we do that. This afternoon we had a telephone call with one production head and said ok, there is a personnel requisition. We are in the benchmark. Explain it. (...) He understood. He reacted. He gave the explanation. (...) Yes.
I: So, with the example benchmarking: what I think is that one YEAR ago, or one and a half years, if you would have asked the heads of production; you would have probably got NO statement back. I am not sure; I float this impression to you. So this is an example that there IS A MOVEMENT in our organization.
B2: Yes, the production heads do know the benchmarking for example.
B3: Yes.
B2: But not the how-to-do.
B4: And this is no criticism, but we learn currently a lot. That means that the benchmarking is improving. Or we use it better.
B3: Correct.
B4: And so there are more opportunities for discussion.
B3: That is the point.
B4: Yes. So today we also said, ok how we/ the external temps, why are there 0.1 in January and 0.5 in February?
B3: Mhm (affirmative).
B4: And now we have understood that this is planning. And then we have to, that means each time that we use it, we can/
B3: Yes.
B4: /improve it and advance ourselves.
B3: This is absolutely right.

B4: Mhm (affirmative). (...) But communication is the key, yes.
(Translated from German transcript document "Group Discussion", paragraph 467-482)

The transcript passage refers to the internal Microsoft Excel cost analysis tool *benchmarking* (s. chapter 6.1.1) which MeFiCom applies to control the costs of goods sold and which is used to communicate and to discuss related business performance results on a monthly base. The following MAXQDA screenshot shows the code system structure of this passage:

Figure 24: Example for an ICS Code in MAXQDA

A code memo (Kuckartz, 2014, p. 144; Barbour, in Flick, 2014, p. 502) was attached to every code included in the MAXQDA code system in order to give a brief explanation of the code and to distinguish between an inductively or deductively derived code. The following MAXQDA screenshot shows an example for the inductive code *Real Life Practice*, being a child of the parent code *Core Values*, within the *Belief System* main category:

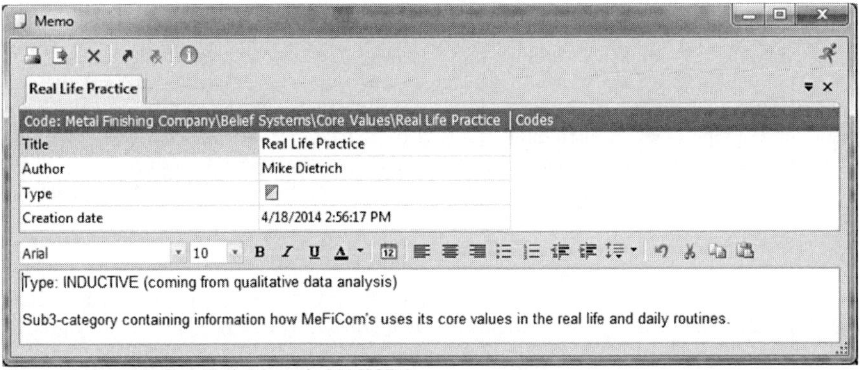

Figure 25: Example for a Code Memo in MAXQDA

To ensure the quality of the analysis, I applied a second coding cycle in which I checked if the established coding system from the first coding circle was valid for all transcripts (Kuckartz, 2014, p. 88).

The content-structuring qualitative content analysis resulted in the code system attached in appendix A.10. A total of 1,070 transcript passages have been assigned to the code system split into categories as presented in the following MAXQDA screenshot:

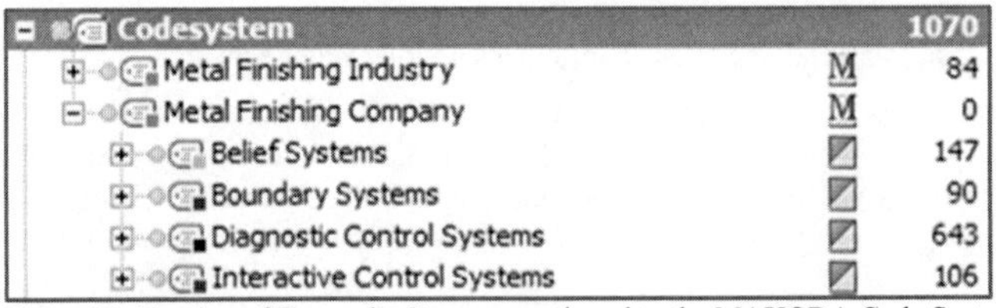

Figure 26: Number of Transcript Passages assigned to the MAXQDA Code System

A total of 84 transcript passages have been assigned to codes within the MFI as area of interest. The MeFiCom area of interest includes 986 transcript passages assigned to the different code categories. The majority of all coded transcript passages is assigned to the MeFiCom main category *Diagnostic Control System* representing BPM within the LOC framework (s. chapter 2.3.3) as it is in the focus of this research project.

6.6 Results – MFI Cause and Effect Model as Base for BPM

After having condensed the transcripts in the content-structuring qualitative content analysis, and after having displayed this condensed data in the code system, I continue with drawing conclusions and verifications (Miles et al., 2014, p. 12) as the final qualitative research process step. I decided to graphically draw and display the results of the qualitative data analysis within a cause and effect model following three perspectives:

- Qualitative research methods:
  Displaying the results of qualitative research graphically is an established academic qualitative method: Saldaña (2010, p. 201) refers to this graphical method as "Operational Model Diagram" in which the "*researcher's analytic synthesis*" is combined with "*the reader's visual grasp of the study*". Saldaña applied such an operational model diagram in order to illustrate his findings from a three year ethnographic study including "*literature reviews, fieldwork, data analysis, and writing*" which is methodological comparable to my research project. Miles et al. (2014, p. 236) advocate the application of "Causal Networks" in order to organize "*field study data in a coherent way*".
- BPM research methods:
  Creating a cause and effect model is an academically accepted design step in the modeling process of BPM systems: the third issue of the 29th Journal of Management Studies was dedicated to "Cognitive Mapping". According to the issue's editorial (Eden, 1992, p. 262), cognitive mapping represents an alternative way of communication in order to display elements of thinking in a map-like diagram.
  Neely and Bourne (2000, p. 4) advocate the creation of a "Success Map", i.e. a cause and effect model as key for the design of a good measurement system. Such a success map shall explain *"the organization's strategy and the manager's theory about how the business operates."*
  "Strategy Maps" (Kaplan and Norton, 2003, 2004; Simons, 2005, p. 86) are a further example of cause and effect models being used for the design of BPM systems especially in the design process of BSCs (s. chapter 2.4.2).
  By analyzing such strategy maps de Salas and Huxley (2013) have identified this method as rather inappropriate for SMEs. With their "Strategy-to-Process Maps" they introduced a cause and effect method which would provide clear links between

organizational strategy elements and those processes which would be most important for the strategy execution.
- Practitioner oriented approach:
  With a graphic approach, I set about presenting the results of the qualitative data analysis in a comprehensible way in order to allow the experts to falsify or approve the results (s. member checking in chapter 6.7). I further assumed that showing the experts solely the CAQDAS code system without interrelations of the codes would mean a shortcoming in modeling an integrated and holistic MFI-BPM system.

Various terms exist for the graphical interpretation of qualitative results within a BPM model process as presented in the different academic perspectives above. I continue by using the term *cause and effect model* (CEM) within my dissertation referring to the umbrella term applied by Neely and Bourne (2000), as it explicitly combines the organization's strategy with the manager's view of the organization – a principle which is also reflected in Simons' (1995) LOC-framework (s. chapter 2.3.3).

This approach embodies the most detailed level in my analyses of the MFI referring to research question *RQ.1c: How can the MFI be described and differentiated from other industry classes?* Displaying the qualitative codes as BPM model elements and their interactions is necessary for presenting the model in its integrality. The aim is to develop an explanatory model in order to "*generate the behavior of the real system by similar structure and elements*" (Bossel, 2007, p. 25).

I applied Microsoft Visio Premium 2010 as software for the transformation of the CAQDAS code system into the MFI CEM, which took place in May 2014. To enhance the CEM I applied a "stocks and flows" notation based on system dynamics (Sterman, 2000, p. 191). Enhancing the quality of strategy maps by applying a system thinking approach is described by Kunc (2008), highlighting the positive effects of including reinforcing and balancing causal loops in the BPM system creation process.

As a first step, the basic concept of MFCs was modeled accordingly, resulting in the following MFI core process (s. chapter 3.2):

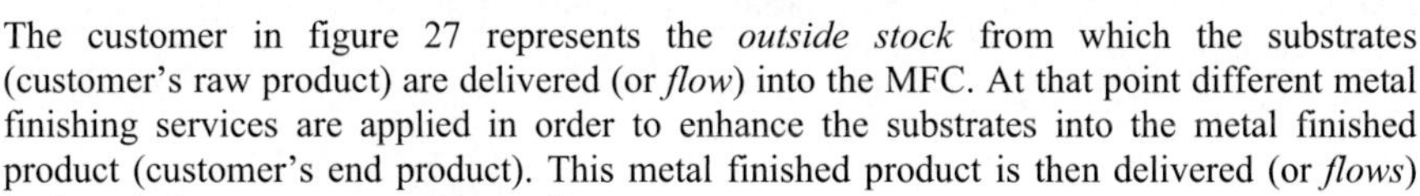

Figure 27: MFI Core Process adopted in System Dynamics Notation

The customer in figure 27 represents the *outside stock* from which the substrates (customer's raw product) are delivered (or *flow*) into the MFC. At that point different metal finishing services are applied in order to enhance the substrates into the metal finished product (customer's end product). This metal finished product is then delivered (or *flows*) back to the customer as outside stock.

Having formed the main categories in the coding system, the LOC-framework (Simons, 1995) is also the leitmotif in the CEM by representing quadrants within the MFI model as presented in the following figure:

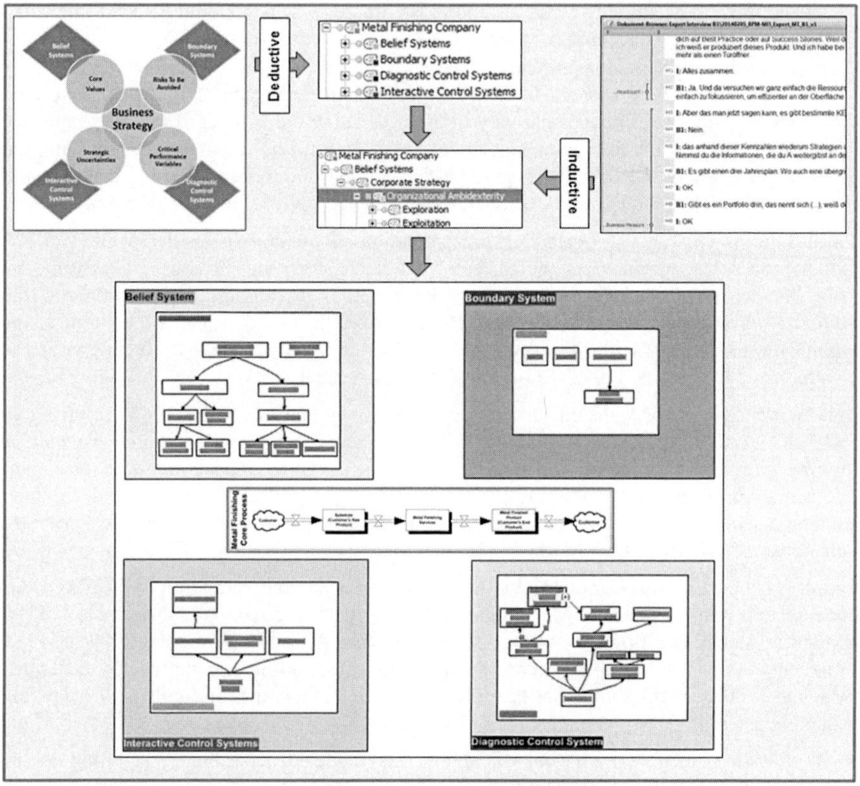

Figure 28: From the CAQDAS Code System to the Cause and Effect Model

These quadrants serve as aggregated containers (Sterman, 2000, p. 213) to be filled with the coding system sub-categories. A fifth container is added to the CEM including the CAQDAS codes related to the MFI as the research area of interest (s. previous chapter). As shown in figure 28, I adhered to the same color scheme as I already applied in the CAQDAS coding system (s. previous chapter).

The transformation process from the CAQDAS code system into the CEM is exemplarily presented for the interaction between the success factors *Service Focus* and *Quality Focus* within the *Diagnostic Control System*:

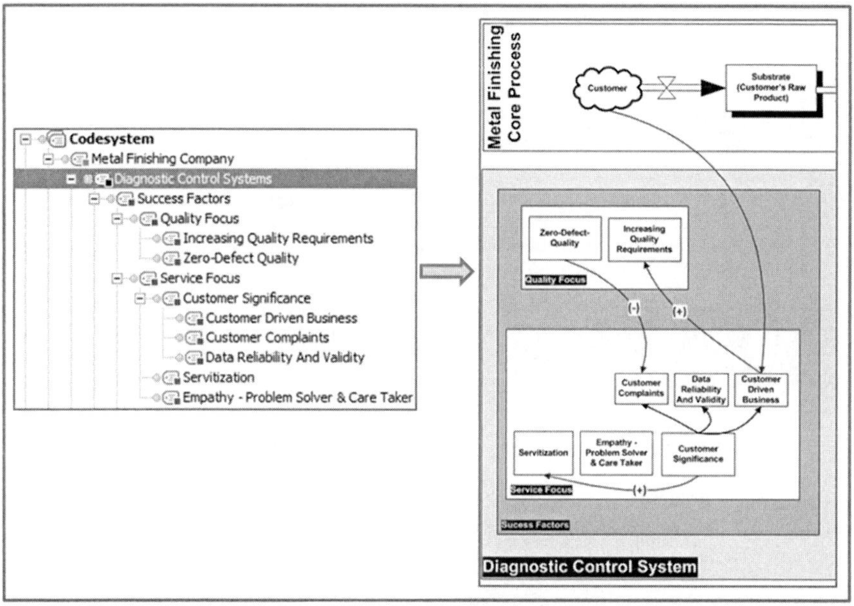

Figure 29: Example for the CAQDAS Code System Transformation

Element connections, particularly their reinforcing (+) and balancing (–) polarities (Sterman, 2000, p. 142), have been added. I followed the advice of Miles et al. (2014, p. 239) to use one-way arrows (colored red) to connect the elements in order to *"provide a temporal order to the diagrammed research story"*. In figure 29 a *Zero-Defect-Quality* approach has a balancing (negative) effect on customer complaints, i.e. delivering best possible quality results in reduced customer complaints. The *Customer Driven Business* on the other hand has a reinforcing (positive) effect on the *Increasing Quality Requirements* coming from the customer.

As delineated above the CAQDAS code system is the main source for the MFI CEM. However the CEM is also influenced by the past three years of my research, referring to the literature reviewed, the statistical data analyzed, and the knowledge produced during my ethnography at MeFiCom (s. chapter 6.1.1). As an example I created the following brief memorandum on April 12th, 2014:

> To be clarified: In case a new BPM report or tool is created, which, if any, already existing, old BPM report or tool gets removed? (Translated from a German ethnographic brief memorandum, created on April 12th, 2014)

This brief memorandum was created after a conversation with a colleague from MeFiCom's IT department who questioned the sustainability of BPM reports and emphasized the risk of information overload which he related to a large amount of ERP reports. This indication strengthens the CEM container *Diagnostic Control Systems / Information Supply / Challenges / Information Systems Design* related to the *Streamlining of Existing Systems* as shown in the following MFI CEM detail:

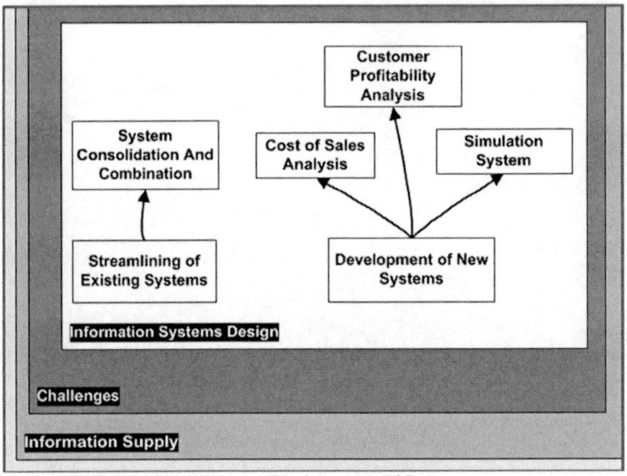

Figure 30: Information Systems Design at MeFiCom

The resulting MFI CEM includes the core process, all aggregated containers, related code elements and element connections as presented in the following overview:

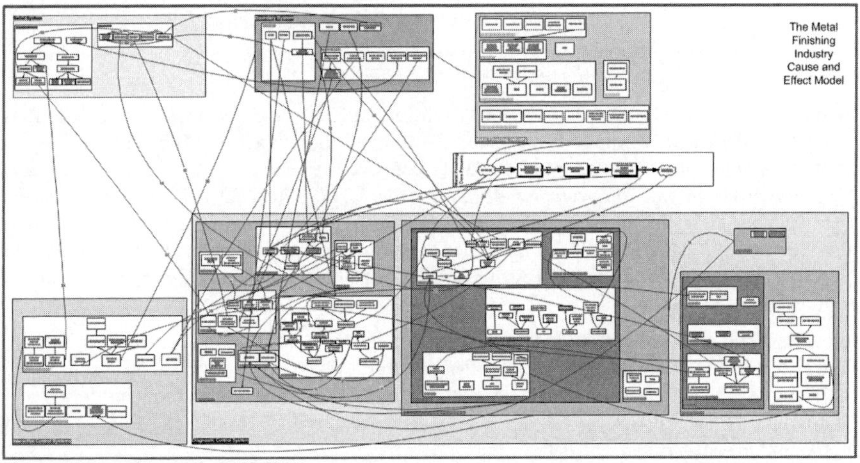

Figure 31: Overview on the MFI CEM

The entire MFI CEM can be downloaded as an Acrobat PDF file (MFI CEM, 2015) as its readability was difficult to include in its entirety due to the paper format of this thesis. In the next chapters I have split the CEM according to the different aggregated containers following the LOC-framework which allows a more detailed view and opportunities for discussion of the results. The connections between the aggregated containers are not shown in the overview, but are described in the text. I derived several propositions from these results in each chapter which I have marked with a tag (***PROP:***) at the beginning of each proposition.

Referring to the result of ethnographic research, Creswell (2013, p. 96) notes:

> "This final product is a *holistic cultural portrait* of the group that incorporates the views of the participants (emic) as well as the views of the researcher (etic)."

In conclusion this holistic cultural portrait corresponds to the cause and effect model of the MFI developed at MeFiCom. To validate and enhance this ethnographic single case portrait, a triangulation was performed within the German and UK MFI (s. chapter 7).

6.6.1 CEM – Metal Finishing Industry

The following figure illustrates the MFI CEM container:

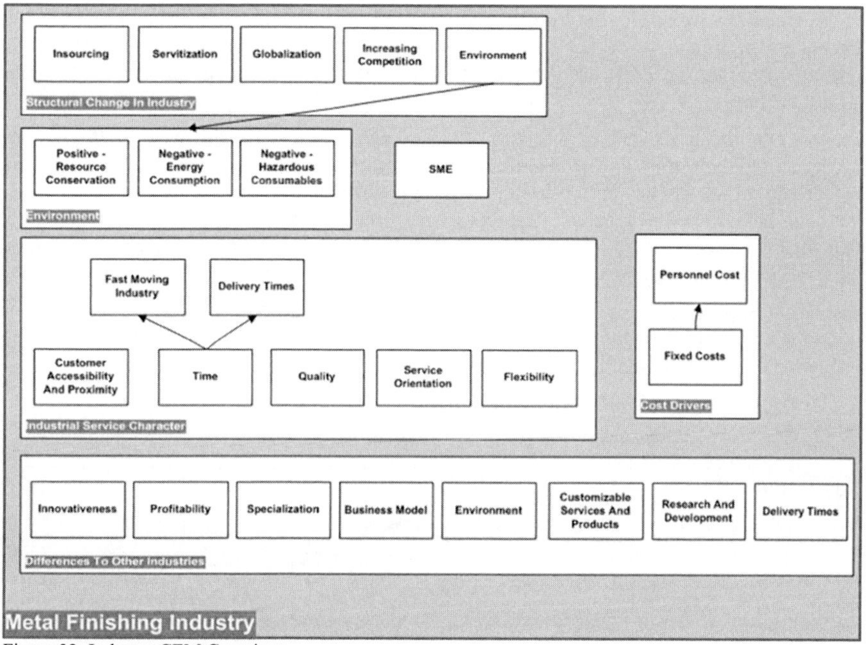

Figure 32: Industry CEM Container

This container triangulates the established MFI characteristics from chapter 3 such as the high share of SMEs that form the MFI:

PROP: *SMEs form the majority of MFI enterprises.*

The high proportion of fixed costs, mainly personnel costs, within MFCs cost calculations is also mentioned by the experts.

Service orientation and the focus on quality and reliable delivery times are elements forming the *Industrial Service Character* container. Additional elements are the accessibility and proximity to the MFC's customers, combined with a high degree of flexibility. It is the latter that refers to a difference with other industries related to customizable services and products: MFC's have to be flexible and innovative to react to changing customer products on which the metal finishing services have to be applied.

Besides innovativeness the corresponding specialized MFI business model requires the implementation of research and development processes.

PROP: *Being flexible is a prerequisite for economical successful MFCs.*

As an additional distinction of the MFI to other industries, the significance of the environment is also an element of the *Structural Change in Industry* container and is related to the legislation requiring some MFC's to alter their operations. This indication conforms to the environmental characteristics illustrated in chapter 3.3. But the qualitative research revealed an additional facet: besides the negative environmental influences like high energy consumption or the application of hazardous consumables, the MFI positively influences the environment by enabling the conservation of resources. A metal finished drill bit for example, which has a ten times higher endurance capability compared to an uncoated drill bit, conserves the raw material and energy otherwise required for the production of nine additional uncoated drill bits.

As the two major drivers of structural change within the MFI, the experts foresee an increase in globalization and related increasing competition as well as an increasing trend in insourcing. Some larger MFI customers, whose businesses have grown over the past years and who have therefore reached a critical mass of metal substrates to be coated, reflect on their make or buy decisions and tend to insource the metal finishing process by investing in corresponding finishing technology and workforce.

The MFI CEM container is connected with the customer in the MFI core process as well as with the MeFiCom Boundary Systems CEM container, as the industry influences for example the environmental legislative boundaries (s. chapter 3.3) of the MFC.

6.6.2 CEM – MeFiCom's Belief Systems

The following figure illustrates the Belief Systems CEM container:

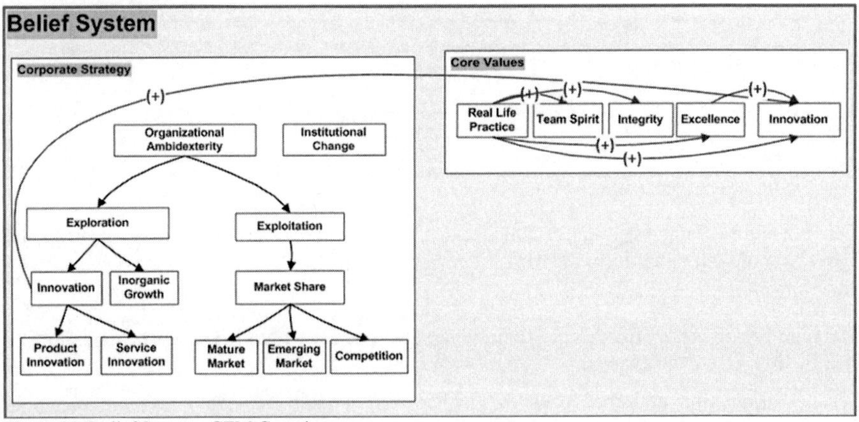

Figure 33: Belief Systems CEM Container

The sub-categories *Corporate Strategy* and *Core Values* have been assigned to the *Belief Systems*.

The results show a balanced influence of exploration and exploitation regarding the corporate strategy. Therefore I suggest that organizational ambidexterity (March, 1991; Simsek et al., 2009) has a significant influence on the MFI. Being ambidextrous refers to organizations' ability to be *"aligned and efficient in their management of today's business demands while simultaneously adaptive to changes in the environment"* (Raisch and Birkinshaw, 2008, p. 375). The significance of this research paradigm for BPM in the MFI centers on the one hand in MFC's ability to fulfill the regular customer requirements like short delivery times, and high quality standards to sustain the customer base. Conversely, MFCs have to react to changes like environmental legislation (s. REACH in chapter 6.6.3) by finding innovative ways to alter their production processes. Consequently, an MFI-BPM system has to include related measures with which both the exploration and exploitation foci can be controlled.

PROP: *Organizational Ambidexterity is a characteristic of economical successful MFCs.*

The experts mentioned the necessity of *Institutional Change* as MeFiCom operates in a mature market in which it becomes more and more difficult to secure profitable margins. This factor therefore has influence on *Business Model Innovation* in the *MeFiCom Boundary Systems* CEM container with which it is connected. In this context the experts mentioned *Inorganic Growth* as one component of *Exploration* which reinforces the related element in the *Strategic Uncertainties* main container: *Inorganic Growth* bears the risk that the expected opportunities during the acquisitions phase do not turn out in reality, as soon as the acquired organization is fully merged into the acquiring company.

With respect to the core values, all experts stated that MeFiCom's values are practiced in the daily work routines. It was seen that this real life practice is basic to the success of the business which is illustrated with reinforcing connectors that link the real life practice with each core value. *Innovation* as core value is reinforced by the *Exploration* portion of the *Organizational Ambidexterity* CEM element, emphasizing the importance of innovation for MeFiCom. Innovation is also connected with *Research and Development* in the *Variable Costs* as an element of the *Internal Controls* in the DCS main container. R&D thus is a driver for innovation but produces internal costs for MeFiCom. *Service Innovation* reinforces *Servitization* as an element of *Service Focus* in the DCS *Success Factors* container. Also related to Service Focus is the core value *Integrity*, by reinforcing the *Empathy – Problem Solver & Care Taker* element. The *Team Spirit* core value reinforces *People Focus* as a major *Success Factor* in the DCS main container.

6.6.3 CEM – MeFiCom's Boundary Systems

The following figure illustrates the Boundary Systems CEM container:

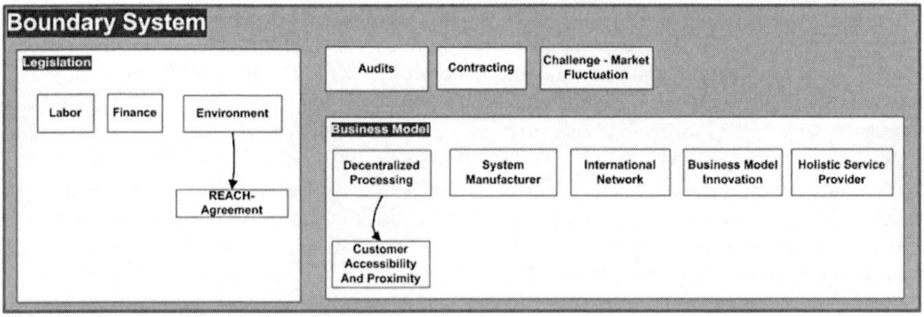

Figure 34: Boundary Systems CEM Container

The sub-categories *Business Model* and *Legislation* have been assigned to the *Boundary Systems*. Within the *Business Model* category *Decentralized Processing* and *Holistic Service Provider* have the strongest impact:

The results show that *Customer Accessibility And Proximity* is essential for the MFI business. Being located close to the customers ensures a minimum of logistics problems and therefore enhances the delivery times. MFI-BPM consequently has to control for example, the customer contribution margin in order to allow management decisions concerning expansion alternatives (should a new factory be built near to customer A or customer B?). This facet is reflecting in MeFiCom's business model by having more than ten production sites all over Germany (s. chapter 6.1.2).

PROP: *Larger MFCs are decentralized organized.*

Holistic Service Provider means that a MFC should act as a problem solver and care taker for its customers and is therefore connected to the *Service Focus* in the *DCS Success Factor* container. It is not sufficient solely to offer a service, but to align its own production and services with changing customers' requirements. A related MFI business performance measure could be the number of sales calls within a certain time frame.

Environment has the strongest impact within the *Legislation* category. Metal finishing services regularly require large amounts of hazardous chemicals and energy (s. chapter 3.3). New legislation like the European REACH (2006) agreement has therefore strong impacts on the MFI, for example by restricting required consumables like Chromium VI for some metal finishing processes (Thiery and Wojczykowski, 2013). *Environment* therefore has a balancing effect on *Hazardous Consumables* in the *Production Focus* as an element of the DCS *Success Factors.*

PROP: *Environmental uncertainties are relevant for MFCs.*

Labor, referring for example to labor protection laws, and *Finance*, referring to fiscal laws, are further elements of the *Legislation* category. *Labor* is connected to *Production Planning* as a *Production Focus* item in the DCS *Success Factors*, as related laws include the maximum workload in hours for employees, which has to be reflected in shift schedules.

Audits check if MeFiCom operates in the boundaries set. They are related to *Bureaucracy* in the DCS Success Factors following the experts' argumentation to this point.

The *Boundary Systems* main container reinforces the *Strategic Uncertainties* in the *Interactive Controls Systems* main container as described in the example above.

Environmental legislation bears the risk for some MFCs to end their operations as they may not afford the necessary investments in new technology to substitute Chromium VI.

6.6.4 CEM – MeFiCom's Diagnostic Control Systems

The sub-categories *Set Standards*, *Business Performance Measurement – Measure Output*, *Information Supply* and *Success Factors* have been assigned to the *Diagnostic Control Systems*.

The following figure illustrates the DCS *Business Performance Measurement – Measure Output* CEM container:

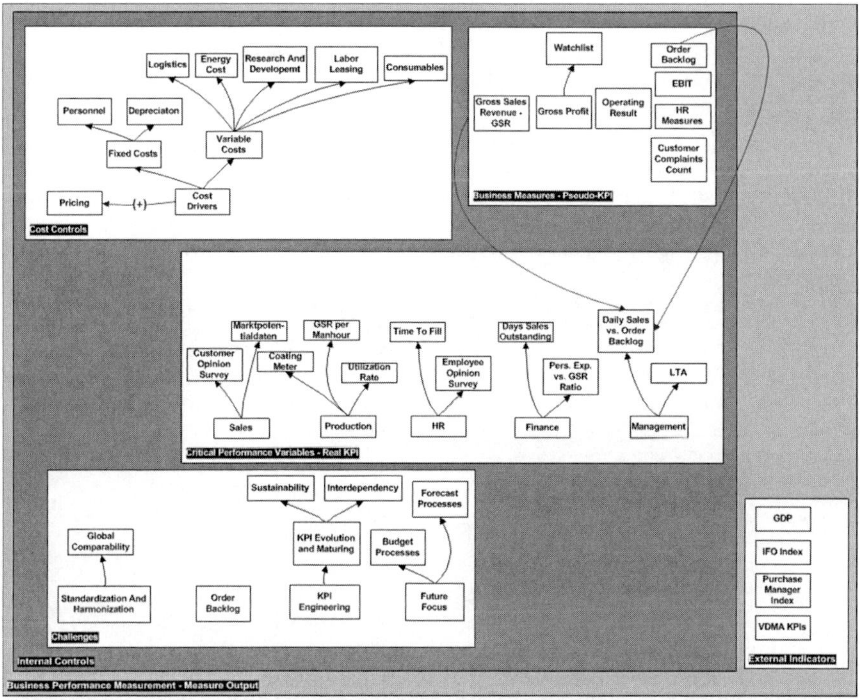

Figure 35: Diagnostic Control Systems – BPM – Measure Output CEM Container

The *Business Performance Measurement – Measure Output* container is split into the categories *Internal Controls* and *External Indicators*. *Internal Controls* includes sub-categories related to *Critical Performance Variables* as well as *Business Measures*.

MeFiCom's managers use a set of *Internal Controls* and *External Indicators* to monitor *Critical Performance Variables* (compare to Simons, 1995, p. 63). A BPM system does not yet exist at MeFiCom in which the business performance is measured. MeFiCom being a proxy model for the MFI (s. chapter 6.1.2) leads to the following proposition:

PROP: *The majority of MFCs are not using a BPM-system.*

And by reflecting the hitherto findings that a BPM system in general does not exist for the MFI leads to the following proposition related to MFCs applying a BPM system:

PROP: *BPM-systems in MFCs are based on generic BPM system models.*

Focus in the *Internal Controls* is confined to the single-dimensional *Business Measures*, namely *Gross Sales Revenue, Gross Profit* and *EBIT*, by comparing actual results against budgeted values. Subordinated elements are the multi-dimensional *Critical Performance Variables* like the *Utilization Rate* of the coating equipment or the internally derived MeFiCom sales market analysis data. These are partly reflected in MeFiCom's quality and environment management systems (s. chapter 6.2). However besides single-dimensional business measures focused on financial aspects, MeFiCom applies non-financial measures and indicators referring for example to the results of *Employee Opinion Surveys* or to the number and reasons for occurred lost time accidents (*LTA*) i.e. an occupational injury resulting in a loss of working time.

It was surprising that the majority of the MeFiCom colleagues treated single-dimensional business measures as being more important than multi-dimensional KPIs; a circumstance that is reflected in the number of transcript passages assigned to single-dimensional business measures (77 passages) compared to the rather low number of coded KPI transcript passages (25 passages) as shown in detail in the following CAQDAS code system screenshot:

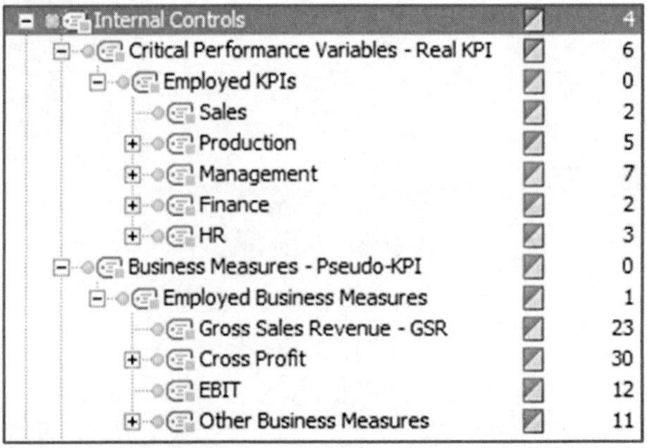

Figure 36: CAQDAS Code System – Business Measures vs. KPIs

Including single-dimensional business measures is therefore at least equally important as including multi-dimensional KPIs to measure the business performance of MFCs. With reference to *External Indicators* MeFiCom regularly analyzes publicly available data like the German *GDP* or the *Purchase Manager Index* reflecting the economic cycles in order to prepare the organization for economic up- or downtrends.

Furthermore *Cost Controls* and *Challenges* are parts of the BPM *Measure Output* container. The experts identified future-oriented processes, like forecasting or budgeting, as challenges for sound internal controls. They also refer to those operations as sources for uncertainty and therefore linked with the corresponding element in the *Interactive Control Systems* CEM main container. In addition, engineering multi-dimensional key performance indicators was seen as major task for the future. Such processes help to improve and sustain internal controls for example in order to assure a global comparability to other MFCs.

Fixed Costs, especially *Personnel* cost, are the major drivers of expenses. This is not surprising as the MFC regularly does not purchase raw materials. The customer provides the raw materials for the metal finishing services in order to receive the value-added product back (s. chapter 3.2). This is a characteristic that conforms to international statistical MFI data related to a high share of personnel costs at the MFCs (s. chapter 3.5). Within MFI-BPM, controlling the break-even point is of high significance as additional gross sales revenue will result in disproportionately high gross profit. This circumstance becomes vital in the sense that a sales collapse combined with high *Fixed Costs* could lead to severe economic problems:

PROP: *A high share of fixed costs is a MFC characteristic.*

However, *Variable Costs* are also of interest as *Energy Cost* is another *Cost Driver* at MeFiCom – and also for many other MFCs (s. chapters 3.3 and 3.5.1). *Logistics* is interrelated with the MFI core process in- and outflows between the customer and the MFC, as these process steps result in logistic costs for MeFiCom that organizes the logistics for its customers.

All these *Cost Drivers* have of course direct influence on the product *Pricing*, being interrelated with the *Customer* as outside stock in the MFI core process. If the price for the metal finishing operation is too high, the customer will likely attempt to substitute the metal finish process with alternatives or switch to a cheaper MFC.

The *Business Performance Measurement – Measure Output* container is interrelated with the *Interactive Intergroup Communication* container, as selected *BPM Results* are communicated at MeFiCom. Such reports for example could be published on bulletin boards or in town hall meetings in order to inform the employees about the company's success.

The following figure illustrates the DCS – *Set Standards* and the DCS – *Information Supply* CEM containers:

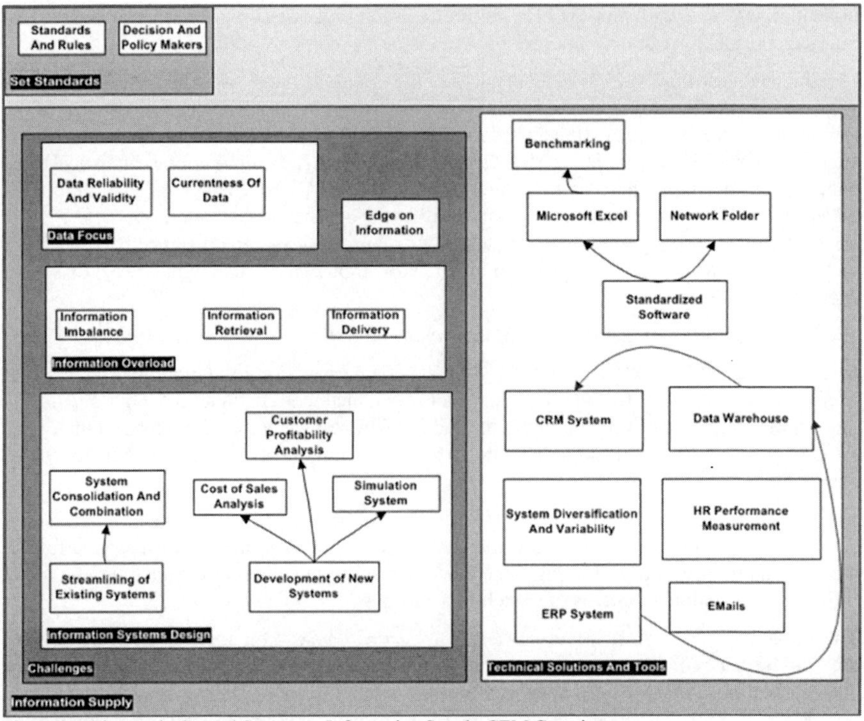

Figure 37: Diagnostic Control Systems – Information Supply CEM Container

In the DCS *Set Standards* container, the experts related that they followed the rules and strategies set by MeFiCom's headquarters. In consequence, I suggest including headquarters representatives within the implementation process of a BPM system in larger MFCs. Moreover, the *Set Standards* container has a reinforcing effect on the *Business Performance Measurement – Measure Output* container, as the *Decision and Policy Makers* directly influence the method of BPM at MeFiCom. Following the same rationale, the *Set Standards* container is also linked to the *Flexibility vs. Bureaucracy* container within the DCS *Success Factors* container.

In the DCS *Information Supply* container the experts highlighted the necessity of an integrated and streamlined information system providing up-to-date, reliable and valid data. Currently they face an *Information Overload*, created by numerous information systems. *System Consolidation and Combination*, including a BPM system for the MFI, could be one way to fulfill this management requirement. The experts discussed the introduction of a *Customer Profitability Analysis* system in order to enhance the already existing *Pricing* tools. The *Production Planning* could be enhanced by the creation of a *Simulation System* which reflects the predicted demands by the customers. This would lead to a production schedule to include the optimum required workforce and demanded coating system capacity. Such is not a trivial undertaking at MeFiCom on several dimensions, like the decentralized organization and the vast number of various customer substrates in differing volumes which have to be reflected in such a system.

As in *Technical Solution and Tools*, the experts referred to the SAP Business *Data Warehouse* (s. chapter 6.2) which retrieves data from MeFiCom's *Enterprise Resource Planning* System and supplies data to the *Customer Relationship Management* system. In addition, *Standardized Software* like *Microsoft Excel* is adopted for BPM at MeFiCom. BPM information and data are shared at MeFiCom via *Email* or by using *Network Folders* to provide data files. Employee portals (Urbach et al. 2011) should be considered in the BPM design as an alternative way for the provision of BPM information throughout hierarchical organization levels.

PROP: *MFCs are using IT software to support BPM processes.*

The following figure illustrates the DCS – *Success Factors* CEM container:

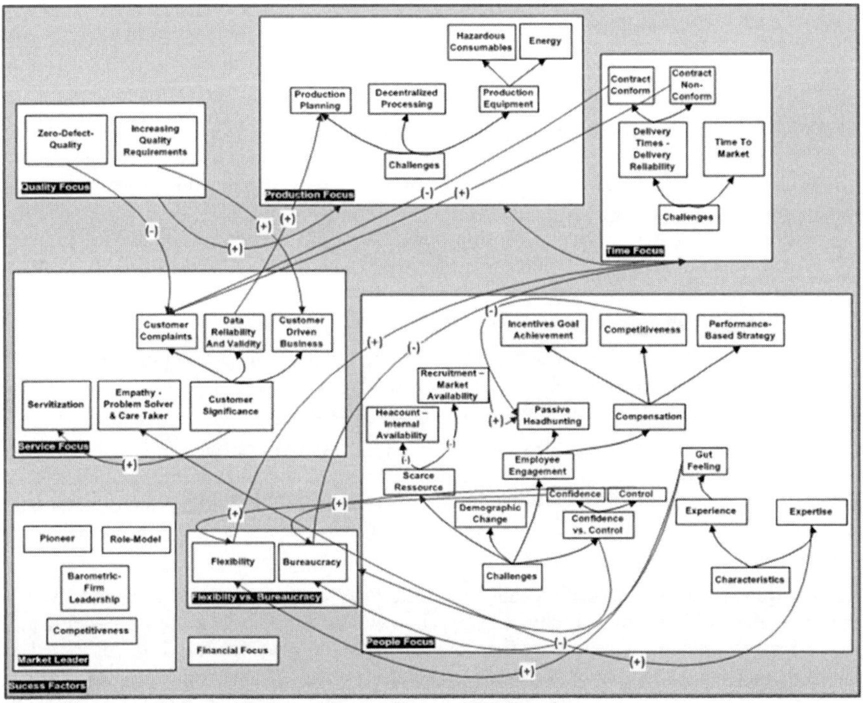

Figure 38: Diagnostic Control Systems – Success Factors CEM Container

People is a major success factor: the experts characterized the MFC employee as a *Scarce Resource* combining *Experience* as well as *Expertise*. *Expertise* is required in order to be seen by customers as integral to *Integrity*. Customers need to deal with *Care Takers* and *Problem Solvers* in order to help improve their products. This circumstance can also be related to the Knowledge-Based View theory (s. chapters 2.2.1 and 2.5.1) in which the level of knowledge of the firm's employees represents a significant competitive advantage.

Reflecting Simons (1995), the experts emphasized the importance of *Employee Engagement* in the sense of *"reconciling self-interest with the desire to contribute"* (Simons, 1995, p. 24). Enhancing the desire to contribute could be established with a competitive (vis-à-vis MFC's competitors) compensation program including incentives for goal achievement. An example of an MFI-BPM related measure for employee engagement could be the number of improvement suggestions within a continuous improvement process in a certain time frame.

Service, Production, Time and *Quality* are further major *Success Factors* which should be reflected in the BPM. *Service* at MeFiCom is *Customer Driven*. The experts agreed in that an MFC has to be a holistic service provider and carefully observe the customers' business in order to react to (customer) product changes while being ready to adapt metal finishing services to those changes. The *Customer Driven Business* element is therefore interrelated with the *Customers'* outside stock within the MFI core process (s. Relational View in chapter 2.2.2).

PROP: *MFC's customers strongly influence MFC's strategy.*

The *Customer Driven Business* element is also interrelated with the *Strategic Uncertainty* elements a) *Commutability* and b) *Insourcing*:

a) The customer requires the metal finishing processes to enhance his products for example by increasing the endurance of cutting tools in increased wear resistance (s. chapter 3.1). In the event the customer finds an alternative for this process resulting in a comparable level of wear resistance, for example by substituting the alloy of the substrate's metal, the metal finishing process would become obsolete as it is not a competitive advantage offered by the MFC anymore (s. Process-Based View in chapter 2.2.3).
PROP: *Technological uncertainties are relevant for MFCs.*

b) An effect which has already been described in the MFI CEM container: as soon as a customer reaches a critical mass of production volume, he may decide to insource the metal finishing processes previously performed by the MFC.

Servitization (Lightfoot et al., 2013) is a topic within the MFI *Service Focus*. In the past years more and more metal treatment services like brushing, blasting or edge rounding have been added to the core metal finishing product portfolio. Providing those additional services, often applied by the customer before and then outsourced to the MFC, is seen as a competitive advantage by the experts.

Finding ways to handle production challenges like *Production Planning* is seen as a success factor, as the order backlog in days is very short (about three to five days) at MeFiCom. Therefore, *Production Planning* is interrelated with the *Substrate* and *Metal Finished Product* stocks in the MFI core process.

PROP: *MFCs have a short order forward book.*

Furthermore the *Production Planning* significantly depends on the customer's input being related to demand of the required volume at which the metal substrates shall be finished (s. also Lange, 2009, p. 84). If the related customer *Data Reliability And Validity* is insufficient the production schedule will be obsolete.

PROP: *Demand uncertainties are relevant for MFCs.*

This leads to another major success factor: *Time*. Short *Delivery Times and Delivery Reliability* are crucial within the MFI business. Especially as just-in-time customers from the automotive industry rely on such exacting delivery times. A strategy to fulfill this demand is that a MFC has to be located closely to its customers and has to offer a *Decentralized Processing* of the customer's substrates. This circumstance is interrelated a) to the *Logistics* in the *Variable Costs* container as MeFiCom's network of production sites results in increased logistic costs and b) to MeFiCom's *Business Model* in the *Boundary Systems* CEM container.

Quality shall be ensured in the MFI company by applying a *Zero-Defects*-Policy, which will assist in reducing quality related *Customer Complaints* in order to enhance the overall service. A further element is *Increased Quality Requirements* by the customer reflecting also the *Customer Driven Business*. *Quality* is interrelated with the *Production* container as the customer's requirements need to be reflected in the MFC's production processes.

Reflecting all the cost types as well as success factors which drive MeFiCom's BPM leads to the following proposition:

PROP: *MFCs are using a mix of financial and non-financial indicators to measure their business performance.*

Additionally, the experts have described the customer's willingness to pay the price for the metal finishing services as being subordinated to fulfilling the aforementioned success factors:

PROP: *Price is subordinated to Quality in MFCs.*

PROP: *Price is subordinated to Delivery Reliability in MFCs.*

The experts delineated a conflict area between *Flexibility* and *Bureaucracy*: related to the *People Focus* they correlated *Flexibility* with *Confidence* in the employees' decisions combined with the manager's gut feeling pertaining to additional steps to be taken. As a note of caution, *Bureaucracy* appears to increase in case the management introduces too many measures in order to *Control* the employees. This container is therefore also interrelated with the DCS *Internal Controls* container. The experts advocated an appropriate relation between *Confidence* and *Control* and demanded a reflection of this in the design of a sustainable BPM system.

6.6.5 CEM – MeFiCom's Interactive Control Systems

The following figure illustrates the Interactive Control Systems CEM container:

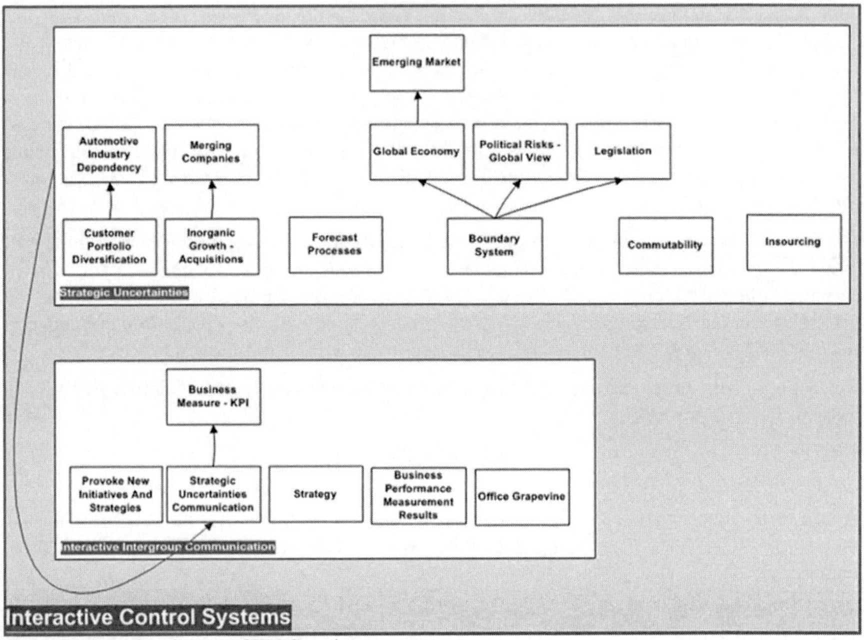

Figure 39: Boundary Systems CEM Container

The sub-categories *Interactive Intergroup Communication* and *Strategic Uncertainties* have been assigned to the Interactive Control Systems.

Major drivers in the *Interactive Intergroup Communication* between the hierarchical business levels are the BPM results as well as the *Strategic Uncertainties Communication* related to *Provoke New Initiates and Strategies* in order to counteract the ever-present strategic uncertainties.

A too low *Customer Portfolio Diversification*, and therefore a strong dependency on single customers or industry branches like the automotive sector, represents a major strategic uncertainty. Controlling the percentage distribution of customers or customer industry branch affiliation should be a further MFI-BPM element. Another major uncertainty, especially for larger MFI companies, is related to inorganic growth by performing acquisitions. Not all planned positive aspects of an acquisition may turn into reality as described above in MeFiCom's *Believes Systems* CEM main container. A related capitalized expenditures controlling could assist and give input to BPM in the MFI.

Further *Strategic Uncertainties* are related to global economic and political issues like the fortification of foreign, emerging metal finishing markets (driven for example by low personnel costs in Eastern Europe – s. chapter 3.5.1) in competition with mature markets like Germany. Including external indicators in MFI-BPM retrieved from statistical databases can help to increase transparency related to MFI growth in those emerging markets.

Propositions related to this CEM main container referring to environment, demand, and technological uncertainties are located in the DCS CEM main container as the experts mentioned these uncertainties within the DCS section.

6.7 Quality Criteria and Assurance – Qualitative Research

> "The basic issue in relation to trustworthiness is simple: How can an inquirer persuade his or her audiences (including self) that the findings of an inquiry are worth paying attention to, worth taking account of? What arguments can be mounted, what criteria invoked, what questions asked, that would persuasive on this issue?" (Lincoln and Guba, 1985, p. 290)

It was not trivial to decide which quality (rigor) measures I should chose for my research project, especially in the light of the academic discussion[43] (Maxwell, 2013, p. 122; Barbour, in Flick, 2014, p. 496) between using traditional (classic) quality criteria related to reliability, validity and objectivity (Yin, 2009, p. 40; Mayring, 2010, p. 116) or using criteria dedicated to qualitative and constructivistic research like credibility, transferability, dependability, confirmability (Lincoln and Guba, 1985, pp. 301-318), and authenticity (Lincoln and Guba, 2013, p. 70).

For my dissertation I followed the recommendations of Miles et al. (2014, p. 311) who have combined both perspectives in five main categories by *"pairing traditional terms with those proposed as more viable alternatives for assessing the trustworthiness and authenticity of naturalistic research"*:

1. **Objectivity / Confirmability of qualitative work:**
   Significant space in this dissertation has been devoted to the detailed reasoning related to various qualitative methods and procedures to allow the reader to follow the sequence of how I collected, processed and analyzed the data within the research setting which shall assist to judge about the confirmability of my qualitative work.

   I reflected possible biases and described various ways I took countermeasures like the group discussion method to reduce researcher bias (s. chapter 6.4).

   The study data, namely the qualitative code system and the MFI cause and effect model, are available online or attached as appendices to this dissertation. Additional detailed data such as transcripts or MeFiCom artifacts (business reports, minutes of meetings, business related field notes, etc.) cannot be made available for re-analysis due to the non-disclosure agreement between EBS and MeFiCom. However my supervisors were granted access to the transcripts in order to audit (Lincoln and Guba, 1985, p. 378) the coding process and related research findings.

2. **Reliability / Dependability / Auditability:**
   The research questions (s. chapter 2.5.4) were derived following the bottom-up approach to improve BPM feasibility (s. chapter 2.5.3). The research methodology (s. chapter 5) followed this bottom-up approach to answer the research questions.

   Within the research setting (s. chapter 6.1) I described my role and status at MeFiCom and showed meaningful parallels making MeFiCom a proxy for the MFI (s. chapter 6.1.2).

[43] This discussion can be referred in German to "methodenangemessene Gütekriterien in der qualitativ-interpretativen Forschung" (Flick, 1987, p. 247).

The second chapter of this dissertation was dedicated to selected basic paradigms leading to the theoretical research framework of this dissertation (s. chapter 2.5.1), as *"reliability depends, in part, on its connectedness to theory"* (Miles et al., 2014, p. 312).

By applying different sources of information like international nomenclatures of economic activities (s. chapter 3.4), international statistical databases (s. chapter 3.5), and MFI literature (s. chapters 3.1, 3.2, 3.3, and 4) I was able to combine these data with the empirically retrieved information in order to find answers to the research questions (s. chapter 2.5.4).

I included a pretest with three EBS professors before applying the interview guide within the expert interviews, in order to allow academic colleague review of the applied research tool. As I was the only field-worker and as I assured confidentiality to the participating experts, it was not possible to apply intercoder checks within the qualitative data analysis; however I checked the data quality by having a review phase of the raw transcripts (s. chapter 6.5.1) as well as a repeated coding cycle within the CAQDAS analysis (s. chapter 6.5.2).

3. **Internal Validity / Credibility / Authenticity – MeFiCom Member Check:**
Although internal validity is not a necessary quality criterion in explorative studies (Yin, 2009, p. 40), it was necessary for my research project to prove the robustness of the qualitative results (Miles et al., 2014, p. 313) as they were the bases for the subsequent quantitative survey within the MFI (s. next chapter). Therefore, I presented both the CAQDAS coding system and the CEM as a member check[44] (Lincoln and Guba, 1985, p. 314; Ybema et al., 2009, p. 62; Lincoln and Guba, 2013, p. 71; Flick, 2014, p. 511) in a joint meeting on May 28th, 2014, to the MeFiCom managers having participated in the expert interviews and the group discussion.

By applying Microsoft Visio's layer properties function I sequentially introduced the different CEM elements starting with the MFI core process and then introduced the different main containers. After having shown the core process and the different containers with CEM elements, I added the interconnections between the different elements including their balancing and reinforcing characters. With this graphic approach I followed the recommendation of Miles et al. (2014, p. 309):

> "As with analysis, matrices and networks work much better than text alone to help participants access the information. They will find it easier to get an overview, *to see how the pieces fit together*."

I presented the CEM containers and elements in English (s. previous chapter) as the participating experts were trained in English. Nevertheless the meaning of some applied terms in the CEM had to be explained to the audience.

I had the impression that the experts appreciated the discussion based on the CEM as it reflected the interactions between the company's key areas of interest, an effect also described by Franco-Santos and Bourne (2005, p. 117). This validation process was fundamental in order to establish trustworthiness (Lincoln and Guba, 1985) in the collected qualitative data by getting feedback from the participants (Hammersley and Atkinson, 1995, p. 280; Miles et al., 2014, p. 309). The member check resulted in suggestions for minor adjustments of the CEM; for example B1 suggested including

[44] In German „Kommunikative Validierung" (Flick, 1987, p. 253).

Customer Opinion Survey in the *Employed KPIs* section of the *Diagnostic Control Systems* main container, whereas B4 spotted a spelling mistake in the term *Gross Profit*. The member check session was audiotaped; corresponding audio files were not transcribed as the demanded changes were minor and therefore were directly and jointly incorporated in the CEM during the member checking session making the CEM the final version (as presented in chapter 6.6). All in all, the experts gave a positive feedback and approved the qualitative results with unanimous assent at the end of the member checking session.

4. **External Validity / Transferability / Fittingness:**
   The transferability of the results, referring to the findings at MeFiCom, is a prerequisite within this research project. Finding solely the recommendations for a BPM at MeFiCom would mean a severe shortcoming of this research project. It would then appear to be a consultancy rather than a research project. Therefore, I have chosen the explorative sequential mixed methods research approach with the ethnographic research at MeFiCom followed by an MFI survey through which I validate the qualitatively derived propositions (s. previous chapter) as well as triangulate the qualitative findings (s. next chapter).

   The transferability of the results, referring to the findings at the MFI, is a topic for future research in order to adapt the hitherto MFI findings to other contract processing industries (s. chapter 9.3). This explicitly includes the adaption of the bottom-up approach to improve BPM feasibility (s. chapter 2.5.3) in future, especially explorative, BPM research studies, which is not limited to contract processing industries.

5. **Utilization / Application / Action orientation:**
   The colleagues at MeFiCom played a central role in my research project as I applied a Mode 2 Knowledge Production policy in the ethnographic research setting (s. chapter 6.1). In order to become *one of them,* it was important for me to apply an overt ethnographic approach, also reflecting ethical issues related to a covert approach (s. chapter 6.1.1). Having an informed consent with the participating experts in the interviews and group discussion, and especially in order to protect these individuals regarding their statements (s. chapter 6.3.1), was also a prerequisite to establish an ethical sound research project.

   Although the whole MFI is in the focus of the research project, MeFiCom receives direct benefits of this research project. The participating experts stated that the interviews and group discussion brought up ideas to improve company internal processes, like the streamlining of existing information services. Furthermore the final and triangulated results of this research project shall enable the implementation of an MFI dedicated BPM system at MeFiCom – somewhat as an industry prototype.

7 BPM in the MFI – Quantitative Research Process

The next step within the research process is a web-based survey (Dillman et al., 2009, p. 195) within the MFI in order to triangulate (Flick, 2011) the hitherto findings following the exploratory sequential mixed methods research design (s. chapter 5).

7.1 MFI Survey

7.1.1 Survey Sample

The MFI survey addressed larger MFC as they are in the focus of this research project (s. chapter 2.5.4). I assumed that those larger MFCs can best be found in MFI associations as they are potentially more willing a) to pay association member fees and b) to provide manpower required in order to take an active part in an association compared to smaller MFCs.

I included the United Kingdom (UK) as one of the European top five MFI countries (s. chapter 3.5.1) into my criterion sampling strategy (Patton, 2002, p. 238), as I wanted to check if the preceding findings, mainly derived from the German MFC single case study (s. chapter 6.1), hold true also for a country other than Germany.

Following this strategy I identified the following associations as being influential for the MFI in Germany and the UK:

- Germany:
  - Zentralverband Oberflächentechnik e. V. (German Central Association of Surface Treatment Professionals) – ZVO
  - Europäische Forschungsgesellschaft Dünne Schichten e. V. (European Society of Thin Films) – EFDS
  - Verband Deutscher Maschinen- und Anlagenbau e. V. (German Engineering Federation), Fachabteilung Oberflächentechnik (Surface Technology) – VDMA
- United Kingdom:
  - Surface Engineering Association – SEA
  - Institute of Materials Finishing – IMF

I contacted all five MFI associations and asked for their support a) in order to retrieve contact information of the associations' members and b) in allowing me to use their logos as supporters of the EBS-SIIE research project expecting a higher response rate within the survey.

Besides the ZVO, which in general does not directly support external surveys involving its members[45], all other associations granted me support: EFDS, SEA and IMF allowed me to use their logos as supporters for instance in the invitation email to participate in the survey (s. next chapter). IMF has sent a copy of their published buyer's guide in which the IMF member organizations are listed. The members of EFDS, ZVO and SEA are listed online (EFDS, 2014; ZVO, 2014; SEA, 2014). The VDMA Surface Technology member organizations are listed in VDMA's industrial sector guide (VDMA, 2014).

I created a Microsoft Excel table for each association in which I copied and structured all association member organizations' contacts. Having all member organizations in one digital

[45] Personnel communication with Dr. Jacob, managing director ZVO, August 12th, 2014.

file allowed me to sort out redundant organizations being members of more than one MFI association.

The following table gives an overview on the resulting MFI sample:

Sample		Size (N)	Country
Zentralverband Oberflächentechnik	ZVO	122	Germany
European Society of Thin Films	EFDS	82	Germany
German Engineering Federation	VDMA	68	Germany
Surface Engineering Association	SEA	201	United Kingdom
Institute of Materials Finishing	IMF	102	United Kingdom
TOTAL Sample Size		**575 = N_{TOTAL}**	
thereof Germany:		272 = N_{GER}	
thereof United Kingdom:		303 = N_{UK}	

Table 19: MFI Survey – Sample Composition

The share of German and British MFCs in the total sample is nearly equally distributed (47.3% German and 52.7% British MFCs). The largest German MFI association is the ZVO followed by the EFDS. The German VDMA Surface Technology members are mainly formed by surface finishing system manufacturers and not autonomous job shops or captive in-house MFCs (s. chapter 3.1); however I included these organizations in the sample a) as they are also part of the MFI and b) as surface finishing system manufacturers are also included in the other MFI association sub-samples. The largest British MFI association is SEA, followed by the IMF with approx. half of the number of member organizations than SEA.

By relating the sub-samples N_{Ger} and N_{UK} (s. table 19) to the total number of German and British MFCs in 2012 (s. appendix A.2) it can be identified that N_{Ger} covers 8.4% of the German MFC population whereas N_{UK} covers 23.9% of the British MFC population. Following the assumption that larger MFCs are more strongly represented in MFI associations, I related N_{Ger} to the total number of German MFCs with more than 49 employees (s. chapter 3.5.1) which leads to a population coverage of above 80%. However, the accurate share of larger survey sample MFCs in the national total MFC population could not be identified, as the number of employees was not disclosed in the associations' member lists.

7.1.2 Design and Operationalization

The following figure gives an overview on the propositions derived from the MFI CEM in chapter 6.6:

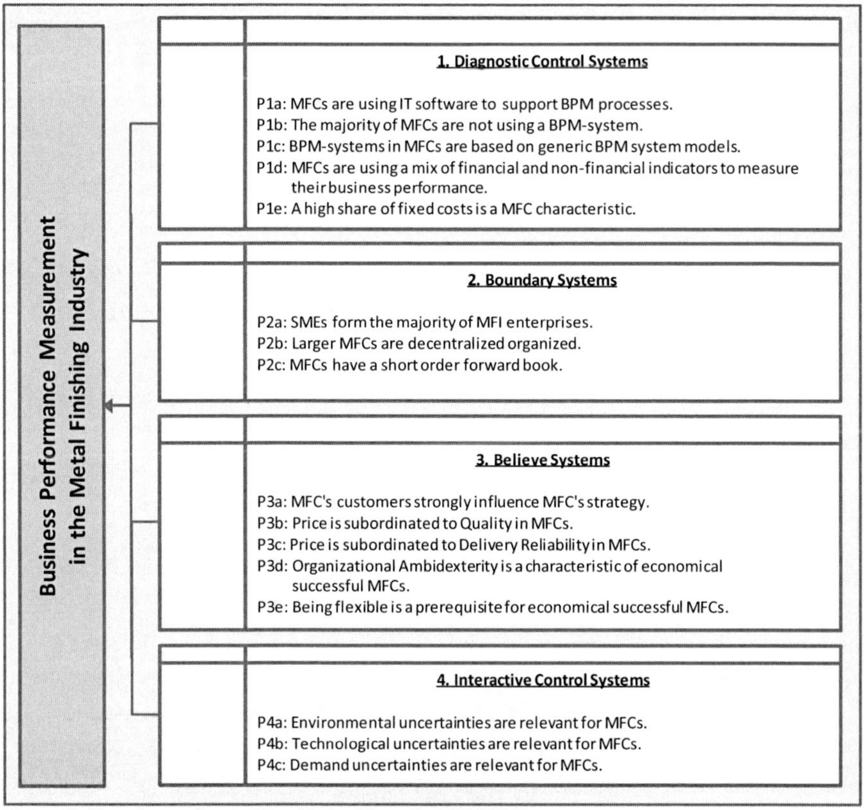

Figure 40: Overview on Propositions Derived from the MFI CEM

In order to validate these propositions within an MFI survey, I adopted existing scales from academic literature (Hair et al., 2010, p. 655) and created additional questions. Long survey questionnaires bear the risk of *"transient mood states such as boredom and fatigue"* (Lindell and Whitney, 2001, p. 118) resulting in a reduced quality of the survey results. Therefore I designed the survey to include only those for the research process necessary scales and questions which were estimated to take in total not more than 15 to 20 minutes to be completed in the survey.

Regarding the content, I divided the survey into the following three parts:

MFI Survey Part I – Statistical Data

The first part is introduced by a summary of the research project goals and an overview on the survey structure. This introduction also shows logos of MeFiCom as well as of three MFI associations (s. next chapter) to emphasize their support.

In addition, I included the official EBS and SIIE university logos into the survey template design throughout all questionnaire sides to emphasize the academic component of my research project as shown in the following survey screenshot:

EBS Business School

SIIE. Strascheg Institute for Innovation & Entrepreneurship

Survey within the Research Project

„Business Performance Measurement in the Metal Finishing Industry“

"Business Performance Measurement in the Metal Finishing Industry" is the dissertation research project of Mr Mike Dietrich, who is a doctoral candidate at the *Strascheg Institute for Innovation & Entrepreneurship (SIIE)* of the *European Business School, Germany*. In addition to the present project, Mr Dietrich has been working as a financial analyst in the Metal Finishing Industry since several years.
The goal of the research is to analyse the Metal Finishing Industry (MFI) in order to build a foundation for a Business Performance Measurement model that is suitable for MFI organizations.

We kindly ask you to complete this survey as it is a fundamental part of the whole dissertation research project. We are convinced that you, as an industry insider, can add valuable insights to our research. We are interested in your personal evaluation and experience in the industry - there are no "right or wrong" answers.

The survey is structured into three parts - each of them will be introduced by a brief description to explain their relevance in the context of our research project:
1. Statistical Data
2. Overall Characteristics of your Organization
3. Business Performance Measurement

Please answer all questions by the date of September 30th, 2014.

You have the opportunity to leave your contact information at the end of the survey in order to receive a short report on the survey results as a reward for your assistance.
Your evaluations and data will be processed and treated anonymously and strictly confidently. The data and information is solely for the research project and will not be passed on to a third party.
Please contact **Mr Mike Dietrich** for further information regarding the survey: E-mail: **mike.dietrich@students.ebs.edu** Phone: **+49** [redacted]

The Research Project is supported by:

[redacted]	**Surface Engineering Association, UK**
European Society of Thin Films	**Institute of Materials Finishing, UK**

Figure 41: Screenshot of the MFI Survey Introduction Site

I discussed the point of including MeFiCom's logo with my academic supervisors as well as with two of my MFI association contacts; the discussion was based on two facets:

1. Including MeFiCom's logo could lead to reluctance of the survey participants as a possible competitor of the participating MFCs is supporting the research project.
2. Not including MeFiCom's logo and not referring to MeFiCom's involvement would indeed reduce the reluctance risk but it would involve an ethically questionable behavior: in this case those respondents who would not wish to take part in a competitor-supported survey would simply not have the chance to opt-out.

I decided to include MeFiCom's logo due to the ethical concerns described above.

The first survey questions are related to the respondent's characteristics referring to:

- his/her function in the organization
- his/her department
- the number of years that he/she is working in the MFI

The next survey questions are related to statistical data related to the respondent's organization (referring to the legal entity in which the respondent is working):

As the association members are not limited to MFCs but also include metal finishing equipment manufacturers or MFI suppliers, I asked the respondents to classify their organization according to the majority of operations it performs. Further questions refer to additional control variables on which the answers were intended to be clustered in the quantitative data analysis referring to:

- the type of MFI process (s. chapter 3.1) the MFC mainly applies
- the number of years the MFC has been in operation
- the country in which the organization is located

Concluding questions of this first survey part refer a) to the organization's SME characteristics (proposition P2a – s. figure 40) with regard to turnover (in 2013) and the number of employees (full time equivalents) and b) to the organization's number of production sites controlling for a possible decentralization of larger MFCs (proposition P2b).

MFI Survey Part II – Overall Characteristics of the MFC

The second survey part starts with an introduction referring to the respondent's organizations characteristics in order to understand the organization's strategy, orientation and behavior:

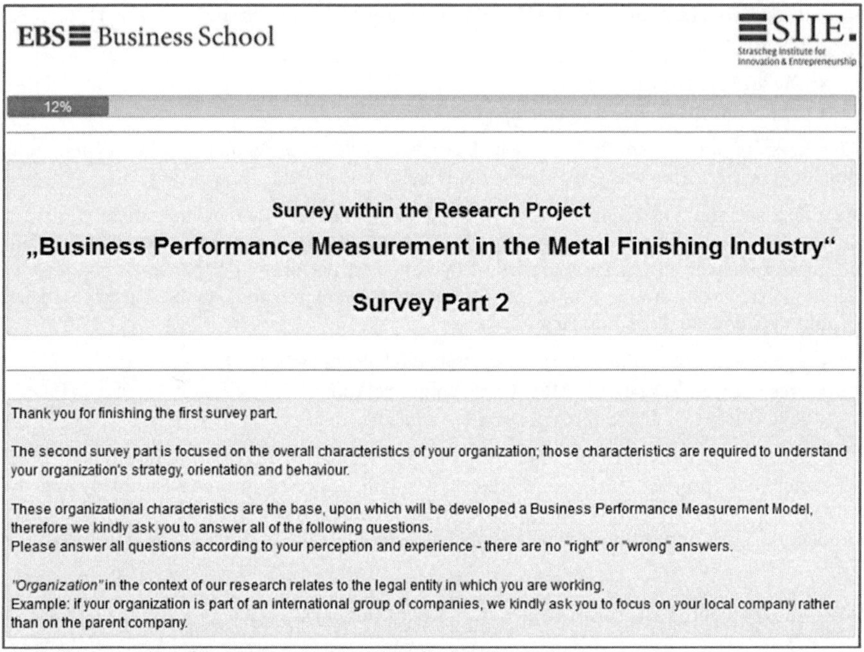

EBS Business School

SIIE. Strascheg Institute for Innovation & Entrepreneurship

12%

Survey within the Research Project

„Business Performance Measurement in the Metal Finishing Industry“

Survey Part 2

Thank you for finishing the first survey part.

The second survey part is focused on the overall characteristics of your organization; those characteristics are required to understand your organization's strategy, orientation and behaviour.

These organizational characteristics are the base, upon which will be developed a Business Performance Measurement Model, therefore we kindly ask you to answer all of the following questions.
Please answer all questions according to your perception and experience - there are no "right" or "wrong" answers.

"Organization" in the context of our research relates to the legal entity in which you are working.
Example: if your organization is part of an international group of companies, we kindly ask you to focus on your local company rather than on the parent company.

Figure 42: Screenshot of the MFI Survey Part 2 Introduction Site

This information is required for triangulating the corresponding qualitative derived MFI characteristics as a foundation for an MFI-BPM model and therefore refers to the first research question cluster (s. chapter 2.5.4 – ***RQ.1:*** *What are the characteristics and specifics of the MFI?*).

To control for the role of the customers and their importance related to the MFC's strategy (proposition P3.a), I applied the "Customer Focus"[46] scale by Chen and Paulraj (2004, p. 123). This scale was created for a supply chain study and involved a seven item scale; by focusing on the customer, *"organizations can outperform their competition by exceeding, not just satisfying, the needs of their customers"*. In addition I applied the "Closer to Customers"[47] scale by Powell (1995, p. 19) created for a total quality management (TQM) study and formed by a seven item scale. Powell relates this scale to one out of twelve TQM factors by *"determining customers' (both inside and outside the firm) requirements, then meeting those requirements no matter what it takes"*. Meeting or even outperforming customer's requirements "no matter what it takes" was the tenor of MeFiCom's experts referring to the strong influence (and involvement) of the customers on the metal finishing business and representing also MFI's industrial service character (s. chapter 3.2).

[46] The authors report a Cronbach's alpha with α=0.86 for the scales reliability.

[47] The author reports a Cronbach's alpha between α=0.78 and 0.90 for the reliability of the scales applied in their study, not referring directly to the "Closer to Customer" scale.

Also referring to MFC's business strategy, I applied the scale "*Competitive Priorities*"[48] by Chen and Paulraj (2004, p. 124) which is formed by a six item scale. This scale is related to propositions P3b and P3c (s. figure 40) and controls for the MFC's priorities related to price, quality, innovation and service including delivery times. In addition and as validation of the competitive priorities, I included a click ranking question (EFS, 2014, p. 186) in which I asked the survey participants to rank the three factors of costs/price, quality and delivery reliability to rank as one (being most important), two and three according to their priority within the MFC. Following the qualitative results I expected that price would have a rather subordinate priority compared to the other competitive aspects.

Referring to the MFC's strategic uncertainties and relating to propositions P4b and P4c, I applied the scales "*Technology Uncertainty*" and "*Demand Uncertainty*"[49] both adopted from Chen and Paulraj (2004, p. 138). Technology uncertainty is measured on a four item scale referring to the MFI characteristic that MFCs have to be prepared to adjust their production and technology according

a) to changing customer products on which the MFC applies its metal finishing processes (s. chapter 6.6.1),
b) to changing legislation which for example forces MFCs to substitute process chemicals like Chromium VI (s. chapter 6.6.3), or
c) to more competitive products and services in order to avoid the risk of commutability (s. chapter 6.6.4).

Demand uncertainty is measured on a five item scale referring to the MFI contract processing characteristic (s. chapter 3.2): as the MFC applies its metal finishing services regularly on customer supplied products, its production is highly dependent on the customer's supply – which in the case of the MFI equals the customer's demand.

In addition and relating to proposition P4a, I asked the survey respondents for their evaluation of the predictability of factors comprising the MFC's external environment which I adopted from Pagell and Krause[50] (2004, p. 632). Nature is only implicitly included by these factors, although one factor refers to the predictability of government regulations controlling the MFI which becomes relevant by relating it to legislation protecting the nature as shown in the US EPA case (s. chapter 3.3).

Related to the organizational ambidexterity (s. chapter 6.6.2) proposition P3d, I adopted scales referring to exploration and exploitation[51] each measured with four items as adopted from Jansen et al. (2009, p. 809). By applying these scales I did not aim to prove whether organizational ambidexterity is a *prerequisite* for economical successful MFCs, but to check if organizational ambidexterity is a *general characteristic* of MFCs. This should avoid an assumption that is solely based on the corresponding MeFiCom characteristic of applying a combination of exploration and exploitation.

[48] The authors report a Cronbach's alpha with α=0.83 for the scales reliability.

[49] The authors report a Cronbach's alpha with α=0.83 for the reliability of the "Technology Uncertainty" scales and an α=0.84 for the reliability of the "Demand Uncertainty" scales.

[50] The factors applied by Pagell and Krause (2004) were originally developed by Duncan (1972, p. 315). Pagell and Krause apply number 4B, 6A, 6B, 7A, and 7B of Duncan's measures. The authors did not report a Cronbach's alpha for the scales reliability.

[51] The authors adopted scales they published in the past (Jansen et al., 2006) for which they reported a Cronbach's alpha with α=0.86 for the reliability of the exploratory scales and an α=0.80 for the reliability of the exploitative scales.

The characteristic of MFCs adapting their production to changing customer products has hitherto been reflected by adopting a technology uncertainty scale as described above. Besides the strategic uncertainty factor I applied another perspective on this aspect, namely flexibility (s. proposition P3e). Therefore I followed the recommendations by Pagell and Krause (2004, p. 634) to measure the multidimensional scale "Flexibility"[52] by adopting a set of five measures related to the importance of the different dimensions, i.e. delivery, volume, mix, changeover, and modification flexibility. I expected volume flexibility, referring to *"the ability to effectively increase or decrease aggregate production in response to customers"* (Pagell and Krause, 2004, p. 635), to have a very high importance for MFCs.

The second part of the survey concludes with a question related to the MFC's order backlog in working days referring to the proposition P2c derived from MeFiCom's production planning characteristic (s. chapter 6.6.4). Following the qualitative results I expected a rather short order backlog of on average less than five working days.

MFI Survey Part III – Business Performance Measurement

The third survey part is dedicated to BPM and is introduced by the definitions of the BPM system, business measures and KPI in order to clarify the terms applied in the subsequent survey questions to avoid ambiguity between the terms' denotation within the research project and the respondent's perception of the terms:

[52] The authors report a Cronbach's alpha with α=0.90 for the scales reliability.

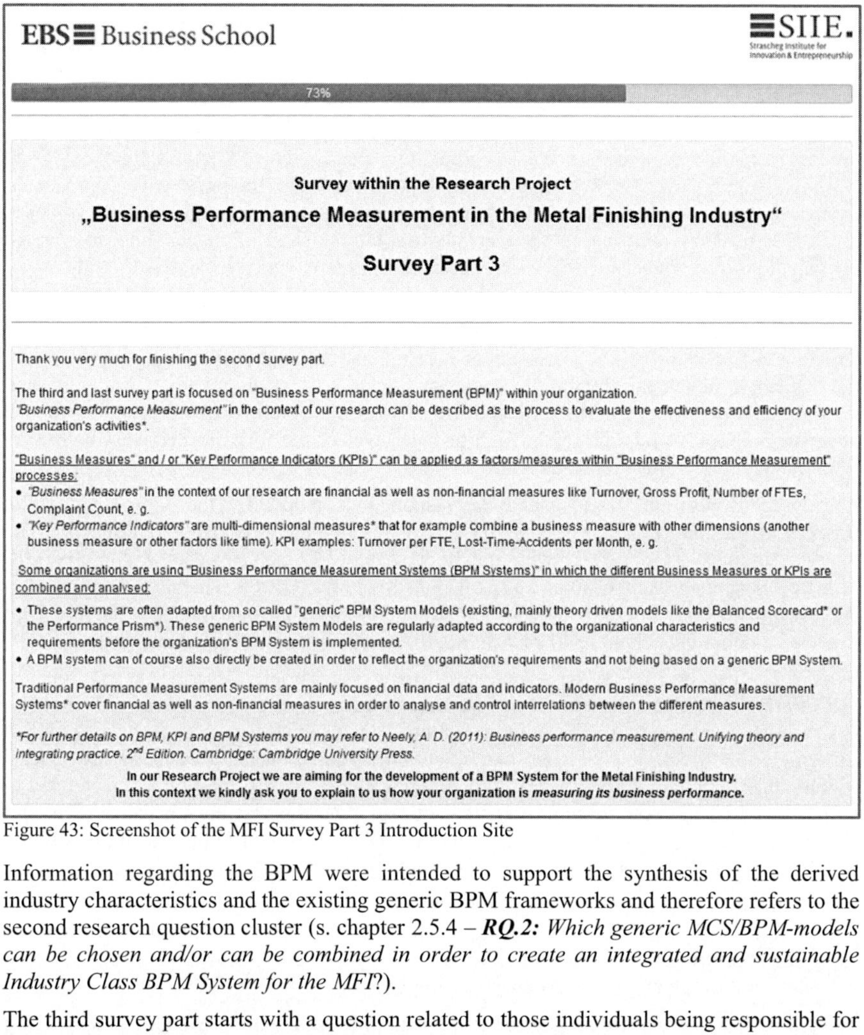

EBS Business School

SIIE. Strascheg Institute for Innovation & Entrepreneurship

73%

Survey within the Research Project

„Business Performance Measurement in the Metal Finishing Industry“

Survey Part 3

Thank you very much for finishing the second survey part.

The third and last survey part is focused on "Business Performance Measurement (BPM)" within your organization.
"Business Performance Measurement" in the context of our research can be described as the process to evaluate the effectiveness and efficiency of your organization's activities*.

"Business Measures" and / or "Key Performance Indicators (KPIs)" can be applied as factors/measures within "Business Performance Measurement" processes:

- *"Business Measures"* in the context of our research are financial as well as non-financial measures like Turnover, Gross Profit, Number of FTEs, Complaint Count, e. g.
- *"Key Performance Indicators"* are multi-dimensional measures* that for example combine a business measure with other dimensions (another business measure or other factors like time). KPI examples: Turnover per FTE, Lost-Time-Accidents per Month, e. g.

Some organizations are using "Business Performance Measurement Systems (BPM Systems)" in which the different Business Measures or KPIs are combined and analysed:

- These systems are often adapted from so called "generic" BPM System Models (existing, mainly theory driven models like the Balanced Scorecard* or the Performance Prism*). These generic BPM System Models are regularly adapted according to the organizational characteristics and requirements before the organization's BPM System is implemented.
- A BPM system can of course also directly be created in order to reflect the organization's requirements and not being based on a generic BPM System.

Traditional Performance Measurement Systems are mainly focused on financial data and indicators. Modern Business Performance Measurement Systems* cover financial as well as non-financial measures in order to analyse and control interrelations between the different measures.

**For further details on BPM, KPI and BPM Systems you may refer to Neely, A. D. (2011): Business performance measurement. Unifying theory and integrating practice. 2nd Edition. Cambridge: Cambridge University Press.*

In our Research Project we are aiming for the development of a BPM System for the Metal Finishing Industry.
In this context we kindly ask you to explain to us how your organization is *measuring its business performance*.

Figure 43: Screenshot of the MFI Survey Part 3 Introduction Site

Information regarding the BPM were intended to support the synthesis of the derived industry characteristics and the existing generic BPM frameworks and therefore refers to the second research question cluster (s. chapter 2.5.4 – ***RQ.2:*** *Which generic MCS/BPM-models can be chosen and/or can be combined in order to create an integrated and sustainable Industry Class BPM System for the MFI?*).

The third survey part starts with a question related to those individuals being responsible for setting BPM rules and guidelines as "*Set Standards*" as one core process within the LOC Diagnostic Control Systems (Simons, 1995, p. 179; s. also chapter 2.3.3). This group of individuals being responsible for setting the standards for BPM in a MFC should certainly be included in the design process of a MFC BPM system as they are supposed to be the recipients or at least the stakeholders for BPM information. This idea refers to the rationale of cybernetic regulating loops (s. chapter 2.3.1) in which an assessor needs to have a preset standard on which it can reconcile a current situation in order to react appropriately.

As propositions P3d and P3e refer to *economically successful* MFCs, I included the control variable "Performance"[53] by Boyer et al. (1997, p. 345) which is formed by two items related to growth (market share growth and sales growth) and three items related to profit (return on investment (ROI), growth in ROI and return on sales (ROS)). The respondents were asked to evaluate all five items regarding their position (focusing on their major product or service) with respect to their competitors measured on a 7-point Likert scale from 1 = significantly lower to 7 = significantly higher. Applying this subjective approach of comparing one's performance with those of their competitors is recommended by Boyer et al. (1997, p. 337) referring to the *"reluctance on the part of respondents to divulge confidential information regarding performance"* in a more objective question setting.

Referring to *Measure Outputs* as a further LOC DCS core process, I adopted eight performance measures as items as applied by Neely et al. (1994, p. 146) in 1992 in a survey among 858 SMEs in the UK. These eight measures are related to the four perspectives of time, quality, price and flexibility. The respondents were then asked to evaluate each measure regarding its importance for the MFC. By adopting this approach I aimed for the understanding and validation of the significance level of the four perspectives for MFCs, again referring to propositions P3b and P3c.

Related to proposition P1e, I asked the respondents to indicate the approximate ratio between fixed and variable costs to the total costs of their organization. Reflecting the hitherto findings associated a) with the high share of personnel costs within the MFI statistical data analysis (s. chapter 3.5) as well as b) at MeFiCom (s. chapter 6.6.4) linked to the low raw material cost due to the MFI contract processing characteristic (s. chapter 3.2), I expected a rather high share of fixed costs significantly above 50% of the total costs.

In an open question I asked the respondents to name and briefly describe the significance and information value of up to five of the most important key performance indicators or business measures applied in their organization. I included this open question in order to triangulate the finding at MeFiCom that single-dimensional business measures are at least equally important than multidimensional KPIs in MFI-BPM. This question is also related to proposition P1d in order to control for the mix of financial and non-financial KPIs and business measures. Besides this internally focused BPM, I also asked the respondents if and which external business and market indicators are relevant for their organization.

As information services play an important role in BPM (s. chapter 2.5.1) and in relation to proposition P1a I asked the respondents to indicate the IT systems or software they are using for measuring their organization's business performance, also having the option to indicate that they are not using any such systems.

The following questions are all directed to a potential BPM system that the MFC is applying including alternative answering modes as presented in the following figure:

[53] The authors report a Cronbach's alpha with α=0.93 for the reliability of the "Profit" items. They did not report a Cronbach's alpha for the reliability of the "Growth" items.

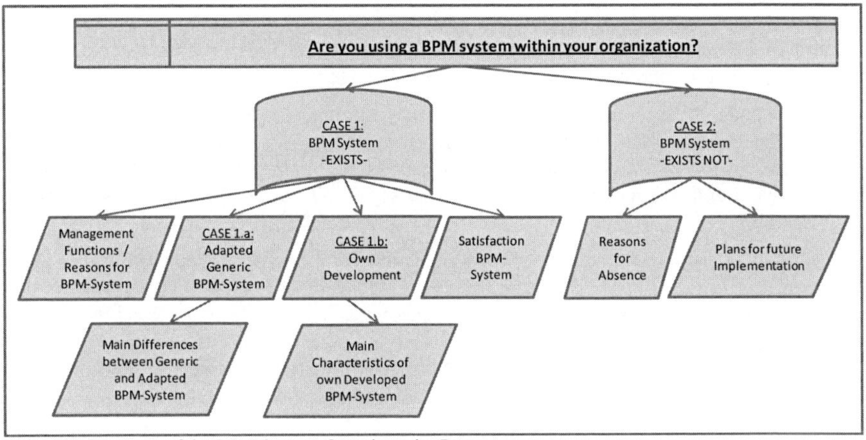

Figure 44: Structure of the BPM System Questionnaire Part

In case 1 the respondents indicated that their organization applies a BPM system:

The first question in this case is directed to the functions for which the MFC uses the BPM system including the cybernetic cycle with plan, do, check, and act items as well as further functions related to interactive communication and employee evaluation (s. chapter 2.3.3). The respondents got the opportunity to add functions that were not listed as items.

In reflecting proposition P1c I asked the respondents if their BPM system was based on a generic BPM model (s. chapter 2.4):

- If they indicated yes, I asked the respondents on which generic BPM model their BPM system was based and asked them to describe the main alterations their organization has performed in order to transform the generic BPM model into the organization's BPM system. Identifying such a process already existing in a MFC could have served as the best practice approach to triangulate the rational within this research project (s. chapter 2.5.3).
- If they indicated no, I asked the respondents to indicate the main characteristics of the organization's BPM system focusing on the application of financial and non-financial indicators as well as on the users of the BPM system. The respondents could add further characteristics to this selection.
- In both cases, generic BPM foundation or not, I asked as a final question if the respondents were satisfied with their BPM system. If they answered no, I asked for their suggestions to improve their BPM system to understand potential shortcomings of existing BPM systems applied by MFCs.

In case 2, the respondents indicated that their organization does not apply a BPM system:

Following proposition P1b, I expected that the majority of MFCs were not using a BPM system thus, those respondents indicating case 2 would support this proposition.

I asked the respondents to indicate the reasons for the absence of a BPM system in their organization, including items referring to the efforts required and the consequences of measurement (Bourne et al., 2002, p. 1305). As to one item, I asked for the reasons an already implemented BPM system was not used anymore. The participants were allowed to indicate multiple responses and to add further reasons.

To understand the requirements of the survey participants related to BPM, I asked them to indicate the characteristics of a *potential* BPM system. And as a final question, I asked the respondent's if they plan to implement a BPM system in their organization within the next six months. This was intended in order to identify potential candidates for future research.

The survey questionnaire concluded by giving the respondents the opportunity to leave comments or thoughts regarding the survey topic or of the survey itself. They could voluntarily leave an email contact address in case they were interested in receiving a short report on the survey results as an incentive for having taken part in the survey. On the final survey slide I thanked the respondents for their survey participation and assured them that their evaluations and data would be processed and treated anonymously.

Operationalization

A web-based compared to other survey modes has several advantages (Dillman et al. 2009, p. 417):

- The survey status, including response and completion rates, can be monitored in real-time via the survey administration software.
- Performing a web-based survey is intended to be cheaper and not as time consuming as other modes. In particular, a manual transfer of paper & pencil survey results into analysis software is avoided as the web-based results are already in digital form that allows transferring the survey data easily from the survey administration into the data analysis software.
- A web-based survey allows applying plausibility checks which encourage the respondent to completely answer the survey questions.
- Eliminating the risk of interviewer bias (Braunsberger et al., 2007, p. 759) as the survey is conducted independently from human interviewers.

Performing a web-based survey was also recommended by my academic supervisors as the method of choice in case an international survey should take place. However the response rates for web-based surveys are expected to be lower compared to other modes (Manfreda et al., 2008, p. 97). Furthermore data security and confidentiality have to be considered in the choice of the web-based survey software (Dillman et al., 2009, p. 418).

I applied the web-based survey software Questback EFS Survey to operationalize the survey model (Dillman et al., 2009, p. 195). As I possessed an international survey sample (s. previous chapter), I created a German and English version of the survey questionnaire (s. appendices A.15 and A.16).

In order to translate the original scales adopted from English literature into German, I applied the *back-translation for cross-cultural research* method (Brislin, 1970). Schaffer and Riordan (2003, p. 189) have analyzed 210 cross-cultural studies published in academic journals and identified back-translation as the best practice technique in cross-cultural organizational research *"before administering an instrument to respondents who speak a different language from the one in which the instrument was originally developed and validated"*. The following figure illustrates the applied back-translation process for one of the "Exploitative Innovation" scale items adopted from Jansen et al. (2009, p. 809):

0. EN - Original	**Literature**	**We frequently make small adjustments to our existing products and services.**
1. 1st DE - Translation	Researcher	Wir machen häufig kleine Anpassungen an unseren existierenden Produkten und Services.
2. 1st EN - Back Translation	Bilingual Test Pers. 1	We frequently make small adaptations to our existing products and services.
3. 2nd DE - Translation	Bilingual Test Pers. 1 & Researcher	**Wir machen häufig kleine Anpassungen an unseren existierenden Produkten und Services.**
4. 2nd EN - Back Translation	Bilingual Test Pers. 2	We often perform small adjustments to our existing products and services.
5. 3rd EN - Back Translation	Bilingual Test Pers. 3	We frequently make small adaptions to our existing products and services.
6. FINAL DE-Translation	**Researcher**	**Wir machen häufig kleine Anpassungen an unseren existierenden Produkten und Services.**

Figure 45: Example for the Back-Translation Process of Scale Items

As a first step I translated each original scale item from English into German, and later a bi-lingual test person translated each item back into English. I discussed these first round English translations with the bi-lingual test person regarding the differences to the original English scale items. I adjusted my German translation in case the semantic sense was ambiguous. These adjusted items were sent to two other bi-lingual test persons who independently from each other translated the adjusted German items back into English. I adopted the adjusted German translation in case these back-translations were congruent to the original English items. In less than five cases the second and third round English back-translations differed significantly from the original English items; I adjusted these few items again in order to be congruent with the original scales.

As I combined scales from different literature sources, I had to align them in order to establish a harmonized MFI survey which included generally addressing the survey respondents in the sense that he/she answered the questions related to his/her organization. Where applicable I employed a non-comparative scaling technique using 7-point Likert scales (Bradburn et al., 2004, p. 126) in the questionnaire.

As a further quality measure I performed a pretest of the web-based survey instrument before the invitation to participate was sent out to the international survey sample. The pretest sample included:

- three academics including my supervisors
- one native English speaker
- a small sample of the survey population (Bradburn et al., 2004, p. 317) including
    - two representatives of the supporting MFI associations (s. previous chapter) and
    - two colleagues from MeFiCom[54].

I asked the pre-testers to indicate any unclear as well as ambiguous wording or phrasing of the survey questions. Furthermore the participants were supposed to report any problems related to the web-based survey design, including testing the different filter alternatives. The pre-test resulted in minor adjustments referring mainly to the wording of some questions.

In order to invite the potential respondents to the survey I created an invitation email (s. appendices A.11 and A.12) which I structured according to the recommendations of Dillman et al. (2009, p. 277) including

- an appeal for help,

[54] Employees of MeFiCom took part only in the pre-test, not in the final survey as the data for MeFiCom had already been retrieved during the qualitative research process.

- the reasons why the respondent was selected referring to his/her organization's membership in an MFI association,
- the usefulness of the survey,
- confidentiality statement,
- contact information, and
- the time frame (in this case 15 to 20 minutes) required to answer the survey.

In this email I attached two Acrobat PDF files:

- an invitation letter (s. appendices A.13 and A.14) with an EBS-SIIE company letter head signed by the SIIE director Prof. Dr. Ronald Gleich and me to emphasize the academic character of the survey.
- a copy of the Transactions of the IMF (Larson, 2014) editorial[55] calling MFCs to support the EBS-SIIE MFI survey.

The invitation, including a hyperlink to the survey questionnaire, was sent via email to the survey sample on August 31st, 2014. A German version was sent to those participants whose organization was located in Germany; while an English version was sent to the participants whose organization was located in other countries.

I applied the email addresses included in the MFI association member lists, mainly covering a general organization email address (like info@xy.com). This is also the reason why the email was not addressed to particular individuals but treated as an anonymous invitation. I asked the recipients to forward the email to the chief executive officer (CEO), general manager or chief financial officer (CFO) of the organization being the target respondents of the MFI survey.

After I had sent out the invitation emails I retrieved several mailer-daemon emails stating that the invitation email could not be delivered to the recipients. This was the case either when the email address retrieved from the MFI associations' member lists included spelling errors, or when the organization listed in the member list no longer existed anymore. In the latter case I eliminated the address record from the survey sample or otherwise corrected the email address and sent it again to the organization.

For the EFDS and SEA members, I did not send the invitation email directly to the sample participants, but forwarded a German version of the invitation to Dr. Brzezinski (managing director of EFDS) and an English version to Mr. Elliott (chief executive of SEA, UK). Both were willing to forward the MFI survey invitation mail directly to their associations' members, combined with a brief comment in order to encourage the members to take part in the survey. I decided to employ this process in order to enhance the response rate of the survey. Referring to possible ethical issues arising from the associations' role in this process, I should clarify that the addressed association members had no obligation to respond to the survey and that the members were not forced by their associations to take part. Their choice to participate in the survey or not was not reported to the associations as the survey was held anonymously.

However this approach had the disadvantage of not applying a wave analysis technique (Armstrong and Overton, 1977) by analyzing early and late respondents in order to address a potential non-response bias (Atif et al., 2012). Due to different reasons, EFDS and SEA

[55] Publicly available also on the Transactions of the IMF website (IMF, 2014).

did not forward the survey invitation emails to their members on August 31st, 2014 but some days later. As the respondents could not be assigned to the different types of invitations (coming either directly from myself, or being forwarded from EFDS or SEA) due to the anonymity of the respondents, the wave analysis technique was not applicable.

After three weeks of having the survey in the field I sent an email reminder to all of the participants (Dillman et al., 2009, p. 279) in which I thanked those who had already taken part in the survey and encouraged those who had not yet taken part to participate.

7.2 Quantitative Data Analysis and Results

The following table gives an overview on the survey response and completion rates:

No. of total survey visitors:	**82**
% of total sample size (Response Rate):	**14.3%**
Thereof chosen German as survey language:	43
% of total German sample size:	15.8%
Thereof chosen English as survey language:	39
% of total UK sample size:	12.9%
No. of completed questionnaires:	27
No. of excluded responses due to unreliable data (uniform answering style):	-1
Final number of responses to be analyzed:	**26**
% of total visitors:	31.7%
% of total sample size (Completion Rate):	**4.5%**
No. of responses from Germany:	**10**
% of total German sample size:	3.7%
No. of responses from UK:	**14**
% of total UK sample size:	4.6%
No. of responses from other countries (Switzerland & Netherlands):	**2**

Table 20: Response and Completion Rates in the MFI Survey

Out of the initial sample, 82 visitors accessed the MFI survey (response rate of 14.3%), out of which 27 have completed the whole questionnaire. I had to delete one data record from the completed questionnaires as it contained unreliable data due to a uniform answering style. The resulting 26 responses form the final sample to be analyzed. This represents a completion rate of 4.5 %[56], including 10 from Germany, 14 from UK, 1 from the Netherlands, and 1 from Switzerland.

These rates are at the lower end of the response rate range for surveys conducted in organizational research (Baruch and Holtom, 2008, p. 1149). The low absolute number of

[56] The response rate is comparable to Brummet and Arnett's (1967, p. 2) MFI survey in the mid-1960s in which they had sent a questionnaire to 485 MFCs receiving 42 responses. However their survey was performed at a time where the overall response rates can be supposed to have been higher (Baruch and Holtom, 2008, p. 1150).

completed questionnaires, however, does not allow multivariate data analysis like structure equation modeling as the minimum sample size is not reached (Hair et al., 2010, p. 662).

Furthermore, the low response rate does not allow a reliable calculation of the Cronbach's coefficient alpha (Cronbach, 1951; Cortina, 1993). The minimum sample size – not only for clinical studies – should ideally be higher than N=300 (Charter, 2003, p. 118). Although the minimum sample size can be calculated according to the desired confidence interval precision or coefficient power (Bonett and Wright, 2015), Charter (2008, p. 249) recommends an N=30 as lower limit for the Cronbach's alpha calculation. In the MFI survey the total N=26, already below this lower limit, needs to be further divided into the German (N=11) and English (N=15) language fractions to calculate each coefficient alpha; following Charter (2008) the results of such calculations could not be treated as reliable. The confirmability of the adapted MFI survey scales' reliability is therefore limited to the Cronbach's alpha stated in the corresponding original publications from which the scales were taken.[57]

On average, the respondents (N=26) completed the survey within 24.7 minutes (Min=12.2; Max=46.7), with 16 participants requiring less than 25 minutes[58], which is slightly higher than the projected time frame of 15 to 20 minutes estimated during the design of the MFI survey (s. previous chapter).

Within the next chapters I will analyze the 26 responses univariate descriptively due to the low response rate using IBM SPSS.

7.2.1 MFI Survey Part I – Statistical Data

From the completed sample, all respondents have indicated that they are working at a management level within their organization. The highest share of the participant's function is formed by chief executive officers and general managers each with eight participants, followed by four chief financial officers as presented in the following table:

Participant's Function	Frequency	Percent
Chief Executive Officer (CEO)	8	30.8
General Manager	8	30.8
Chief Financial Officer (CFO)	4	15.4
Head of Sales	1	3.8
Other (as indicated by the respondents):		
Department Head	1	3.8
Business Administration Manager, Authorized Officer	1	3.8
Managing Director	1	3.8
Owner of the Business	1	3.8
Previously Operations Director for the group, now semi-retired and working as a Director of our Holdings company.	1	3.8
Total	**26**	**100.0**

Table 21: Composition of the Survey Sample – Focus on Participants' Function

[57] Please refer to appendix A.18 for the calculation and discussion of the MFI survey scales alphas.

[58] Please note that these values also include possible breaks or interruptions during answering the survey. The mean of the German respondents (N=10) is with 25.7 minutes slightly higher compared to the British respondents (N=14) with 24.4 minutes.

This composition suits the survey target recipients of CEOs, general managers and CFOs as described in the previous chapter. Regarding their experience measured in years of work within the MFI the respondents indicated as follows:

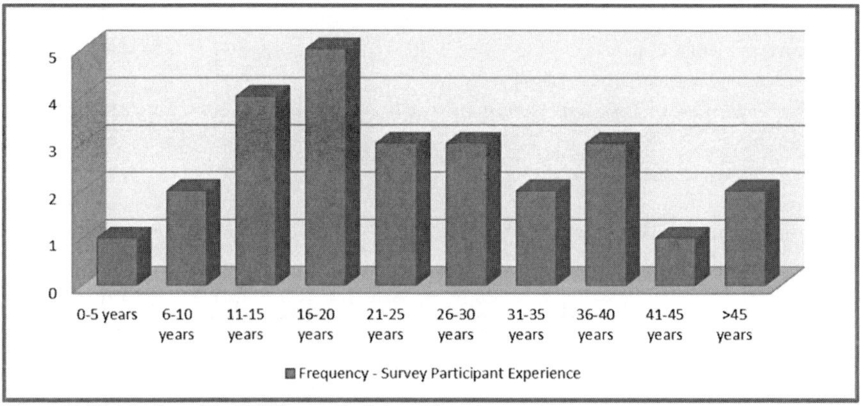

Figure 46: Distribution of the MFI Survey Participant Experience

The majority of participants (53.8%) have been working in the MFI for more than 20 years. Only one participant indicated working in the MFI for less than six years. The MFI experience is therefore expected to be high in the sample. This is a positive indicator related to the information quality referring to one result from the qualitative data transcription, where the richest information came from those experts with the highest seniority (s. chapter 6.5.1).

The following table refers to the type of metal finishing company represented in the sample:

MFI Organization Type	Frequency	Percent
Metal finishing company	17	65.4
Manufacturing company with metal finishing department	1	3.8
Supplier for metal finishing companies	2	7.7
Metal finishing system manufacturer	1	3.8
Other (as indicated by the respondents):		
Research & Development Service Provider	1	3.8
Research Institution	1	3.8
Heat Treater	1	3.8
Heat Treatment	1	3.8
Coat Thickness Measuring	1	3.8
Total	**26**	**100.0**

Table 22: Composition of the Survey Sample – Focus on Organizations

As the two heat treatment organizations can be added to the metal finishing company share (s. chapter 3.4), the total number of MFCs in the sample is 19 (73.1%). I retained the seven other organizations in the sample as they are also MFI organizations. The respondents related the following metal finishing processes to their organizations (multiple answers were allowed):

Metal Finishing Processes	Quoted	Percentage
Electro Plating (incl. chrome plating)	12	19.7
Pre- / Post-treatment	12	19.7
Galvanization (incl. hot-dip galvanization, phosphate conversion)	9	14.8
Heat treatment (incl. nitriding)	9	14.8
Paint finishing (incl. wet and powder painting)	7	11.5
Chemical / Physical Vapour Deposition (CVD / PVD)	4	6.6
Anodizing	4	6.6
Thermal Spraying (incl. plasma spraying)	1	1.6
Electropolishing	1	1.6
Research and Development	1	1.6
Coat Thickness Measuring	1	1.6
Total	**61**	**100.0**

Table 23: Composition of the Survey Sample – Focus on Metal Finishing Processes

The percentage of metal finishing processes related to the 26 British, German, Swiss, and Dutch MFI organizations can be compared to the distribution of processes in the German MFI in 2005 (s. figure 12).

Only three of the 26 MFI organizations have been in operation for less than 20 years, whereas ten organizations have been operating for more than 45 years as presented in the following figure:

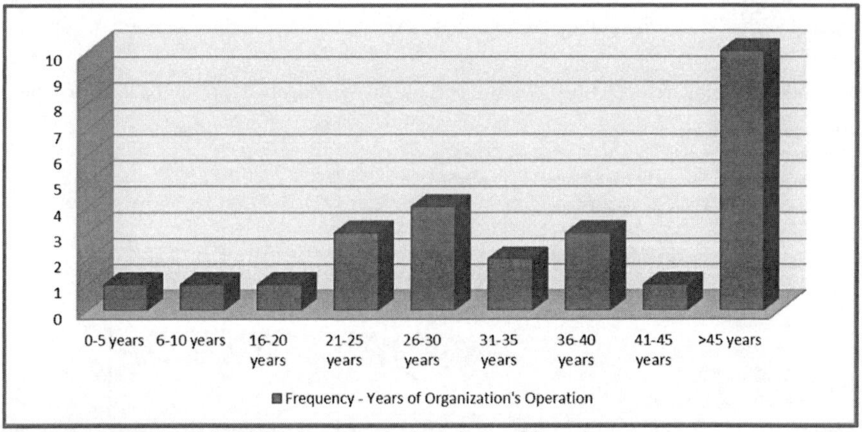

Figure 47: Distribution of the MFI Survey Organization's Years of Operation

This result illustrates the maturity of the MFI (s. chapter 3.1 for MFI's origins). Under the assumption that the sample organizations have not significantly changed their metal finishing operations during their existence, this is also an indicator of stability in the MFI.

The survey data regarding the 26 organization's turnover and number of employees is attached in table A.17.1 of appendix A.17. The following table is an extract of these data and is structured according the SME definitions in chapter 3.5.1:

Enterprise Class	Germany			United Kingdom			Switzerland			Netherlands			Total		
	Frequ.	Turn.* [kEUR]	Empl.	Frequ.	Turn. [kEUR]	Empl.	Frequ.	Turn. [kEUR]	Empl.	Frequ.	Turn. [kEUR]	Empl.	Frequ.	Turn.* [kEUR]	Empl.
Large	0	0	0	3	189,226	1,340	0	0	0	1	130,000	980	4	319,226	2,320
Medium	6	78,200	515	2	13,132	205	0	0	0	0	0	0	8	91,332	720
Small	4	10,200	113	8	19,214	195	0	0	0	0	0	0	12	29,414	308
Micro	0	0	0	1	419	5	1	408	3	0	0	0	2	827	8
Total	**10**	**88,400**	**628**	**14**	**221,990**	**1,745**	**1**	**408**	**3**	**1**	**130,000**	**980**	**26**	**440,798**	**3,356**

* Excluding one German case with no value in turnover.
Frequ. = Frequency // Turn. = Turnover // Empl. = Number of Employees

Table 24: Turnover and Employees of Sample Organizations per Country and Enterprise Class

The sample organizations account for a total turnover of 440.8 million Euros in 2013, of which the 14 British organizations have a share of 50.4% and the nine German organizations[59] have a share of 20.1%. The single Dutch enterprise has a share of 29.5% of the overall turnover showing the significance of the MFI enterprise classes' mix. The twelve large and medium sized organizations, being the target group of larger MFCs for BPM in the MFI (s. chapter 3.5.1), have a share of 46.2% in the survey sample. However these twelve larger MFCs account for 93.1% of the overall turnover and 90.6% of the overall employees in the survey sample. This illustrates the disproportionately high portion of turnover and employees attributed to larger MFCs as presented in the statistical data analysis in chapter 3.5.1. Eighty-four point six percent of the sample organizations belong to the SME cluster (EC, 2003, p. L124/39; s. chapter 3.5.1), a circumstance that is congruent with the analyzed statistical data (s. chapter 3.5) and therefore supports proposition P2a.

In the following table I compare the survey sample turnover for 2013 and the number of employees for June 2014 with the corresponding European statistical data for 2012 (s. appendix A.2):

Data Source	Germany		United Kingdom		Netherlands	
	Turn.* [kEUR]	Empl.	Turn.* [kEUR]	Empl.	Turn.* [kEUR]	Empl.
MFI Survey 2013/14	88,400	628	221,990	1,745	130,000	980
Eurostat 2012	8,299,100	71,918	2,029,500	20,182	1,382,100	7,713
Ratio	1.1%	0.9%	10.9%	8.6%	9.4%	12.7%

* Excluding one German case with no value in turnover.
Data for Switzerland not available from Eurostat. Turn. = Turnover // Empl. = Number of Employees

Table 25: Turnover and Employees of Sample Organizations Compared to Overall MFI Population

Comparing data from different sources and different time periods can never lead to accurate results. However table 25 gives an indication of the survey population related to each nation's total MFI turnover and number of employees:

- United Kingdom's is represented with approximately 10% of all national MFCs
- Germany is represented with approximately 1% of all national MFCs[60]
- The Netherlands is represented with approximately 10% of all national MFCs by one Dutch enterprise

[59] N=9 as one German organization did not indicate its turnover.

[60] Please note that MeFiCom with an annual turnover of above 50 MEUR (s. chapter 6.1.2), representing more than 0.5% of the German MFI turnover, is not included in the survey as it represents the single case to be studied in this research project.

Fifty-eight point three percent of the twelve larger organizations have reported operating solely one production site within their country (s. table A.17.1 of appendix A.17) which does not support proposition P2b regarding the decentralization of larger MFCs.

7.2.2 MFI Survey Part II – Overall Characteristics of the MFC

On the next pages I will analyze the results of the applied scales. Therefore I will present descriptive statistics results generated in SPSS for each MFI survey scale reflecting the whole N=26 sample. In addition I have included the descriptive statistics for the German (N=10) and the UK (N=14) sub-samples as well as the results of *independent-samples t-tests* (Student, 1908; Box, 1987) for these two sub-samples in appendix A.17.2. In the t-tests, I compare the item responses of both sub-samples as grouping variables in order to spot any significant statistical variance (below a five percent significance level) that may be driven by cultural differences (Trompenaars and Hampden-Turner, 2012, p. 193). Although located in Western Europe, both Germany and UK show cultural differences within organizations as presented by Hofstede et al. (2010, p. 314) who refer to *"Mintzberg's Five Preferred Configurations of Organizations"*:

- UK gets referred to its organizational market model aiming for *"mutual adjustments"* and *"ad-hoc negotiations"* reflecting the national characteristic of low uncertainty avoidance.
- Germany by contrast, showing a high degree of uncertainty avoidance, is characterized for its *"standardization of skills"*, referring to the high standards of employee qualification, and its *"professional bureaucracy"*.

These two different levels of uncertainty avoidance have implications on the national attitudes regarding planning, controlling and accounting processes (Hofstede et al. 2010, p. 316) as major MCS and BPM elements (s. chapter 2.3.1): related to its low uncertainty avoidance level, UK focuses on strategic planning whereas Germany by contrast focusses on detailed, operational planning and control processes[61].

Besides these cultural differences, having the survey in two different languages may play a role within statistical variances between the German and the British sub-samples: although applying the back-translation method (s. chapter 7.1.2) for the adaption of the original scale items into German, I cannot fully negate slight semantic differences remaining in the two survey questionnaires versions (s. appendices A.15 and A.16) leading to statistical variances between the two sub-samples.

Furthermore I have analyzed significant statistical differences between the organizational sizes classes (s. table 24) applying independent-samples t-tests. Therefore I grouped the N=26 sample into the two sub-samples "Larger MFCs" (with more than 49 employees – N=12) and "Smaller MFCs" (with less than 50 employees – N=14). The detailed result of this analysis is attached in appendix A.17.3.

As described above I will present the descriptive statistics for the whole N=26 MFI sample in subsequent pages. I will include information of those items showing significant statistical differences (below a 5% significance level in the t-tests) between the MFC country sub-samples and the MFC size class sub-samples into the discussion of the results.

[61] MeFiCom, being the German MFC in my ethnographic single case study, also applies a sophisticated and detailed monthly forecast process in order to reduce uncertainty, what suits to the above descriptions by Hofstede et al. (2010) for German organizations.

The following table shows the descriptive statistics related to the "Customer Focus" scale adopted from Chen and Paulraj (2004, p. 123):

Customer Focus	N	Minimum	Maximum	Mean	Std. Deviation
	Statistic	Statistic	Statistic	Statistic	Statistic
We anticipate and respond to customers' evolving needs and wants.	26	5	7	6.31	.788
We emphasize the evolution of formal and informal customer complaints.	26	3	7	5.85	1.047
We follow-up with customers for quality/service feedback.	26	2	7	5.73	1.282
We interact with customers to set reliability, responsiveness, and other standards.	26	2	7	5.77	1.275
Satisfying customer needs is the central purpose of our business.	26	5	7	6.46	.647
Customer focus is reflected in our business planning.	26	4	7	6.19	.849
We produce products and/or offer services that satisfy or exceed customer expectations.	26	5	7	6.42	.643

Table 26: Descriptive Statistics – Customer Focus Scale

The items were measured on a 7-point Likert scale from 1 = Strongly Disagree to 7 = Strongly Agree. The means of all seven items are above 5.7 showing the importance for MFCs to focus on their customers and therefore supporting proposition P3a.

The first item shows a statistically significant difference between Germany and UK with $t(22)=2.28$, $p=0.033$, where the German sub-sample MFCs ($M=6.70$, $SD=0.48$) more strongly agrees to anticipate and respond to customers' evolving needs and wants than the British MFCs ($M=6.00$, $SD=0.88$).

The third item shows a statistically significant difference between larger and smaller MFCs with $t(24)=2.43$, $p=0.023$, where the larger MFCs ($M=6.33$, $SD=0.78$) more intensively follow-up with customers for quality/service feedback than the smaller MFCs ($M=5.21$, $SD=1.42$).

The following table shows the descriptive statistics related to the "Closer to Customers" scale adopted from Powell (1995, p. 19):

Closer to Customers	N	Minimum	Maximum	Mean	Std. Deviation
	Statistic	Statistic	Statistic	Statistic	Statistic
We always try to increase our organization's direct personal contacts with customers.	26	4	7	6.04	0.774
We are actively seeking customer inputs to determine their requirements.	26	3	7	5.92	1.129
We are using customer requirements as the basis for quality.	26	5	7	6.27	0.778
Involving customers in product/service design is crucial for our organization.	26	2	7	5.65	1.413

Table 27: Descriptive Statistics – Closer to Customers Scale

The items were measured on a 7-point Likert scale from 1 = Strongly Disagree to 7 = Strongly Agree. The means of all four items are above 5.9 further emphasizing the importance of the direct contact connection of the MFC with its customers, also supporting proposition P3a.

The first item shows a statistically significant difference between Germany and UK with t(22)=2.46, p=0.022, where the German sub-sample MFCs (M=6.50, SD=0.53) has a stronger emphasis to always attempt to increase direct personal contact with customers than the British MFCs (M=5.79, SD=0.80). Furthermore, the fourth item shows a statistically significant difference between Germany and UK with t(20.07)=2.25, p=0.036, where the German MFCs (M=6.30, SD=0.82) show stronger indications to involve their customers in the product/service design than the British MFCs (M=5.14, SD=1.66).

The fourth item also shows a statistically significant difference between larger and smaller MFCs with t(17.53)=2.65, p=0.017, where the larger MFCs (M=6.33, SD=0.65) see the involvement of their customers into product and service design as more crucial for their organization than the smaller MFCs (M=5.07, SD=1.64).

The following table shows the descriptive statistics related to the "Competitive Priorities" scale adopted from Chen and Paulraj (2004, p. 124):

Competitive Priorities	N	Minimum	Maximum	Mean	Std. Deviation
	Statistic	Statistic	Statistic	Statistic	Statistic
Our strategy cannot be described as concentrating solely on offering products with the lowest price.	26	1	7	6.35	1.325
Our strategy is based on quality performance rather than price.	26	5	7	6.54	0.647
We place greater emphasis on innovation than price.	26	3	7	5.73	1.116
We place greater emphasis on customer service than price.	26	3	7	5.96	1.076
Our strategy places importance on delivering products/services with high performance.	26	3	7	6.31	1.123
We emphasize launching new products/services quickly.	26	2	7	5.00	1.414

Table 28: Descriptive Statistics – Competitive Priorities Scale

The items were measured on a 7-point Likert scale from 1 = Strongly Disagree to 7 = Strongly Agree. The means of items one to five are above 5.7 showing that the price has a lower competitive priority compared to quality, innovation and customer service. This result supports proposition P3b.

The second item shows a statistically significant difference between Germany and UK with t(20.12)=2.76, p=0.012, where the German sub-sample MFCs (M=6.90, SD=0.32) more strongly emphasize that their strategy is based on quality rather than on price compared to the British MFCs (M=6.36, SD=0.63).

The following table shows the results of the click ranking question (EFS, 2014, p. 186) which validates the above shown MFC's competitive priorities results:

Competitive Priorities	**Quality**	**Delivery Reliability**	**Costs/Price**
Rank 1	**24**	1	1
Rank 2	2	**19**	5
Rank 3	0	6	**20**
N=	*26*	*26*	*26*

Table 29: Descriptive Statistics – Competitive Priorities Click Ranking

Within the click ranking question, the experts were asked to rank the three factors *cost/price, quality*, and *delivery reliability* according to their priority within the respondent's organization starting with "1" having the highest priority.

Twenty-four of the 26 respondents ranked *quality* as having the highest priority in their organization with none ranking it to the subordinated rank 3. Nineteen respondents have chosen *delivery reliability* to be at priority rank 2. Twenty respondents have ranked *costs/price* to have a subordinated priority compared to *quality* and *delivery reliability*. These results support propositions P3b and P3c as well as being congruent with the industrial service nature of the MFI being based on contract processing (s. chapter 3.6). The superordinate priority of quality and delivery reliability compared to costs and price can be explained with the MFI core process (s. chapter 3.2) which is regularly one of the last process steps within the value added chain of a product. Therefore the quality of this metal finishing process determines the overall quality of the product. A low quality of the metal finishing process will lead to reduced metal finishing benefits, like a reduced wear protection. In a worst case scenario, an unacceptable quality will lead to the scrapping of the product, which would include the risk of warranties for the MFC. Likewise, a reliable delivery is required by MFC's customers to plan their own shipping and delivery processes (in case of tools as metal finished products) or assembly (in case of components being part of assembly groups).

The following table shows the descriptive statistics related to the "Environmental Uncertainty" scale adopted from Pagell and Krause (2004, p. 632):

Environmental Uncertainty	N	Minimum	Maximum	Mean	Std. Deviation
	Statistic	Statistic	Statistic	Statistic	Statistic
Actual users of our products/services are ...	26	1	7	3.69	1.784
Competitors for the supply of raw materials are ...	26	1	6	4.31	1.436
Competitors for our customers are ...	26	1	7	3.77	1.531
Government regulations controlling our industry are ...	26	1	7	3.12	1.558
The public's political views and attitudes toward our industry are ...	26	1	7	3.50	1.530

Table 30: Descriptive Statistics – Environmental Uncertainty Scale

The items were measured on a 7-point Likert scale from 1 = Never Predictable to 7 = Always Predictable. The respondents heterogeneously answered all four items leading to means that have a central tendency of being slightly below the midpoint (higher uncertainty), excepting the second item which is slightly above the midpoint (lower uncertainty). Proposition 4a is therefore not supported as environmental uncertainty is not relevant for the majority of MFCs in the sample.

The t-tests showed no significant statistical differences between the MFC country sub-samples and between the MFC size class sub-samples.

The following table shows the descriptive statistics related to the "Technology Uncertainty" scale adopted from Chen and Paulraj (2004, p. 138):

Technology Uncertainty	N	Minimum	Maximum	Mean	Std. Deviation
	Statistic	Statistic	Statistic	Statistic	Statistic
Our industry is characterized by rapidly changing technology.	26	1	7	3.96	1.637
If we do not keep up with the changes in technology, it will be difficult for us to remain competitive.	26	3	7	5.65	1.294
The rate of process obsolescence is high in our industry.	26	1	6	3.38	1.577
The production technology changes frequently and sufficiently.	26	1	6	3.23	1.366

Table 31: Descriptive Statistics – Technology Uncertainty scale

The items were measured on a 7-point Likert scale from 1 = Strongly Disagree to 7 = Strongly Agree. Comparable to the environmental uncertainties the respondents heterogeneously answered all four items leading to means that have a central tendency of being slightly below the midpoint, excepting the second item. Having means below the midpoint represents in this case a lower technology uncertainty as the respondents rather disagree about rapidly changing technology or process obsolescence within the MFI. However the second item indicates that MFCs have to keep up with technology changes to remain competitive. Proposition 4b is not supported as technology uncertainty is not relevant for the majority of MFCs in the sample.

The fourth item shows a statistically significant difference between Germany and UK with $t(22)=2.59$, $p=0.017$, where the British sub-sample MFCs ($M=2.64$, $SD=1.08$) strongly disagreed with the statement of frequently and sufficiently changing production technology compared to the German MFCs, whose answers where neutral on this item ($M=4.00$, $SD=1.49$).

The following table shows the descriptive statistics related to the "Demand Uncertainty" scale also adopted from Chen and Paulraj (2004, p. 138):

Demand Uncertainty	N	Minimum	Maximum	Mean	Std. Deviation
	Statistic	Statistic	Statistic	Statistic	Statistic
Our master production schedule has a high percentage of variation in demand.	26	2	7	5.35	1.648
Customer demands fluctuate drastically from week to week.	26	1	7	5.08	2.058
Our internal supply requirements (for example for consumables) vary drastically from week to week.	26	1	6	3.08	1.468
We keep weeks of inventory of critical materials to meet changing customer demands.	26	1	7	4.65	2.097
The volume and/or composition of customer demands are difficult to predict.	26	1	7	5.38	1.577

Table 32: Descriptive Statistics – Demand Uncertainty scale

The items were measured on a 7-point Likert scale from 1 = Strongly Disagree to 7 = Strongly Agree. The means of items one, two, and five are above 5.0 indicating an existing but not distinct demand uncertainty, which however does not show an effect on the internal supply requirements with a mean of 3.1. These results indicate support for proposition P4c.

The fourth item shows a statistically significant difference between larger and smaller MFCs with $t(18.61)=2.65$, $p=0.016$, where the larger MFCs ($M=5.67$, $SD=1.073$) tend to maintain weeks of inventory of critical materials to meet changing customer demands compared to smaller MFCs ($M=3.79$, $SD=2.39$).

The following tables show the descriptive statistics related to the organizational ambidexterity scales "Exploratory Innovation" and "Exploitative Innovation" adopted from Jansen et al. (2009, p. 809):

Exploratory Innovation	N	Minimum	Maximum	Mean	Std. Deviation
	Statistic	Statistic	Statistic	Statistic	Statistic
Our organization accepts demands that go beyond existing products and services.	26	1	7	5.04	1.777
We commercialize products and services that are completely new to our organization.	26	1	7	4.27	1.733
We frequently utilize new opportunities in new markets.	26	2	6	4.81	1.021
Our organization regularly uses new distribution channels.	26	1	5	2.65	1.355

Exploitative Innovation	N	Minimum	Maximum	Mean	Std. Deviation
	Statistic	Statistic	Statistic	Statistic	Statistic
We frequently make small adjustments to our existing products and services.	26	1	7	5.08	1.917
We improve the production efficiency for our products and/or the provision efficiency for our services.	26	2	7	5.85	1.190
We reduce costs by having efficiency effects channelled through the volume increase of existing products.	26	1	7	4.50	1.817
Our organization expands services for existing clients.	26	3	7	5.42	1.027

Table 33: Descriptive Statistics – Exploratory and Exploitative Innovation Scales

The items of both scales were measured on a 7-point Likert scale from 1 = Strongly Disagree to 7 = Strongly Agree.

Regarding the exploratory scale, the means of items one to three lie slightly above the Likert scale midpoint, not showing a statistically high or low significance of exploratory orientation. The fourth item, referring to the utilization of new distribution channels, has a mean of 2.65 and a maximum value of 5, and thus was rejected by the respondents. I suspect this circumstance is due to the contract processing character of MFCs mainly being located near their main customers and therefore not requiring new distribution channels (s. chapter 3.6).

The means of the exploitative items are all above the Likert scale midpoint and also higher compared to the exploratory items. These results show a tendency of the sample MFCs to be exploitative rather than exploratory oriented.

Organizational ambidexterity refers to an organization's ability to simultaneously reflect exploration and exploitation orientation (s. chapter 6.5.2). Therefore I countered the exploratory items relative to the exploitative items for each respondent within appendix A.17.4. The results show that six organizations are slightly more exploratory than exploitative oriented (negative variances between the means). Organizational ambidexterity plays only a minor role for seven organizations for which the mean variance is above a value of 2. These organizations showed a clear exploitative orientation without emphasizing the exploratory perspective.

The item means of economically successful MFCs[62] showed no significant difference to the means of all other sample organizations. The results therefore do not support proposition P3d referring to organizational ambidexterity being a characteristic exclusively for economically successful MFCs. However a certain degree of organizational ambidexterity can be expected for the MFI as 19 respondents showed a low mean variance between the exploratory and exploitative items (variances between -1.75 and +1.75).

[62] I selected those 15 out of the 26 organizations for which the respondents have indicated to have a significantly higher economic performance compared to their competitors within the last two years (item values >= 5 – s. performance scale by Boyer et al. (1997) in the next chapter).

The t-tests showed no significant statistical differences between the MFC country sub-samples and between the MFC size class sub-samples within the exploratory and exploitative scales.

The following table shows the descriptive statistics related to five flexibility dimensions adopted from Pagell and Krause (2004, p. 634):

Flexibility Dimensions	N	Minimum	Maximum	Mean	Std. Deviation
	Statistic	Statistic	Statistic	Statistic	Statistic
Delivery flexibility: The ability of an organization to effectively respond to changes in planned delivery dates.	26	5	7	6.31	0.736
Volume flexibility: The ability of an organization to effectively increase or decrease aggregate production in response to customers.	26	2	7	5.77	1.336
Mix flexibility: The ability of an organization to effectively produce a wide variety of different products or offer a wide variety of different services.	26	2	7	5.96	1.341
Changeover flexibility: The ability of an organization to effectively handle additions and subtractions to the product or service mix over time.	26	2	7	5.69	1.320
Modification flexibility: The ability of an organization to effectively implement minor changes in current products or services that result from corrective actions or changing customer requirements.	26	4	7	6.04	1.038

Table 34: Descriptive Statistics – Flexibility Dimensions

The items were measured on a 7-point Likert scale from 1 = Extremely Unimportant to 7 = Extremely Important. The means of all five items are above 5.7 showing the importance of all presented types of flexibility dimensions for MFCs.

No significant difference in the means of economically successful MFCs to all other sample organizations could be identified. Proposition P3e is therefore not supported, as flexibility indicates not only an exclusive prerequisite for economic success but to be a prerequisite for all MFI organizations.

The significance of delivery flexibility is emphasized as the statistical minimum equals five, which means that all participants have ranked the importance of the flexibility dimension above the Likert scale midpoint. This illustrates the high priority of delivery reliability indicated by the respondents within the click rank question. Volume and mix flexibility address those MFCs, offering their services and processes on a high variety of substrates requiring also a higher number of temporary or leased employees adjusting to these variations, compared to other manufacturing industries (s. chapter 3.5.2). The importance of modification flexibility can be linked to the MFI characteristic of customizable services and products, i.e. the adaption of metal finishing processes to new or modified customers' substrates (s. chapter 6.6.1).

The t-tests showed no significant statistical differences between the MFC country sub-samples and between the MFC size class sub-samples.

The following table shows the results related to the MFC's order backlog in working days:

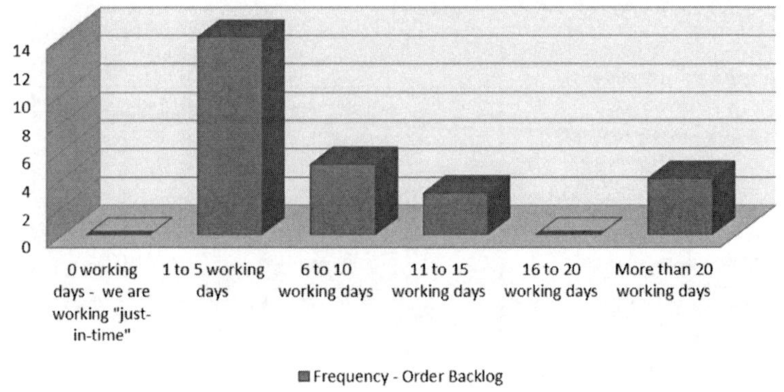

Figure 48: Regular Order Backlog at the MFI Survey Sample MFCs

The majority of the survey respondents have indicated that their MFCs have a short order backlog of one to five working days which supports proposition P2c.

7.2.3 MFI Survey Part III – Business Performance Measurement

The first question in the third MFI survey part is related to the responsible individuals setting the rules and guidelines for BPM in the MFC. The results are presented in the following table:

Responsibility BPM Rules and Guidelines Setting	Frequency	Percent
Corporate Headquarters	4	12,5
Executive Board	8	25,0
Supervisory Board	1	3,1
Owner	6	18,8
Local Management	8	25,0
Local Department Heads	2	6,3
Other (please indicate):		
Head of Institution	1	3,1
Managing Director	1	3,1
The responsibility for setting BPM rules/guidelines in our organization is NOT defined.	1	3,1
	32	**100,0**

Table 35: Descriptive Statistics – Responsibility BPM Rules and Guidelines Setting

Only one out of the 26 respondents indicated that the responsibility for setting BPM rules and guidelines is not defined in his/her organization. This result indicates that for the majority of MFI organizations the responsibility for setting BPM standards (Simons, 1995, p. 179; s. also chapter 2.3.3) is defined and dedicated to the organization's management, mainly either to the executive board, to the owner and/or to the local management of the MFI organizations. Six respondents indicated a combination of two responsible parties for BPM standards-setting in their organizations. One of these six respondents indicated a combination of corporate headquarters and executive boards being responsible, whereas the other five respondents named their local management as responsible, combined either with their corporate headquarters, executive board, owner or local department heads (indicated twice). The results document the heterogeneity of different BPM standard-setting responsible parties in MFI organizations. Therefore a general rule to include either individuals from the executive boards, the business owner, or the local management cannot be given for the MFI. In fact the results document the importance – and perhaps the difficulty – in identifying the *right* individuals being responsible for BPM standards-setting in each MFI organization as it approaches the customizing and implementation of an MFI-BPM system. The task of these responsible individuals is to foster and support the implementation of a BPM system in their organization, not only by setting the standards, but by ensuring the sustainability of the BPM system through management commitment (Webb, 2004) and management integration (Gleich, 2011, p. 330).

The following table shows the descriptive statistics related to the "Performance" scale adopted from Boyer et al. (1997, p. 345):

Performance	N	Minimum	Maximum	Mean	Std. Deviation
	Statistic	Statistic	Statistic	Statistic	Statistic
Market share growth	26	3	7	4.77	1.032
Sales growth	26	3	7	4.88	0.993
Return on investment (ROI)	26	3	6	4.92	0.891
Growth in ROI	26	3	7	4.92	0.977
Return on sales (ROS)	26	3	6	4.92	0.891

Table 36: Descriptive Statistics – Performance Scale

The items were measured on a 7-point Likert scale from 1 = Significantly Lower to 7 = Significantly Higher. The respondents were asked in this scale to evaluate the economic performance of their organization in comparison to their competitors within the past two years. The means of all items lie in a narrow range between 4.77 and 4.92, which indicates that the sample organizations slightly improved their economic performance compared to their competitors. The main purpose of this scale was to identify the economically successful organizations in the survey sample as described in the previous chapter.

The following table shows the descriptive statistics related to eight "SME Performance Measures" adopted from Neely et al. (1994, p. 146):

SME Performance Measures	N	Minimum	Maximum	Mean	Std. Deviation
	Statistic	Statistic	Statistic	Statistic	Statistic
Unit product costs	26	1	7	4.42	2.212
Costs versus budget	26	1	7	4.96	1.843
Incoming customer parts quality	26	3	7	6.12	1.177
In process quality	26	5	7	6.65	0.689
Field failure under warranty	26	1	7	5.35	1.875
On-time delivery to customer	26	6	7	6.58	0.504
Delivery lead time	26	4	7	6.27	0.919
Product/machine change over times	26	1	7	4.81	1.980

Table 37: Descriptive Statistics – SME Performance Measures

The items were measured on a 7-point Likert scale from 1 = Extremely Unimportant to 7 = Extremely Important. The respondents indicated all measures to be important for their organizations by having means above 4.4. The results furthermore support propositions P3b and P3c as they are congruent to the hitherto quantitative findings that measures related to quality performance (items 2 – 5) and delivery performance (items 6 and 7) are more important than financial measures (item 1 and 2). This mix of financial and non-financial measures which is applied by MFI organizations, supports proposition P1d.

The following figure presents the results regarding the distribution of fixed and variable costs in the overall MFC's costs:

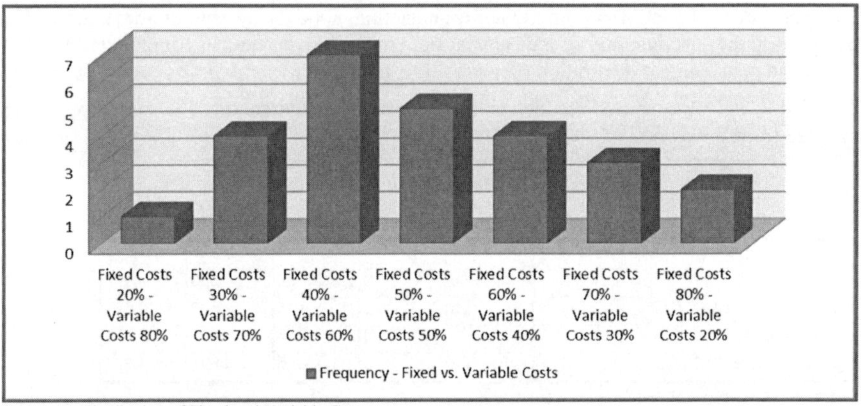

Figure 49: Distribution of Fixed and Variable Costs

The majority of the survey respondents indicated that the share of fixed costs relative to the total costs of their organization equals or is less than 50 percent. This result is surprising due to the MFI's contract processing character (s. chapter 3.2) referring to a share of fixed costs being supposed to be significantly higher than 50%. Removing the seven organizations that are not MFCs (s. chapter 7.2.1) from the sample does not influence the result. Thirteen of these 19 MFCs indicated that they have equal or less than 50 percent fixed costs. This finding does not support proposition P1e, which is based on the single case findings at MeFiCom as well as on the statistical database analysis (s. chapter 3.5.1). The latter showed a significantly high share of personnel costs, having a fixed costs character, in their overall costs indicating the service character of the MFI.

Asked for a listing of up to five of the most important KPIs and/or business measures to assess their organization's performance, the 26 respondents named a total of 82 items. I have clustered these financial and non-financial measures into 16 subject areas. Some of these areas are split into further dimensions as presented in the following figure:

BPM in the MFI Survey
Key Performance Indicators and Business Measures

FINANCIAL		44
Sales		**17**
	Amount	8
	Per Employee	3
	Per Month	2
	Per Production Hour	1
	Per Customer	1
	Per Sales District	1
	Target vs. Actual	1
Costs		**9**
	Energy	3
	Personnel Cost in Sales	2
	Consumables Costs	1
	Unit Labor Cost	1
	Unit Materials Cost	1
	Quality Cost (for Customer Defects)	1
Gross Profit		**7**
Net Profit		**5**
	Amount	2
	Before Tax	1
	Per month	1
	Per employee	1
Cash Flow / Liquidity		**3**
EBIT		**1**
Stock Value		**1**
Actuals vs. Budget		**1**

NON-FINANCIAL		38
Production Process		**9**
	Backlog	2
	Capacity Utilization	1
	Machine Utilization	1
	Material Utilization	1
	Throughput Time	1
	Units Produced per Period	1
	Working Hours per Order	1
	Workshop Orders per Day	1
Human Factor		**7**
	Absenteeism / Sick Days per Empl.	2
	Productivity	2
	HSE - Accident Statistics	1
	Empl. Satisfaction	1
	Staff Performance	1
Quality		**7**
	Failure Count / ScrapRate	5
	Customer Complaints per Month	2
Delivery Reliability		**5**
	On Time Delivery	3
	Reliability	1
	Performance	1
Organization		**4**
	New Applications	2
	Product portfolio	1
	Audit Performance	1
Customer		**3**
	Focus	1
	Satisfaction	1
	Service	1
Market		**2**
	Growth Rate vs. Competitors	1
	Current Market Situation	1
Environment		**1**
	Water Utilization	1

Figure 50: MFI Survey – KPIs and Business Measures

The values in figure 50 equal the frequency of KPIs and business measures per subject area, as well as an itemization per subject area dimension. Forty-four of the 82 items have a financial character whereas 38 are non-financial measures. Comparable to the results at MeFiCom (s. chapter 6.6.4), the majority of MFI survey organizations apply single-dimensional financial business measures, namely *Sales* and *Gross/Net Profit* figures, within their BPM. Further financial measures and indicators refer to the organization's costs, especially to energy and personnel costs being two of the main MFI cost drivers (s. chapter 3.5).

The non-financial measures and indicators mainly refer to the subject areas of the production process, the human factor, quality, and delivery reliability. The importance of the latter two subject areas has already been described in the previous chapters of the MFI survey data analysis. As the metal finishing production process embodies the competitive advantage (s. chapter 2.2.3) of being the enabler for contract processing (s. chapter 3.2), the MFI organizations emphasize the different process dimensions including controlling for order backlog as well as different types of utilization. The efficiency of the production, referring to units or orders produced within a time period, is an additional factor that would be analyzed by some of the sample organizations.

Although the environmental impact of the MFI organizations is undoubted (s. chapter 3.3), the respondents only named energy cost and water utilization as environmental related business measures.

Two non-financial measures refer to the MFI organization's market situation, also compared to its competitors. This external BPM focus is continued within the next survey question in which I asked the respondents to indicate the external indicators being applied in their organization's BPM. The results are presented in the following table:

Applied External BPM Indicators	**Frequency**	**Percent**
GDP - Gross Domestic Product	3	10,7
PMI - Purchasing Managers Index	2	7,1
Other (please indicate):		
Automotive Production Index	1	3,6
CBI Output Expectations	2	7,1
Galvanizing Industry Data	1	3,6
Our organization is NOT using external business / market indicators for Business Performance Measurement.	19	67,9
	28	**100,0**

Table 38: Descriptive Statistics – Applied External BPM Indicators

The majority of the respondents opted not to apply external business or market indicators within their BPM. This result indicates a weakness of current MFI-BPM processes as the inclusion of external market and business cycle indicators is recommended to reduce uncertainties (Simons, 1995, p. 94) in order to early react to economic down- or upturns. This would necessitate internal capacity adjustments (for example by reducing or increasing the number of temporary staff, e.g.). I expect that following such bellwethers is important for the MFI being somewhat dependent on the prosperity of other industry branches (s. chapter 3.1).

Regarding information services applied in BPM processes, the respondents indicated as follows:

MFI BPM Information Services	Frequency	Percent
Standardized Office Software (for example Microsoft Excel or Access, OpenOffice)	19	47,5
Enterprise Resource Planning (ERP) Software	9	22,5
Reporting based on Data Warehouses (for example IBM Business Warehouse)	2	5,0
Customer Relationship Management (CRM) Software	3	7,5
Other (please indicate):		
Own Developed Software	3	7,5
Accounting Software	2	5,0
We are NOT using IT Systems / Software for measuring our organization's business performance.	2	5,0
	40	**100,0**

Table 39: Descriptive Statistics – BPM Information Services

Twenty-four of the 26 respondents indicated to use IT systems or software to measure their organization's business performance; this result supports proposition P1a. Nineteen respondents were receptive to apply standardized office software like Microsoft Excel or Open Office. Eleven respondents indicated to apply more than one IT systems; six of whom apply a combination of standardized office and enterprise resource planning software to measure their organization's business performance. This result demonstrates the predominant role of standard office software in BPM as described by Singh (2013) who refers to the application of Microsoft Excel in planning and forecasting processes especially in small organizations, which also form the majority of MFI organizations (s. chapter 3.5).

Out of the 26 sample organizations, seven (26.9%) indicated to use a BPM system. This result supports proposition P1b, as the majority of the sample organizations are not using a BPM system. Asked for the functions for which the seven organizations use their BPM system, the respondents answered as follows:

MFI BPM Functions	Frequency	Percent
Organizational Strategy Setting ("Plan")	4	11,8
Organizational Strategy Implementation ("Do")	2	5,9
Organizational Strategy Control ("Check")	3	8,8
Organizational Strategy Alignment ("Act" - in case of strategy deviations)	4	11,8
Personal Target Setting	3	8,8
Personal Feedback Processes (including incentives, gratifications and penalties)	4	11,8
Information Supply (Management Information System, e.g.)	4	11,8
Communication (strategy, target achievement, e. g., are communicated via the BPM-System)	4	11,8
Operational Management of the Organization	6	17,6
Other	*0*	*0,0*
	34	**100,0**

Table 40: Descriptive Statistics – BPM Functions

All seven respondents indicated multiple functions for which their organizations are using BPM:

Six respondents indicated to use their BPM system as a cybernetic regulating loop (function 1 to 4; s. chapter 2.3.1) in which one respondent uses all four functions, two respondents use three functions and three respondents use only one function (either *Plan* or *Act*). Four respondents anticipate a need to use their BPM system for human relation purposes (at least for one of the functions 5 and 6). And five respondents intended to use their BPM system also as a communication tool (at least for one of the functions 7 and 8; s. Interactive Intergroup Communication in chapter 6.6.5). Nearly all sample organizations use their BPM system for the operational management of their organization. No respondent included additional functions as those offered within the survey question. The results indicate that an MFI-BPM model should not be solely narrowed to follow the rationale of a cybernetic regulating loop, but also include functions referring to human resource management (like incentives and penalties management; Simon, 1995, p. 51) and communication.

Out of the seven organizations, three applied a BPM system based on the Balanced Scorecard as the generic BPM framework. Proposition P1c is therefore not supported as the survey results indicate that existing MFI-BPM systems are not in general based on generic BPM frameworks. The other four organizations applied self-developed BPM systems which have the following main characteristics:

MFI Customized BPM Systems	**Frequency**	**Percent**
Includes financial indicators	4	28,6
Includes non-financial indicators	3	21,4
The management uses the BPM-System	4	28,6
The department heads are using the BPM-System	2	14,3
The finance department uses the BPM-System	1	7,1
Other	*0*	*0,0*
	14	**100,0**

Table 41: Descriptive Statistics – Customized BPM System Characteristics

Three of the four customized BPM systems include a mix of financial and non-financial indicators. The BPM systems mainly address the organization's management as recipients of BPM information.

When asked if they were satisfied with their BPM system, only one of the seven respondents answered in the negative. As reasons for his/her dissatisfaction the respondent stated: *"We are not utilizing its potential, still too many locally created spread sheet reporting."* I found this also described by the respondent at MeFiCom relating to the streamlining of existing systems (s. chapter 6.6) when referring to a variety of locally created Microsoft Excel spreadsheets in MeFiCom departments. This result also emphasizes the important role of suitable and harmonized information services within BPM (s. chapter 2.5.1).

The 19 organizations which are not using a BPM system indicated the following reasons for the absence of such a system:

Reasons for BPM System Absence	**Frequency**	**Percent**
Implementation too time consuming	5	19,2
Implementation too expensive	1	3,8
Sustaining too time consuming	5	19,2
Sustaining too expensive	2	7,7
Costs disproportional to benefit	11	42,3
Other (please indicate):		
Satisfied with the successful results of current budgetary management and efficiency control measures.	*1*	*3,8*
Satisfied with the way information is gathered and processed.	*1*	*3,8*
	26	**100,0**

Table 42: Descriptive Statistics – Reasons for BPM System Absence

The results show that the majority of the 19 organizations do not see the benefits of a BPM system that justify corresponding costs. However the majority of the respondents did not indicate that implementing or sustaining such a BPM system is too expensive. A more important factor seems to be the amount of time that has to be spent on a BPM system referring to both, the implementation and the sustainment. Therefore the degree of BPM automation (for example automated report compilation) plays a significant role in the acceptance and sustainability of a BPM system by reducing the required time to be spent on it. No respondent indicated that an existing system was no longer in use.

The 19 respondents whose organizations did not use a BPM system were asked for some optimal characteristics of a potential BPM system for their organization. Fourteen respondents answered as follows:

Potential MFI BPM System	**Frequency**	**Percent**
Should include financial indicators	14	26,4
Should include non-financial indicators	14	26,4
The management should use the BPM-System	11	20,8
The department heads should use the BPM-System	7	13,2
The finance department should use the BPM-System	6	11,3
Other (please indicate): *Should be transmitted to factory floor (cell) level.*	*1*	*1,9*
	53	**100,0**

Table 43: Descriptive Statistics – Potential BPM System Characteristics

All 14 respondents indicated that a potential BPM system for their organizations should include a mix of financial and non-financial indicators. The majority of the respondents referred to their management in using the BPM system. The department heads as well as the finance department play a subordinated role. This result demonstrates the role of BPM as an integral element of MCS (s. chapter 2.5.1).

Eighteen of the 19 respondents whose organization did not operate a BPM system indicated that they did not plan to implement such a system within the next six months[63].

Chapter 7 completes the empiric research process and leads to a synthesis of the attendant research results and MFI-suitable generic BPM frameworks into an MFI-BPM model.

[63] The respondent who indicated that his organization has planned to implement a BPM system left his email address in order to receive a short-report on the survey results as an incentive to participate. Therefore I was able to contact the person in order to receive further information related to the introduction of the BPM system in his MFC. However the respondent refused to give more detailed information (personnel communication with the respondent, April 8th, 2015).

8 BPM in the MFI – Synthesis

> "Performance improvement does not simply come from measuring. Performance comes from the processes and practices and how effectively people execute them. Measurement just keeps the score. So, to improve, you need to change what you do, or do it more effectively, something that measurement may encourage, but measurement alone does not create value." (Bourne, 2008, p. 76)

Following the chalice-shaped bottom-up approach to improve BPM feasibility (s. chapter 2.5.3) I have accessed MFI-related international nomenclatures of economic activities in order to analyze statistical databases leading to the first MFI characteristics in chapter 3 of this dissertation thesis which form the general characteristics clusters of the MFI (s. chapter 3.6).

These industry class characteristics were further analyzed in detail within the empiric qualitative research process in chapter 6 leading to the holistic cultural portrait of the MFI, i.e. the MFI cause and effect model as the base for the MFI-BPM design. This MFI CEM was triangulated by an MFI survey, presented in the previous chapter.

Chapter 8 synthesizes the accumulated statistical data, research findings, and survey implications into an MFI-BPM model as an approach towards a sustainable and integrated management system for contract processing businesses. Chapter 8.1 links the MFI characteristics clusters to suitable generic BPM frameworks. Chapter 8.2 delineates the MFI-BPM model design and presents MFI relevant performance indicators and business measures. Chapter 8.3 concludes with further BPM recommendations for the MFI.

8.1 Adaption of Generic BPM Frameworks and Models

Following steps 3 and 4 of the chalice-shaped bottom-up approach to improve BPM feasibility (s. figure 10), I will choose generic BPM frameworks and models suiting to the MFI characteristics clusters (s. chapter 3.6), and then collapse these models into their unique elements to choose those which shall form the MFI-BPM system. This chapter therefore answers research questions RQ.2a and RQ.2b (s. chapter 2.5.4).

In order to identify suitable generic BPM frameworks and models, I followed the analysis of important and generally accepted BPM frameworks and models performed by Gleich (2011, p. 122). In his analysis, Gleich compares eleven criteria (2011, p. 23) out of which contemporary BPM models (s. chapter 2.4.2) should be formed regarding their peculiarities within a) seven generic BPM frameworks derived from an academic or consulting background, as well as b) four BPM models derived from business practice. These are presented in the following figure:

	Generic BPM Frameworks							Business Practice BPM Models			
Framework / Model Criterion	Data Envelopment Analysis	**Results and Determinants**	**Balanced Scorecard**	Tableau de Bord	SMART Pyramid	Quantum BPM Concept	**SCOR-Model**	Hewlett-Packard Concept	DART	FESTO ISC	Skandia Navigator
Connection to Vision and Strategy	◔	◕	●	◕	◕	●	◔	◕	●	●	◑
Stakeholder related Goal Differentiation	◔	◑	◑	◔	◑	◕	◑	◑	●	◑	◑
Consideration of Different Performance Levels	◑	◑	◑	◕	●	●	●	◑	◕	◕	◑
Measures and Indicators Management	◔	◑	◑	◑	◑	◑	◕	◕	◔	●	◑
Measurement of Modalities	●	◔	◑	○	◑	◑	◑	●	◔	○	○
Performance Eevaluation and Variance Analysis	●	◔	◕	○	◑	◑	◕	●	○	◕	◑
Consideration of Incentives	◔	◕	●	○	◔	◑	◔	◔	○	○	○
Concept Reporting	◑	◑	●	◕	◕	◕	○	◕	○	◔	◕
Institutional Framework	◑	◑	◕	◔	◕	◕	◕	◕	◕	◑	◑
Application of BPM Tools	◑	◑	◕	◔	◑	◑	◑	◑	◑	◔	○
Integration of Management Control	◑	◔	◕	◕	◑	◕	◑	◕	◕	◔	◔
MFI Characteristics Cluster		**Industrial Service Nature**	**A Job Shop Industry**				**A Contract Processing Industry**				

Figure 51: Generic BPM Framework Comparison (adopted in modified form from Gleich, 2011, p. 125)

Out of the seven generic BPM frameworks and related to research question RQ.2a, I have chosen the following three frameworks (s. chapter 2.4.2) as the foundation for the MFI-BPM model:

- the Results and Determinants Framework, representing a BPM framework for service businesses, for being related to the MFI's industrial service nature,
- the Balanced Scorecard, representing a BPM framework suitable also for SMEs, for being related to the MFI's job shop characteristic, and
- the SCOR-Model, representing a supply chain BPM framework, for being related to the MFI's contract processing characteristic.

As a generic BPM framework I did not choose

- the Data Envelopment Analysis as it was developed for non-for-profit organizations and as it shows a limited practicability due to its methodical complexity (Gleich, 2011, p. 70),
- the Tableau de Bord, as it can be conceptually compared to the Balanced Scorecard (s. chapter 2.4.2), but its degree of realization in an organizational context is subordinated to the Balanced Scorecard (Gleich, 2011, p. 84),
- the SMART Pyramid, as its design follows a top-down approach (s. chapter 2.5.3) which does not suit the chalice-shaped bottom-up approach applied in my research project. In addition, the SMART pyramid lacks a connection to rewards and incentives (Gleich, 2012, p. 54), and

- the Quantum BPM Concept, as a fast and economical implementation is unreliable due to the following framework's weaknesses (Gleich, 2012, p. 54):
    - the operationalization of the performance indicators management,
    - the performance evaluation, and
    - the execution of variance analyses.

The four business practice BPM models in figure 51 cannot be taken fully into account for the development of an MFI-BPM model as they are developed in different industry contexts than the MFI. However these business practice BPM models share common elements with an MFI-BPM model. Two examples:

- the Skandia Navigator (s. chapter 2.5.3) is based on the employee's motivation embodying the human capital of an organization. This can be linked to the important role of the employees as success factors within the MFI (s. chapters 6.6.4 for employees being a major success factor at MeFiCom, and 8.2.1 for the implementation of this aspect within the MFI-BPM model)
- an organization's employees are also one of the three target groups within the DART BPM model (s. chapter 2.4.2; Gleich, 2011, p. 112). Additionally, DART, compared to the ten other frameworks and models within figure 51, has the strongest peculiarity when it comes to the consideration of stakeholders – this consideration also has a significant relevance within BPM in the MFI (s. chapters 6.6.3 for legislation being part of the MFI boundary system indicating the MFI's relation with and placement in the society, and chapter 8.2.3 for the implementation of this aspect within the MFI-BPM model)

However none of the eleven BPM frameworks and models in figure 51 is explicitly related to environmental BPM. A corresponding BPM concept is therefore added in this chapter in order to represent the fourth MFI characteristics cluster, *MFI's Environmental Impact*. This environmental concept as well as the three chosen generic BPM frameworks and their limitations for the corresponding MFI characteristics clusters, are presented in more detail subsequently related to research question RQ.2b.

8.1.1 Metal Finishing – A Contract Processing Industry

An MFC, as contract processing *metal finishing process provider*, is a supply chain element as illustrated in the following figure[64]:

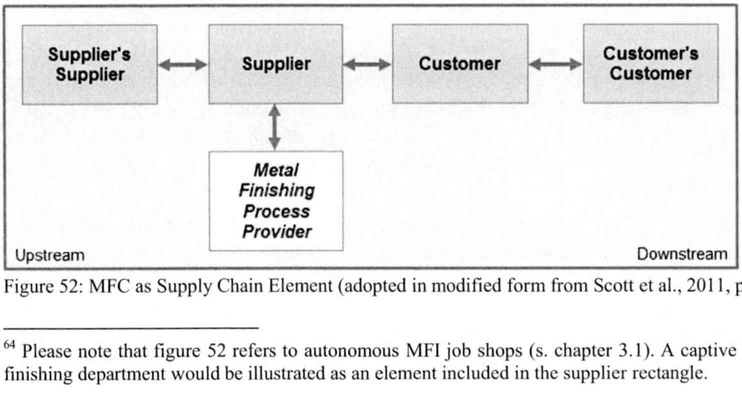

Figure 52: MFC as Supply Chain Element (adopted in modified form from Scott et al., 2011, p. 5)

[64] Please note that figure 52 refers to autonomous MFI job shops (s. chapter 3.1). A captive in-house metal finishing department would be illustrated as an element included in the supplier rectangle.

The Supply Chain Operations Reference (SCOR)-Model (SCOR, 2010), as a supply chain BPM design instrument (Gleich, 2011, p. 97), represents the first generic BPM model chosen for the adoption in an MFI-BPM model. This generic BPM framework focuses on the relations of an organization within its market environment in order to enhance the overall supply chain performance (SCOR, 2010, p. 2). It can therefore be akin to the Relational View (s. chapter 2.2.2) as vital interorganizational relations are the tenet within this strategic theory of the firm leading to competitive advantages. The SCOR model basically consists of five management processes (Gleich, 2011, p. 98):

- Plan – processes coordinating demand and resources with purchases, production and sales
- Source – processes coordinating purchases according to the demand
- Make – processes which convert raw material in finished products
- Deliver – processes which provide the customer with the products
- Return – processes related to the goods returned

The organizations within a supply chain are connected in this SCOR-Model by deliver-source-relationships (SCOR, 2010, p. 6). In order to include a MFC into the supply chain (s. figure 52), the model has to be adjusted for the MFI as presented in the following figure:

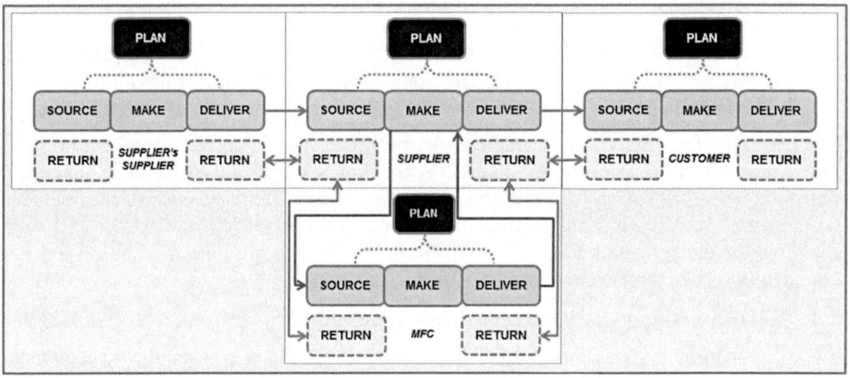

Figure 53: SCOR-Model Adapted for the MFI

Within this adapted SCOR-Model, the MFC becomes directly connected with the make-process of its customer (connections marked orange within figure 53), although logistics processes are included within the transfer of metal substrates to the MFC and metal finished products returned to the customer. The rationale for not connecting the MFC via a deliver-source relationship is grounded in the MFI contract processing characteristic (s. chapter 3.2), as an MFC extends its customer's production ("make") with the metal finishing processes.

This adapted SCOR-Model serves as the explanatory model for the MFI-BPM subsystem which reflects the MFCs relations to its customer (s. chapter 8.2.3). As presented in the previous chapters, a delay or downtime within the MFC's production cycle has a direct influence on the MFC's customer's production in such events where the MFC is not producing on stock to buffer these negative effects. Furthermore the MFC has to handle changing customers' demands by being flexible (s. chapters 6.6.1 for the importance of flexibility at MeFiCom and 7.2.2 for the importance of different flexibility dimensions derived from the MFI survey). This circumstance is directly related to the MFC's production utilization rates (s. production process in figure 50 being reflected in chapter 8.2.2) as an already achieved maximum production capacity reduces the degree of flexibility in case the MFC's customer increases his/her demand. A vital relation to its customers is therefore of great importance for the MFC in order to have corresponding information on demand increases early to be able to react on these changes. This is accomplished either by increasing the production capacity (temporary for example with leased laborers, or for longer periods with the purchase of additional coating equipment) or by increasing the efficiency of the currently installed production equipment. The latter is recommended in cases where other MFC's customers would reduce their demands.

8.1.2 Metal Finishing – A Job Shop Industry

Giannopoulos et al. (2013) have analyzed the usage of Balanced Scorecards (BSC) in British and Cypriot SMEs. They found that the majority of SME's were not aware of BSC as a BPM model, but that they already apply performance measures that scope the four perspectives of a BSC (s. chapter 2.4.2). Garengo and Biazzo (2012) have shown the positive aspects of the implementation of a Balanced Scorecard within an SME. These two studies indicate that the Balanced Scorecard (BSC) is a further generic BPM framework whose elements should be considered for the creation of an MFI-BPM model, as the MFI is mainly formed by SMEs.

The basic structure of a BSC allows the inclusion of

- objectives which get quantified by using
- measures that are compared to
- targets reflecting
- strategic initiatives

for each of the four perspectives

- financial (reflecting an intended shareholder's perception of the organization),
- customer (reflecting an intended customer's perception of the organization),
- internal business process (that has to be excelled in order to satisfy shareholders and customers), and
- learning and growth (sustaining an organization's ability to change and improve)

in order to fulfill an organization's vision and strategy (Kaplan and Norton, 1996, p. 8).

The BSC is a very popular BPM framework (s. chapter 2.4.2), however it shows some weaknesses as described by Gleich (2012, p. 60) related to the contemporary BPM framework criteria presented in figure 51:

- Target differentiation exists only for customers, neglecting other stakeholder groups.
- BSC lacks a detailed description regarding the management of business measures and performance indicators.
- Modalities like performance targets or measuring cycles have to be set individually.
- Concentration solely on an organization's top-management.
- BSC has a strategic rather than an operational orientation.

Reflecting those weaknesses, Gleich (2012, p. 63) recommends that a BPM system should be tailored to the targeted practical area of application (in my case the MFI) including not solely strategic but also operational BPM elements. I will follow this recommendation in the modeling process of the MFI-BPM model (s. chapter 8.2) by having a BPM structure that is in line with the MFC's competitive advantages influencing the MFC's business strategy (s. figure 55), and by including business measures and performance indicators with short measuring cycles, thus reflecting the current operational state, for example of the MFC's production processes (s. chapter 8.2.2).

8.1.3 MFI's Industrial Service Nature

Fitzgerald et al. (1991, p. 2) have identified the following four key differences of services as opposed to manufacturing on which they grounded their *Results and Determinants BPM Framework* (s. chapter 2.4.2):

- Intangibility: most services produce intangible outputs, such as a helpful cabin crew when travelling by air.
- Heterogeneity: the quality of service outputs varies, especially when the services include a high amount of labor, leading to heterogeneous outputs related to different employees performing the services.
- Simultaneity: production and consumption of services occur simultaneously, like *"having a haircut or taking a rail trip"*, which limits quality inspections in advance before the services are delivered to the customers.
- Perishability: most services cannot be stored unlike stocks in manufacturing. This missing stock buffer limits a service organization's scope of action regarding demand fluctuations – *"therefore scheduling operations and controlling quality are key management problems in services, which are made more difficult by the presence of the customer in the service process"*.

Analyzing these four service characteristics for the MFI illustrates the twilight zone between manufacturing and services in which the MFI is located (s. chapter 3.5.1):

- Intangibility does not hold true for the MFI as MFCs produce tangible products, i.e. coating layers which increase the added-value of a metal substrate (s. chapter 3.1).
- Heterogeneity can be seen as a characteristic of the MFI: although many metal finishing processes are executed on automated coating equipment, it still needs a high share of labor (for example in the composition of chemicals for cleaning or galvanizing processes), resulting in a certain degree of heterogeneous metal finishing services output.
- Simultaneity does not hold true for the MFI, as the metal finished processes, like any other manufacturing processes, get performed before the finished products are delivered to the customers.
- Perishability again is an MFI characteristic for those MFCs which do not have the chance to produce the customers' metal finished substrates on stock. At MeFiCom the main share of metal finished products cannot be produced on stock, as the customers have the requirement to receive these products back in the shortest possible time in order to deliver the finished products to their customers. It is this missing stock buffer which emphasizes the relevance of a sound production planning in the MFI (s. chapters 6.6.4 and 7.2.2 for the short regular order backlog at MeFiCom and at the MFI survey sample MFCs).

At first sight the *Results and Determinants BPM Framework* does not seem suitable for BPM in the MFI, as the MFI does not meet all of the four service characteristics. However the performance dimensions related to the framework's results and determinants are well suited to the hitherto MFI-related findings as presented in the following comparison:

Results and Determinants BPM Framework			BPM in the MFI		
Type	BPM Dimension	Source in Generic Framework	BPM Dimension	Related to	Chapter in Dissertation
Results	Competitiveness	Fitzgerald et al., 1991, p. 17	Global Competition	MeFiCom	6.6.1
			Competition Human Resources	MeFiCom	6.6.4
			Competition in Mature Markets	MeFiCom	6.6.5
			Competitive Priorities	MFI Survey	7.2.2
			Competitive Measures	MFI Survey	7.2.3
	Financial Performance	Fitzgerald et al., 1991, p. 19	Critical Financial Performance Variables	MeFiCom	6.6.4
			Financial Measures	MFI Survey	7.2.3
Determinants	Quality of Service	Fitzgerald et al., 1991, p. 35	Quality Management System	MeFiCom	6.2
			Quality as Success Factor	MeFiCom	6.6.4
			Customer Quality Feedback	MFI Survey	7.2.1
			Competitive Priorities	MFI Survey	7.2.2
			Quality Performance	MFI Survey	7.2.3
	Flexibility	Fitzgerald et al., 1991, p. 55	Industrial Service Character	MeFiCom	6.6.1
			Flexibility vs. Bureaucracy	MeFiCom	6.6.4
			Flexibility Dimensions	MFI Survey	7.2.2
	Resource Utilization	Fitzgerald et al., 1991, p. 77	Utilization Rate of Coating Equipment	MeFiCom	6.6.4
			Human Resources as Success Factor	MeFiCom	6.6.4
			Non-Financial Measures	MFI Survey	7.2.3
	Innovation	Fitzgerald et al., 1991, p. 101	Organizational Ambidexterity	MeFiCom	6.6.2
			Competitive Priorities	MFI Survey	7.2.2
			Organizational Ambidexterity	MFI Survey	7.2.2

Table 44: The Results & Determinants Framework and BPM in the MFI

Fitzgerald et al.'s service classification (1991, p. 12) differentiates according to the number of customers being processed per day between:

- Professional Services – with a low number of customers and an accordingly high peculiarity of contact time and customization.
- Service Shops – with a moderate number of customers and a medium peculiarity of contact time and customization.
- Mass Services – with a high number of customers and standardized processes, which only allow a low customization procedure.

According to this differentiation, Fitzgerald et al. suggest corresponding performance measures (1991, p. 121). In order to select the appropriate MFI service performance measures, I have located MFCs on the intersection between professional services and service shops. The degree to which an MFC tends more to be a professional service provider or a service shop depends on its orientation being a) either a service provider that has matched its production processes according to the requirements of very few (one) customers or b) by offering standardized processes for more customers. I will reflect service performance measures related to Fitzgerald et al.'s determinants and results dimensions within the MFI-BPM modeling process (s. chapter 8.2) according to their MFI relation as presented in table 44.

8.1.4 MFI's Environmental Impact

The circumstance of reflecting natural environmental aspects within an organization's management control processes should not be marketing-driven[65]. It should be a necessity which also enables one to react early on upcoming legislation as for instance happened in recent years in the MFI relating to the European REACH agreement (s. chapter 6.6.3).

The application of Corporate Environmental Performance (CEP; Trumpp et al., 2015) systems assists an organization in the assertion of its environmental impact, to evaluate countermeasures to reduce an organization's negative impacts on the natural environment.

Larson and Brown (1997) suggest establishing and classifying quantitative CEP measures by their point of measurement as presented in the following figure:

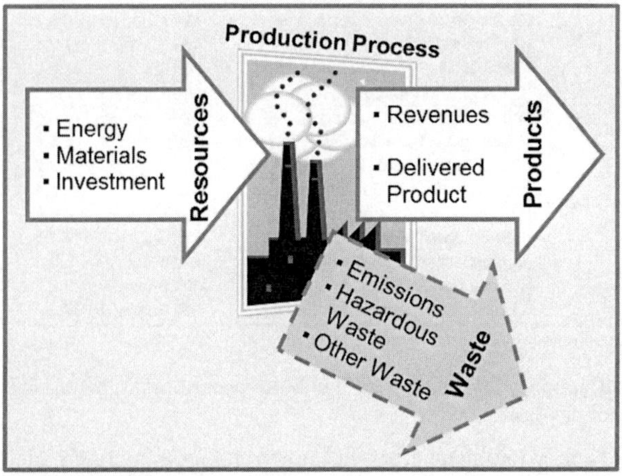

Figure 54: Classifying Quantitative CEP Measures (adopted in modified form from Larson and Brown, 1997, p. 84)

According to the above figure, Larson and Brown (1997, p. 83) differentiate two types of quantitative CEP measures:

- Measures *"that monitor the flow of resources, products, and wastes"*. Such measures are related to the sources of an organization's environmental impacts.
- Measures *"that can measure direct impacts on local environmental systems, including air, water, soil, wildlife, vegetation, and even employee health"*. These measures are related to the outcome of an organization's environmental impact.

Following the three main opportunities for environmental improvements within the MFI (EPA, 2006, p. 55; s. chapter 3.3), business measures should be considered that are related to

- managing and minimizing toxics and waste,
- reducing air emissions, and

[65] Also known as green washing; Walker and Wan (2012) found that green-washing has a negative effect on an organization's financial performance. The authors argue *"that green-washing would be more likely to be perceived, less effective, and finally more likely to be punished"* (2012, p. 238).

- conserving water.

The chalice-shaped bottom-up approach to improve BPM feasibility (s. figure 10) explicitly allows to embrace further BPM elements not included in generic BPM systems. In case an MFC applies an environmental management system, related for example to ISO 14000 (ISO, 2015b; s. environmental management at MeFiCom in chapter 6.2), further performance measures retrieved from this system should be included into the MFC's individual BPM, as in the sense of an integrated management system.

8.2 Modeling the MFI Levers of Measurement

After having chosen the suitable MFI generic BPM systems, I will continue with the final step 5 of the chalice-shaped bottom-up approach to improve BPM feasibility (s. figure 10), i.e. modeling the BPM system for the MFI. Therefore I have adapted my theoretical research framework (s. figure 7) to the MFI as shown in the following figure:

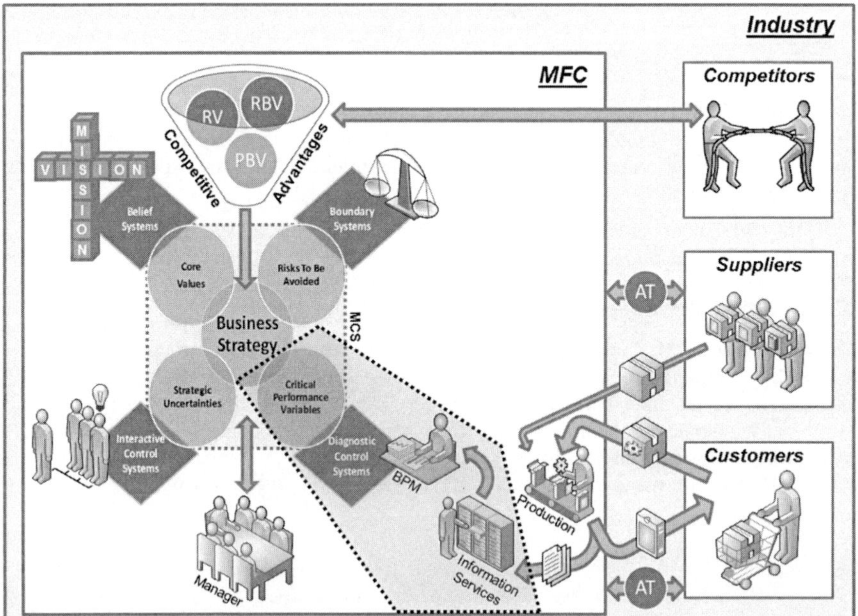

Figure 55: MFC Model following the Research Framework

In comparison to figure 7, an additional arrow symbolizing a material inflow is included in the figure above coming from the customer who delivers the metal substrates into the MFC's production. The material inflow from all other suppliers delivering consumables, energy, etc., is consequently lower as the main customer material inflow and is therefore symbolized by a narrower arrow in figure 55.

Simons' Levers of Control (1995; s. chapter 2.3.3) embody the heart of my theoretical research framework (s. figures 7 and 55). They are also the leitmotif within my dissertation project as shown in the empiric research process. In order to specify this LOC framework regarding the MFI Diagnostic Control Systems (area within a dotted line in figure 55), I designed the MFI-BPM model as presented in the following figure:

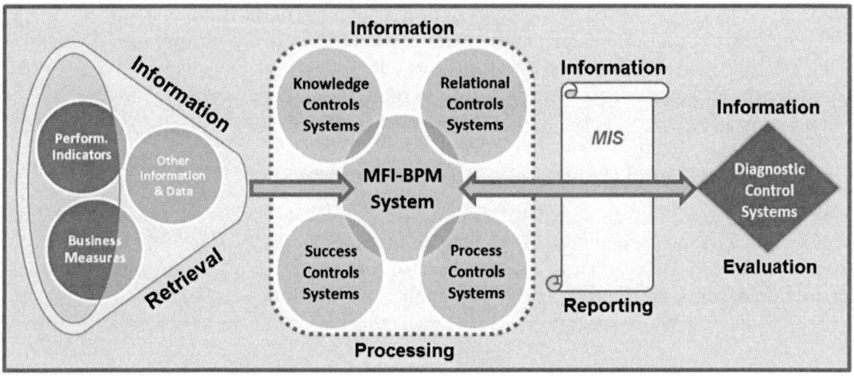

Figure 56: The Levers of Measurement - BPM Model for the MFI

I call this MFI-BPM model *The Levers of Measurement (LOM)* as it serves as a specification of the LOC Diagnostic Control Systems. It is therefore an extension of Simons' LOC framework applied in the MFI (s. chapter 2.3.3 for the expandability of the generic LOC framework).

The MFI-BPM model is centered on the four LOM within the *Information Processing* area in figure 56 which represent BPM sub-systems of the *MFI-BPM System. The Success Controls Systems* include information regarding the MFC's business development, formed mainly by financial performance indicators.

The three other LOM, namely *Knowledge, Relational,* and *Process Controls Systems*, refer to the MFC's competitive advantages, on which the MFC's business strategy is grounded (symbolized by the funnel in figures 7 and 55). These competitive advantages are again based on the strategic theories of the firm which I selected according to the MFI's peculiarities as presented in chapter 2.2. This way BPM becomes integrated into the management control process (s. chapter 2.3.3) as presented in the following figure:

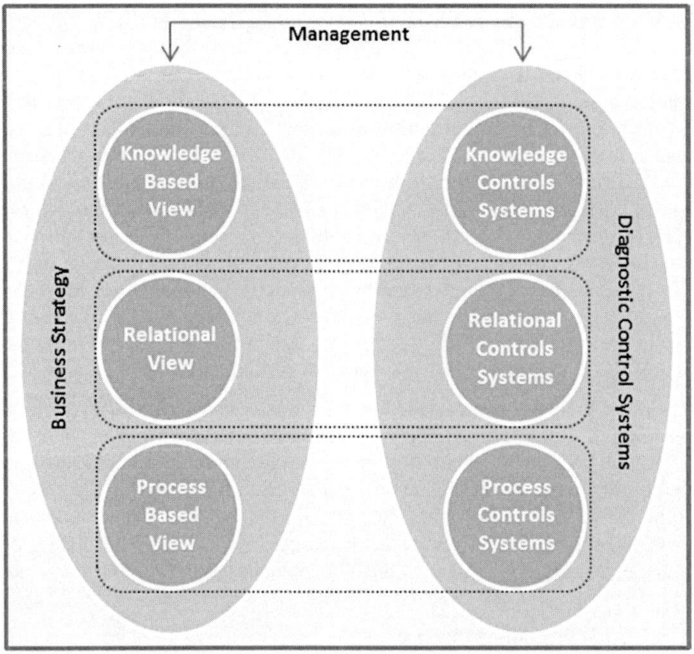

Figure 57: Relation between the Strategic Theories of the Firm and the LOM

The LOM's nomenclature follows the distinction between *control* and *controls* by Drucker (1964, p. 286; s. chapter 2.4.1): the different LOM controls systems include the means to an end, and the end is diagnostic control. Each of the four LOM includes elements from the chosen generic BPM frameworks based on the research process conclusions as shown in the previous chapters. Typical MFI elements are added which are not covered by generic frameworks.

This MFI-BPM model can be clustered into four types of information services which are regularly applied sequentially:

Within the *Information Retrieval* stage, BPM necessary raw data, performance indicators or business measures are retrieved from various data sources. These sources are mainly located in IT software like ERP-systems, databases, log-files from coating systems, etc. But the data can also be retrieved from other non-IT sources like minutes of meetings. However the focus should be concentrated on IT data sources which can be connected to the BPM system via data interfaces in order to allow the automation of BPM processes. The retrieved data is not limited to accounting or other financial figures, but includes all types of information *relevant to diagnostically control* the MFC.

The main part of the MFI-BPM model is the *Information Processing* step in which the retrieved data is processed within each of the four LOM (s. next chapters) and is then integrated into the *MFI-BPM System*.

The *MFI-BPM System* again is the source for the *Information Reporting* as supporting tool for the *Information Evaluation* by the MFC's managers to diagnostically control their business. Aggregated data from the MFI-BPM system should be displayed within a management information system (MIS) according to the managers' requirements regarding relevance, the level of data granularity and data illustration. The resulting MIS reporting shall enable the manager to have a holistic view on his/her MFC business development. The double-sided arrow in figure 56 between the MFI-BPM system and the Diagnostic Control Systems demonstrates the dynamic capability of the MFI-BPM model (s. chapter 2.5.2 for the academic concern regarding static BPM systems). Managers need the opportunity to regularly change the MFI-BPM system in order to respond to the dynamics of the business world. This explicitly includes the reaction on emergent strategies coming from the LOC Interactive Control Systems and the implementation of corresponding BPM performance indicators and measures into the MFI-BPM system. This integration process between the two LOCs is an example of an integrated MFC management system.

The detailed LOM and their elements are presented within the next chapters. A table is presented at the end of each LOM element showing examples for corresponding relevant MFI performance indicators and business measures based on the hitherto presented research findings. Each table column presents one indicator or measure, its purpose as well as its update interval within the LOM. These tables can and should be altered and amended according to each MFC's business objectives as described in chapter 8.3. Target values or limits for the different indicators and measures are not included as these have to be set individually by each MFC.

8.2.1 LOM – Knowledge Controls Systems

The LOM Knowledge Controls Systems (KCS) includes performance indicators referring to the MFC's employees and is related to the Knowledge-Based View within the strategic theories of the firm (s. chapter 2.2.1).

The KCS focusses on the employees as the human capital (Petrash, 1996, p. 366) represented as part of the intellectual capital (Gleich, 2011, p. 46) of the MFC.

The KCS is split into four system elements each being either related to generic BPM frameworks or showing otherwise significance for the MFI.

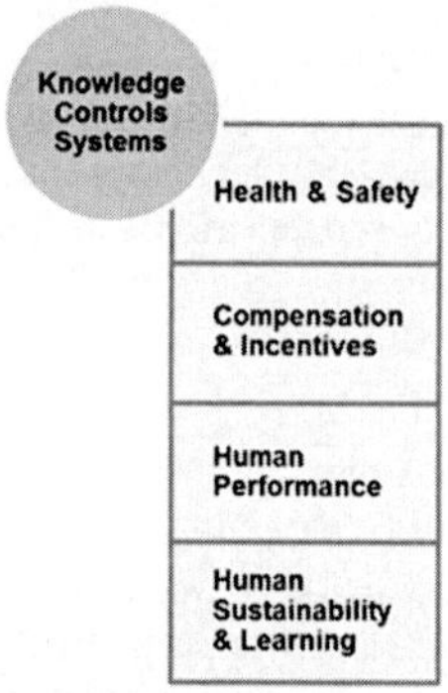

Figure 58: LOM – Knowledge Controls Systems

Health & Safety includes measures indicating corresponding conditions in the MFC. Protecting the employees' health by establishing and maintaining safe working conditions is of special interest to any MFC as their employees regularly work with hazardous chemicals, complex coating systems, or sharp and heavy metal substrates handled in part manually (s. chapter 3.1). Health & Safety within the KCS is not limited to legislative regulations[66] but includes further perspectives assuring a safe workplace. An example would be the consideration of

[66] For German safety and health regulations please refer to the German Occupational Safety Act (ASIG, 2015).

improved ergonomics in production and administrative processes as well as accident prevention.

	MFI-LOM – Knowledge Controls Systems – **Health & Safety**		
Performance Indicator/ Measure	Number of lost-time-accidents (LTA)	Direct investment amount in Health & Safety	Number of training days related to accident prevention
Purpose / Description	Target should be zero. Indicator allows the management to control for this target by analyzing and then avoiding the reasons for LTAs.	Assumption: MFCs need to continuously invest in safe workplaces (like security retrofits of older coating systems, or ergonomic workplace enhancements). Gives an indication over time: decreasing investments may either be a sign for reaching an ideal workplace situation or a sign for an increasing risk for workplace safety deterioration.	Gives an indication regarding raising awareness of threads which lead to accidents and which shall be prevented. Contributes to the target of zero LTAs.
Update Interval	Weekly	Half Year	Monthly

Table 45: MFI-LOM-KCS – Health & Safety Measures

Compensation & Incentives include measures referring to the status and development of salary elements as well as of incentives. Finding excellent employees for the MFI is challenging as they represent a scarce resource. The competition between organizations related to the acquisition of such employees drives MFCs to offer competitive compensation plans (s. people as major success factor in chapter 6.6.4). Incentives to promote employees and acknowledge superior job performance will improve employee engagement (Simons, 1995, p. 24). The incentives, in the form of rewards, should be linked to the achievement of BPM targets (following the idea of a cybernetic regulating loop in chapter 2.3.1) in order to emphasize the importance of BPM within the MFC.

	MFI-LOM – Knowledge Controls Systems – **Compensation & Incentives**		
Performance Indicator/ Measure	Average salary per functional area (production, administration, management)	Short term incentives (STI) achievement	Average incentives amount paid for suggestions within the continuous improvement process
Purpose / Description	Gives an indication of salary increases. In case the salary increase is below the inflation rate, the MFC risks an increasing employee dissatisfaction. Measure can also be compared to the salaries paid by competitors in order to evaluate the MFC's own compensation attractiveness.	Compares the total amount of realized STI's to the maximum possible STI's. Gives an indication if STI's are rather unrealistic to reach (indicator is far below 50% over more than two years) or not challenging enough to reach (indicator is continuously higher than 95%).	Indicates the vitality of the MFC's continuous improvement process which is related to the LOC Interactive Control Systems (idea sharing beyond hierarchical boundaries).
Update Interval	Quarterly	Annual	Quarterly

Table 46: MFI-LOM-KCS – Compensation & Incentives Measures

Human Performance includes measures related to the MFC employees' efficiency within the MFC. The relevance of these measures for the MFI is based on the MFI survey (s. chapter 7.2.3) as well as on the preliminary studies at MeFiCom (s. chapter 6.2). These measures can be linked to other LOM like PCS – for example a growing order backlog could be linked to a decreasing productivity or an increasing absenteeism.

	MFI-LOM – Knowledge Controls Systems – **Human Performance**		
Performance Indicator/ Measure	Absenteeism (sick days per employee)	Productivity I (sales per employee)	Productivity II (output per employee)
Purpose / Description	Important measure which can be referred to capacity shortages leading to an increasing order backlog. An increasing absenteeism may also indicate unsafe working conditions and/or dissatisfied employees. Measure is applied at MeFiCom and at MFCs from the MFI survey sample.	Divides the sales amount through the number of employees. Allows to reflect the overall employee performance under the assumption of stable prices. Measure is applied at MeFiCom and at MFCs from the MFI survey sample. Could be limited to blue collars only in order to focus on the production department. Alternative: sales per production hour	Price effects will influence the Productivity I indicator. The Productivity II indicators therefore divides the number of metal finished substrates through the number of employees (whole MFC or production only).
Update Interval	Monthly	Weekly	Weekly

Table 47: MFI-LOM-KCS – Human Performance Measures

Human Sustainability & Learning includes measures which allow the MFC managers to control for the sustainment of knowledge within the MFC. Having well educated employees who are further trained in their job enables those individuals to be creative and innovative. Maintaining knowledge includes all aspects that allow keeping the organizational know-how within the MFC like

- a sound mix of younger and older employees to counteract the demographic change (so that the MFC's knowledge does not die out over time), and
- the extent of employee training and development plans (especially relevant for larger MFCs in order to promote talents and simultaneously create backup positions for key performers).

The resulting employee satisfaction measured in opinion surveys indicates the success of the MFC's employee related arrangements and affects the human sustainability.

	MFI-LOM – Knowledge Controls Systems – **Human Sustainability & Learning**				
Performance Indicator/ Measure	Share of employees with a development plan in total employees	Average employee age	Number of training days per employee	Cost for training and education per employee	Employee satisfaction measured in employee opinion surveys
Purpose / Description	Refers to the sustainment of the MFC's knowledge related to succession plans of key performers by developing talents.	Indicates the MFC's demography (example: what is the share of employees older than 60 years compared to the share of employees younger than 30 years)	Both measures are related to control the sustainment and enlargement of the MFC's knowledge. However it is necessary to control for both measures as internal training sessions should be reflected in the number of training days which regularly do not result in external cost.		Direct indication of the employee's satisfaction. Such surveys should be combined with the employee's opportunity to express improvement potentials in case they are not satisfied with the MFC.
Update Interval	Quarterly	Half Year	Quarterly	Quarterly	Half Year

Table 48: MFI-LOM-KCS – Human Sustainability & Learning Measures

One should pay attention to the composition of KCS indicators and measures as they include personnel related and sensitive data. It may be necessary to consult the works council or data privacy acts in the event the data can be linked to single individuals (referring especially to measures like absenteeism or productivity that may have a potential negative impact on the individual).

8.2.2 LOM – Process Controls Systems

The LOM Process Controls Systems (PCS) includes performance indicators referring to the MFC's processes and is related to the Process-Based View within the strategic theories of the firm (s. chapter 2.2.3).

The PCS focusses on the MFC's main production processes having the most influence on the MFC's performance, integrating several competitive advantage dimensions like quality and delivery reliability (s. chapter 3.2).

The PCS is split into four system elements, each being either related to generic BPM frameworks or showing otherwise significance for the MFI.

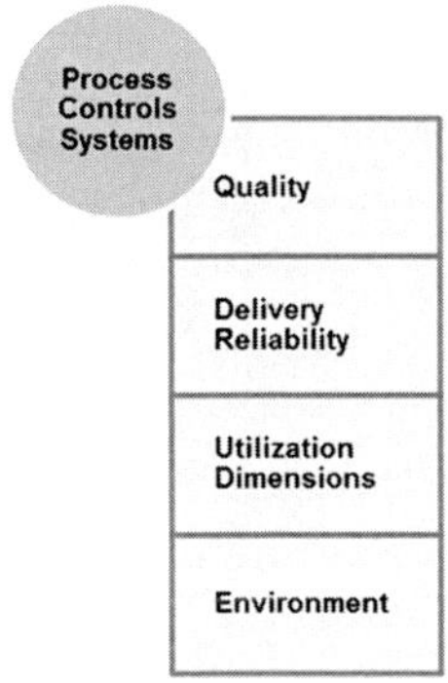

Figure 59: LOM – Process Controls Systems

Quality was ranked as having the highest priority compared to delivery reliability and cost/price within the MFI survey (s. chapter 7.2.2). Following the distinction between hard and soft quality dimensions within a service business-to business relationship (s. chapter 3.1) by Szmigin (1993, p. 10), this LOM element includes hard quality measures indicating the quality of the metal finishing core process (*what* the customer receives). Soft quality measures are included that refer to supporting processes within the relationship of an MFC and its customers such as marketing, sales, or logistics (*how* the customers receives it).

In their study conducted in the MFI regarding service failures, Zhu and Zolkiewski (2015; s. chapter 8.2.3) have gathered the following examples to illustrate hard and soft quality failures, related either to the MFC or to the MFC's customer causing a service failure:

	MFC-related Errors	MFC's Customer-related Errors
Hard Quality	Wrong colour paint	Items unsuitable for metal finishing
Soft Quality	Late delivery	Lack of provision of appropriate information

Table 49: Examples for Hard and Soft Quality in the MFI (adopted in modified form from Zhu and Zolkiewski, 2015)

The distinction between soft and hard quality factors implies that some aspects of a MFC's service management are much more vital than others, for example the wrong paint being a much more serious problem than a late delivery. However the results of the study by Zhu and Zolkiewski (2015) show that, whatever quality dimension is regarded, *"business-to-business customers tend to forgive suppliers' occasional service failure, but have less tolerance towards recurring problems"*. Therefore it is vital for the MFC's management to detect early on the reasons for quality-related service failures in order to establish countermeasures.

	MFI-LOM – Process Controls Systems – **Quality**		
Performance Indicator/ Measure	Process quality and efficiency (number of coating system shutdowns)	Scrap rate (per coating system // per shift // per customer)	Internal quality audit results
Purpose / Description	Measures the number of coating system shutdowns or failures. Each failure results at least in a longer process time and therefore in higher costs per piece. Metal substrates may have to be devarnished in case the coating process would have to be repeated.	The update interval is rather short in order to allow the identification of bad quality patterns. For example, in case the scrap rates of customer A are continuously higher than the scrap rates of customer B although both deliver a comparable substrate may then be related to a worse quality of the metal substrates of customer A. In this case the incoming goods inspection should be intensified.	Reflection of the internal quality audit results which allows the managers to early react on quality risks in case the results deviate from the audit targets set.
Update Interval	Weekly	Daily	Monthly

Table 50: MFI-LOM-PCS – Quality Measures

Delivery Reliability, ranked second by the MFCs in the survey sample regarding the competitive advantages, refers to the soft quality of the MFC's service as presented in the previous paragraph. Besides an on-time delivery, this LOM element also refers to the delivery of the demanded quantity. Further dimensions can be added to this LOM element. MeFiCom for example offers a pickup service to its customers (s. chapter 6.1.2) in which it organizes all the logistics related to the transfer of the metal substrates to MeFiCom and the metal finished substrates back to the customer. Such additional delivery services can be summarized under logistics performance.

	MFI-LOM – Process Controls Systems – **Delivery Reliability**		
Performance Indicator/ Measure	Adherence to delivery dates	Adherence to delivery quantities	Logistics Performance
Purpose / Description	Gives an overview on the number of cases in which the actual delivery date differs from the customer agreed delivery date. Cases in which the delivery took longer as agreed should be reduced. In case the delivery time is shorter as agreed (= delivers is over performed) price renegotiations with the customer may be advisable, or the production schedule could be rearranged focusing on the time critical products.	The measure compares the demanded quantity of metal finished substrates to the actual delivered quantity. As the metal finishing process is regularly one of last production steps in the values-added chain, the MFC's customer depends on the correct delivery of the agreed quantity of metal finished substrates. A under delivery may result in severe problems for the MFC's customer as it may not be able to deliver the demanded quantity to its customers.	In case the MFC offers special logistics services (like pickup services) the quality of these service should be monitored (for example the actual number of pickup tours to the top customers versus the demanded frequency).
Update Interval	Daily or Weekly	Daily or Weekly	Monthly

Table 51: MFI-LOM-PCS – Delivery Reliability Measures

The SCOR-Model (s. chapter 8.1.1) can be applied as an additional source for delivery reliability measures as it includes hierarchical ordered indicators related to the supply chain performance attributes of reliability, responsiveness, and agility (SCOR, 2010, p. 9).

Utilization Dimensions refer to the performance indicators as noted by the MFI survey participants (s. chapter 7.2.3). They include measures related to the utilization of the installed metal finishing capacity as well as to the utilization of required coating materials for the metal finishing processes. In addition they include further important process related indicators like throughput time, order backlog, or units produced per day.

	MFI-LOM – Process Controls Systems – **Utilization Dimensions**		
Performance Indicator/ Measure	Utilization rate of installed metal finishing capacity (in percent of max. capacity)	Utilization rate of coating materials (in quantity / weight per month)	Other production process related measures
Purpose / Description	This operational measure reflects the ratio between the actual usage of the installed metal finishing systems in hours to the technical possible usage (reduced by service intervals). This measure is a flexibility indicator: in case a MFC operates continuously on maximum capacity indicates a low flexibility in case a customer wants to increase his/her demands. On the other side a too low utilization rate indicates an over capacity and reduces accordingly the return on investment related to the installed coating systems.	The utilization rate of coating materials (precious metal targets, e. g.) combined with the utilization rate of the installed metal finishing capacity gives an indication on the coating systems efficiency: In case the consumption of coating materials increased over time but the utilization rate of the metal finishing systems decreases indicates an inefficient coating process. This may be caused either the technical problems during the coating process or to failures during the handling of the coating systems.	Further measures referring to the metal finishing process are: - Number of produced parts (total amount // per coating system // per substrate group) - Throughput time (can be used in combination with the utilization rate in order to evaluate demanded volume increases) - Order backlog (an increasing order backlog with an decreasing utilization rate indicates a decreasing production efficiency) - Finished workshop orders per day (continuous measurement of this indicators reflects the production efficiency under the assumption that all workshop orders are comparable)
Update Interval	Weekly	Weekly	Daily or Weekly

Table 52: MFI-LOM-PCS – Utilization Dimensions Measures

Environment is a key area of MFI performance measurement. As shown in figure 54, the production process of an organization leads not solely to the demanded products but also to undesired side products, i.e. waste. This LOM-PCS element therefore assists the MFC's Management, by virtue of the CEP measurement, to identify sources for natural environmental impacts and to monitor the success of related countermeasures, like the annual reduction of hazardous waste by a certain percentage through more efficient or altered production processes. Such countermeasures can also have positive effects on the MFC's financial performance: MeFiCom for example reduced the costs for certain chemicals by altering the chemical composition of cleaning baths using more environmental friendly chemicals without reducing the quality of the cleaning process.

	MFI-LOM – Process Controls Systems – **Environment**		
Performance Indicator/ Measure	Reduction of energy consumption (measured in kilowatt hour)	Reduction of toxic wastes (measured in tones generated by type of waste)	Reduction of wastes related to water hazard classes (measured in cubic meters)
Purpose / Description	The reduction of electric energy can be related to an energy management system according to ISO 50000. To reduce the energy consumption, a MFC could for example invest in more energy efficient heating systems for metal finishing processes.	Toxic wastes, that have to be disposed as special waste should be reduced as far as technical feasible. Reducing toxic wastes also reduces the risk of potential harm for the MFC's employees.	Especially those liquid wastes that are hazardous to water and that have to be disposed as special waste should be reduced as far as technical feasible. Those liquids bear the risk of significant environmental cost in case of a corresponding environmental accident.
Update Interval	Quarterly	Quarterly	Quarterly

Table 53: MFI-LOM-PCS – Environment Measures

The quarterly update interval for these measures may be increased to monthly or weekly in case the MFC's management wishes to closely observe the current implementation of environmental protection initiatives.

8.2.3 LOM – Relational Controls Systems

The LOM Relational Controls Systems (RCS) includes performance indicators referring to the MFC's relations to its organizational environment and is related to the Relational View within the strategic theories of the firm (s. chapter 2.2.2).

The RCS focusses on the MFC's relation to its customers, competitors, market, and share- & other stakeholders as focal elements of the MFC's organizational environment. Accordingly the RCS is split into these four system elements each being either related to the generic BPM SCOR-Model (s. previous chapter) or showing otherwise significance for the MFI.

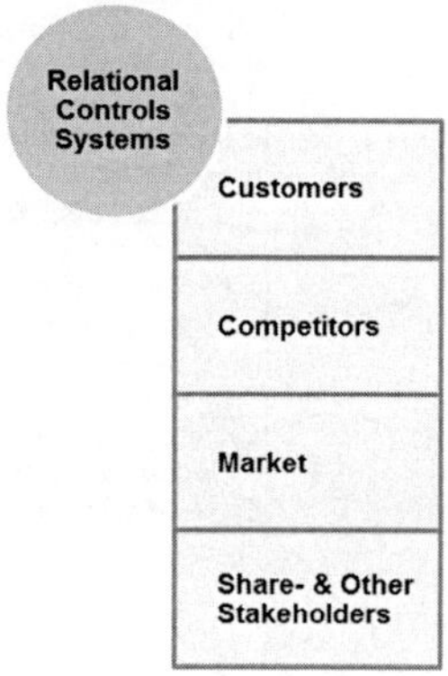

Figure 60: LOM – Relational Controls Systems

Customers have a special role in the MFI as presented throughout the previous chapters. They are not only the customers but also the main material suppliers of the MFC. This special role of the MFC's customers was documented in a recent study: Zhu and Zolkiewski (2015) have explored the reasons and consequences of service failures in business-to-business contexts. This empirical study is based on two British MFI case studies.

The authors have identified a *"Domino Effect"* which *"implicitly dominates business-to-business service failure episodes where negative outcomes cascade downstream and affect the service recipient's customers"*. This domino effect is illustrated in the next figure:

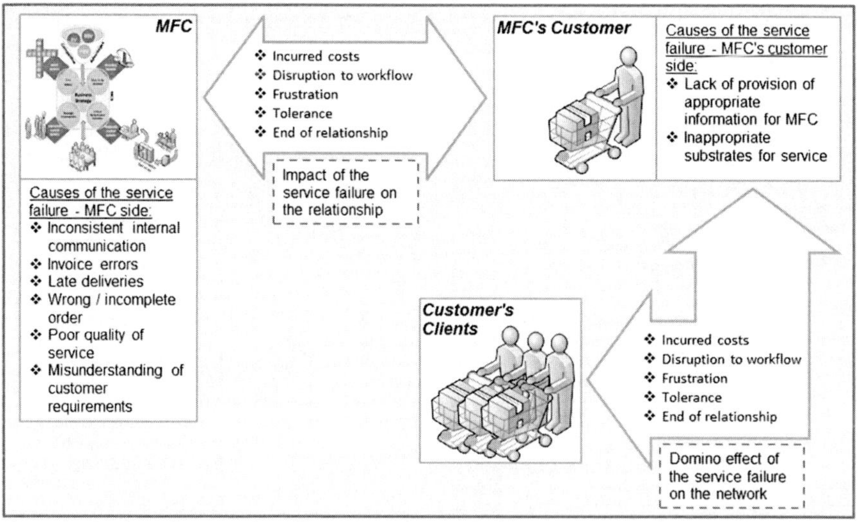

Figure 61: Service Failure Domino Effect (adopted in modified form from Zhu and Zolkiewski, 2015)

According to this study, the reasons for a service failure can either be located on the MFC or the MFC's customer's side. However a service failure, especially related to poor quality of service, delays in delivery, or inappropriate substrates for services, can have spillover (domino) effects on the MFC's customer's clients. The consequences of a service failure on the MFC's customer's relation with its clients are the same as the more direct effects on the relationship of the MFC with its customer. Avoiding service failures are therefore a managerial necessity and related business performance measures should be included within a BPM system *"because customers are concerned about negative impacts on their business performance, such as causing their downstream customers to be dissatisfied and leave them, and losing their profits"* (Zhu and Zolkiewski, 2015).

Two further customer dimensions are adapted from the customer perspective of the BSC (Kaplan and Norton, 1996, p. 68) as generic BPM framework (s. previous chapter) into the LOM-RCS. These dimensions[67] include information regarding:

- Customer Acquisition: gives an indication of new customers being attracted or won related for example to the new introduction of enhanced coating services. However this performance element is of minor relevance for those MFC's that are settled near to their top customers (s. chapters 6.1.2 and 6.6.3). But targets set in this performance element shall reduce the risk of being dependent solely on the few MFC's main customers.
- Customer Retention & Satisfaction: although the BSC divides these two customer perspectives, I combined them as both are linked in the MFI. The qualitative results at MeFiCom illustrate the customer driven business (s. chapter 6.6.4) as MeFiCom

[67] The BSC customer perspective (Kaplan and Norton, 1996, p. 68) lists as further dimensions a) *market share* which is reflected in the LOM-RCS element *market* and b) *customer profitability* which gets reflected in the LOM-SCS element *financials*.

needs to react to changing customer requirements related to the demand for improved coating solutions which fit changing customer substrates. If MeFiCom could not satisfy its customers by creating applicable and customized solutions, it would not be able to sustain and retain existing customer relationships. In conclusion, avoiding service failures combined with the MFC's ability to react on upcoming customer's requirements are necessary prerequisites for customer satisfaction in the MFI.

	MFI-LOM – Relational Controls Systems – ***Customer***		
Performance Indicator/ Measure	Customer complaints count // Service failure count // Number of warranty claims	Customer Fluctuation (measured in orders per customer)	Customer Satisfaction (measured by customer survey, e. g.)
Purpose / Description	An increasing number of customer complaints related to service failures is a risk for the MFC's prosperity. This measures can be differentiated according to customer complaints related to soft and hard quality failures. However the MFC has to respond on the customer complaints.	Gives the management an indication on new customers, customers within increasing demands (winners), customers with decreasing demands (losers), as well as on customers that canceled the business relation with the MFC. The MFC risks to loose profitability in case the number of new customers or winners does not outweigh the lost customers and losers.	Having satisfied customers is of interest of any organization. Knowing the level of customer satisfaction is of significance to retain and maintain the MFC's relation to its customers. Related measures could be the number of sales calls or sales visit indicating a vivid contact between the MFC and its customers.
Update Interval	Weekly	Monthly	Quarterly or Half Year

Table 54: MFI-LOM-RCS – Customers Measures

Competitors include indicators comparing the MFC's own business with those of the key competitors in its MFI market segment. These indicators, in the sense of performance benchmarking, would help the MFC's management to identify the best practice improvement opportunities. As shown before in the LOM-KCS, the MFCs do not only compete for customers but also for highly qualified employees.

However, focusing on the competitor's success may result in negative management decisions as presented in a study by Armstrong and Collopy (1996, p. 188) researching the impact of competitor-oriented behavior. The authors found that firms being directly oriented to profits are more profitable and more likely to survive than competitor-oriented firms.

	MFI-LOM – Relational Controls Systems – ***Competitor***	
Performance Indicator/ Measure	Growth Rate vs. Key Competitors (in percent)	Number of coating systems
Purpose / Description	The reasons should be analyzed by the MFC's management in case the growth rate of the competitor significantly differs from the own growth rate: does the competitor offer innovative metal finishing solutions or does the competitor offer additional service elements? This analysis can also be combined with the MFC's market position (see market share in the LOM-KCS market element).	Knowing the competitor's number of coating systems allows the MFC to evaluate the competitors coating capacity. In case the competitor enlarges the coating capacity of his organization the MFC would face the risk to loose market share.
Update Interval	Quarterly	Quarterly

Table 55: MFI-LOM-RCS – Competitors Measures

Market includes external performance indicators and measures that give an overview on cyclical economic developments. Combining such external indicators (for example GDP; s. chapter 6.6.4) with internal indicators like sales may reveal patterns on which future planning can be based. MeFiCom for example correlates the German GDP with the sales amount of certain product applications and determines that a decreasing GDP indicates a sales reduction for these product applications with a three month time lag.

The market share is reflected as an additional market dimension. Knowing the status quo of this LOM-RCL market element allows reacting to a decreasing market according to the identified reasons for this decline (for example being related to non-competitive pricing, increasing service failures, or a better coating service portfolio of the MFC's competitors).

	*MFI-LOM – Relational Controls Systems – **Market***		
Performance Indicator/ Measure	MFI specific market indicators (future oriented)	Country specific economical indicators (future oriented)	Market share
Purpose / Description	Specific market indicators reflect the market condition in the MFC's target markets: In case the MFC has customers in the automotive industry, the review of the number of car registrations or car manufacturer production indices are of interest (an increasing number of car registrations combined with an increasing car manufacturer production indicates an increasing demand for metal finished car components.	Country specific economical indicators, like Gross Domestic Profit or economical trend indicators (like CBI in UK or PMI in Germany), indicate the prosperity of a country's economy. As the MFI is connected to an ample set of other industries, considering such economical indicators have a significance for MFCs. Combining economical indicators with the MFCs GSR development over time can create early warning indicator in the sense that a downtrend in GDP will result in a (maybe delayed) decreasing sales of the MFC.	A continuous measurement of this indicator allows the calculation of growth rates which can be compared to the competitors (do competitors for example participate more strongly in an increasing market demand situation than the MFC).
Update Interval	Weekly	Monthly	Monthly

Table 56: MFI-LOM-RCS – Market Measures

Share- & Other Stakeholders includes all indicators and measures that are demanded by shareholders and other stakeholders and that are not considered in other LOM. For example, shareholders of larger MFCs may desire information that would not regularly be reflected in a BPM system, as those are not relevant for the operations or strategy implementation. Or the MFC has a partnership with a university and the MFC's management is interested in the number of patents derived from this partnership.

	*MFI-LOM – Relational Controls Systems – **Share- & Other Stakeholders***	
Performance Indicator/ Measure	Measures reflecting the relation to shareholders	Measures reflecting the relation to other stakeholders
Purpose / Description	This LOM-RCS element could include for example indications regarding the MFC's utilization of the shareholder's cash pool (referring to the MFC's liquidity).	Referring to the MFI's environmental impact, reflecting the number of new legal provisions affecting the MFC should be of interest for the MFCs management. In addition performance measures reflecting the success of partnerships to other institutions should be added.
Update Interval	Depends on the measures	Depends on the measures

Table 57: MFI-LOM-RCS – Share- & Stakeholders Measures

8.2.4 LOM – Success Controls Systems

The LOM Success Controls Systems (SCS) includes performance indicators and business measures related to the MFC's business development.

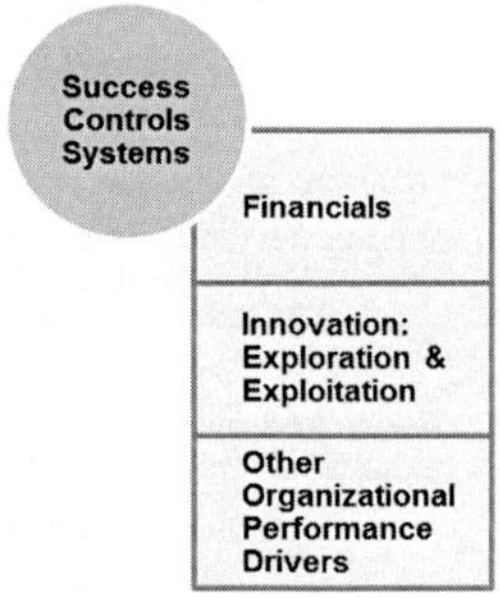

Figure 62: LOM – Success Controls Systems

The SCS focusses on all organizational aspects which are not covered by the other LOM. This includes mainly financial indicators reflecting the MFC's success.

The SCS is split into three system elements each being either related to generic BPM frameworks or showing some significance for the MFI.

Financials include business measures and indicators referring to the MFC's financial success. The research results derived from MeFiCom (s. chapter 6.6.4) and from the MFI survey (s. chapter 7.2.3) show the strong consideration of single-dimensional profit and loss business measures, like sales revenue or gross profit, in order to evaluate the MFC's financial success. Comparing these business measures to financial targets (set for instance in budgets or forecasts) represents the foundation of the LOC-SCS financials.

Accordingly, the following two tables present a selection of business measures that are relevant for MFCs according to the research results. An MFC should compare each of these measures to corresponding financial targets in order to evaluate the success of the MFC's business during the fiscal year. These financial measures and indicators do not exclusively reflect the revenues and costs of goods sold, which were the focus of MeFiCom and the MFI survey participants, but also overhead costs. The consideration of overhead costs is especially relevant for larger MFCs: Gleich et al. (2008) have proposed a BPM tool for process benchmarking derived from an empirical study within the German mechanical engineering industry. They found that *"process cost accounting is an objective tool for measuring and comparing the performance of activities and sub processes in overhead departments"* (Gleich et al., 2008, p. 252). Within process cost accounting, representing the main element of the process benchmarking approach, process-related strengths and weaknesses are compared either internally, with the competitors, with their own industry, or with other industries (Gleich et al., 2008, p. 245). According to the authors, process cost accounting is intended to enhance management accounting processes in order to reduce costs by identifying and subsequently implementing performance improvement initiatives.

	*MFI-LOM – Success Controls Systems – **Financials I***		
Performance Indicator/ Measure	- Sales amount - Sales amount per employee - Sales amount per customer - Sales amount per production hour - Sales amount per coating system - Sales amount per substrate application (piston pins, cutting tools, or forming tools, e. g.) - Sales amount per sales district - Daily sales	- Personnel cost (internal / external) amount - Energy cost amount - Depreciation cost amount - Consumables (incl. coating material) cost amount - Cost per unit produced - Quality Cost - Logistics Cost - Each cost element as share of sales revenue	- Gross profit amount - Customer profitability - Gross Profit amount per coating system - Gross profit amount per substrate application - Gross profit amount per sales district
Purpose / Description	Analyzing the MFC's sales revenue dimensions in order to spot decreasing revenue elements and decide countermeasures	Analyzing the MFC's main cost drivers in order to reduce costs in case on element increases disproportionately	Analyzing the MFC's gross profit dimensions in order to spot and eliminate non-profitable elements
Update Interval	Daily or Weekly	Daily or Weekly	Daily or Weekly

	*MFI-LOM – Success Controls Systems – **Financials II***		
Performance Indicator/ Measure	- R&D costs amount - Purchasing costs amount - Production overhead costs amount - IT costs amount - HR costs amount - Finance costs amount - Management costs amount - Marketing and selling costs amount - Overhead costs amount as share of sales revenue	- Investment amount - Investment amount as share of sales revenue - Asset utilization ratio - Depreciation per coating system - Depreciation as share of sales revenue	- Operational / Net profit amount - EBIT amount - Net working capital development (including stock value, accounts receivable and payable) - Liquidity within the next months
Purpose / Description	Analyzing the MFC's overhead costs in order to spot any unplanned cost increase	Analyzing the MFC's capitalized expenditures in order to analyze the sustainability of the MFC's assets. MFI processes regularly require high investments in related coating equipment resulting in subsequent depreciation costs which have to be reflected in a profit and loss calculation.	Analyzing other MFC's financial measures in order to able to react on financial deviations
Update Interval	Monthly	Quarterly	Weekly (Liquidity) or Monthly

Table 58: MFI-LOM-SCS – Financials Measures

Innovation refers to the status of the MFCs innovation processes. Organizational ambidexterity plays a role in the MFI as presented in chapters 6.6.2 and 7.2.2. Therefore I divided the LOM-SCS innovation element into the two perspectives explorative and exploitative innovation. An MFC can choose measures according to the peculiarity of each innovation perspective. Measures referring to both perspectives would be included in the case of MeFiCom as ambidexterity plays a significant role in MeFiCom's business model.

	MFI-LOM – Success Controls Systems – **Innovation**	
Performance Indicator/ Measure	Measures related to explorative innovation include: - amount of sales related to new metal finishing services - number of new patents - investment amount in new coating equipment - amount of risk capital for opening new market segments - amount of sales acquired through new distribution channels	Measures related to exploitative innovation include: - investment amount in existing coating equipment - number of projects to improve the existing production processes - volume increase of existing substrates to be metal finished - number of additional services for existing customers
Purpose / Description	Evaluate MFC's explorative innovation actions	Evaluate MFC's exploitative innovation actions
Update Interval	Monthly	Monthly

Table 59: MFI-LOM-SCS – Innovation Measures

Other Organizational Performance Drivers leaves free space in the MFI-BPM model which allows the MFC's to add further performance drivers, that a) have either not been considered in the MFI-BPM model, or that b) are special for the MFC. Other performance drivers may include qualitative information, like the results of external audits, or the consideration of the MFC's product portfolio.

	MFI-LOM – Success Controls Systems – **Other Organizational Performance Drivers**	
Performance Indicator/ Measure	Product Portfolio (sales growth / decrease per offered coating type)	External Audit Performance (qualitative information)
Purpose / Description	The MFC's management should check from time to time the MFC's product portfolio in order to spot poor dogs or cash cows: those coating types or metal finishing services should be stopped that are either not profitable or represent less than five percent of the overall production volume in order to streamline the MFC's product portfolio. Such a streamlining process can have positive effects on the production planning process as "exotic" coating processes do not disturb the overall production schedule anymore.	In case the MFC gets audited according to different norms (like ISO 9001 for quality management systems or ISO 14000 for environmental management systems), the communication of the audit results should be a permanent element of the MFC's BPM system. Reflecting the results in the institutional context of a BPM meeting allows the management to discuss possible audit deviations.
Update Interval	Half Year	Half Year or Annual (depending on the audit cycles)

Table 60: MFI-LOM-SCS – Other Organizational Performance Drivers

8.2.5 MFI-BPM System – Integrating the LOM Information

All business performance information and indications from the four LOM are integrated into one MFI-BPM system as presented in figure 55. This integrated BPM system is the source for the information reporting via management information systems, containing aggregated information on the MFC's business.

Therefore, the MFI-BPM system allows for accumulating information. This aggregation is realized by performance indicator engineering (s. chapter 6.5.2), i.e. the combination of single performance indicators and business measures from the different LOM logically to multi-dimensional critical performance variables[68] (CPV; s. chapter 2.3.3) reflecting the MFC's business objectives and strategy. Thus, the MFI-BPM system, as an information

[68] In the context of my dissertation I apply the term *critical performance variable* as a synonym for *key performance indicator* following the LOC-framework nomenclature (s. chapter 2.3.3).

processing service, represents a platform for the analyses of the actual LOM element measures and their variances to preset target values. The level of aggregation within the information reporting equals the composition of a) multi-dimensional CPVs, formed by LOM-elements, and b) the number of single-dimensional measures. It is determined by each MFC itself, depending on the demands of the MFC's management regarding the complexity and scope of the MIS.

In addition to the different LOM sub-system elements, the following three CPVs serve as examples for this performance indicator engineering process:

1) Gross Profit per Substrate Group Volume:
The weekly amount of gross profit is related to the produced volume in parts per week for the MFC's metal substrate groups:

MFI-LOM-KCS	MFI-LOM-PCS	MFI-LOM-RCS	MFI-LOM-SCS
---	**Produced Volume** (number of produced parts per substrate group and week)	---	**Gross Profit** (amount in local currency per substrate group and week)

Table 61: CPV Engineering from LOM-PCS and -SCS

The resulting quotient (equals gross profit per produced part) allows the following trend analyses:

a) For a single metal substrate group: in the event the CPV decreases under a stable gross profit condition, this would be an indication that the sales revenue of this substrate group did not grow proportionally to the costs of goods sold, as a higher volume is required to stabilize the gross profit. As possible countermeasures, the production process of this substrate group should be analyzed (are additional finishing services included in the production process, like blasting, deburring, or polishing, that are either obsolete or not invoiced to the customer?), and / or prices for this group should be increased as far as the market will allow. In the latter case the management has to decide if they accept the decreasing profitability of the substrate group, or if the price needs to be raised above the market selling price, with the risk of losing corresponding orders to the MFC's competitors.
b) For the whole metal substrate portfolio: identifying the gross profit per part for all of the MFC's substrate groups allows ranking them according to their profitability. The MFC's management has to set a certain gross profit limit in order to spot those substrate groups that are below this limit. Same countermeasures apply for these low-profit substrate groups as shown for the single substrate group.

2) Hard Quality Service Failures related to Labor Utilization:
As presented in chapter 8.2.2, service failures related to hard quality issues caused by the MFC should be avoided. Correspondingly, this CPV is a quotient of the number

of such hard quality service failures occurring in the time frame of 1,000 production employee worked hours[69]:

MFI-LOM-KCS	MFI-LOM-PCS	MFI-LOM-RCS	MFI-LOM-SCS
Labor Utilization (1,000 worked hours of production employees)	---	**Hard Quality Service Failures** (number of production related customer complaints in the time frame, for which labor utilization is measured)	---

Table 62: CPV Engineering from LOM-KCS and -RCS

An increasing CPV indicates quality problems which have to be identified and for which containment actions must be taken. Including the labor utilization dimension allows one to identify the reasons for the quality failures by focusing on the employees and not on the coating equipment. An example:

An MFC regularly employs 25 production employees working 8 hours on five days per week. Correspondingly 1,000 worked hours are regularly realized per week. The following three scenarios are given in the event the CPV increases:

a) The number of employees and the number of worked hours per day does not change. An increasing CPV index can therefore be caused a) by employee fluctuation (under the assumption that new employees increase the risk for quality issues), or b) by a decreasing quality awareness of the existing staff. Countermeasures would include in both events a more intensified training of the production employees raising the quality awareness.
b) The number of employees does not change, but the number of worked hours per day increases. The 25 MFC employees' daily working hours are increased from 8 to 10 due to an increased customer demand. In this case the time frame of working 1,000 hours is reduced from 5 to 4 days. An increasing CPV therefore indicates the risk that their production overheats due to overstrained employees. A countermeasure would be hiring temporary workers.
c) The number of employees increases, but the number of worked hours per day does not change. The MFC has hired 6 additional temporary workers (s. chapter 3.5.2) to scope an increased customer demand. In this case the time frame of working 1,000 hours is reduced from 5 to slightly less than 4 days. An increasing CPV indicates that the temporary workers may not be appropriately trained for their new job. A countermeasure would be a more intensified training of the temporary workers or replacing them in the event the training does not return a profit.

3) Market Development:
This CPV serves as an early warning indicator (s. chapter 8.2.3) by comparing the MFC's country's economic indicators, as for instance the projected GDP, with the MFC's projected sales revenue on a monthly base:

[69] Based on 25 production employees working 8 hours per day on 5 days a week. This can be changed according to the number of the MFC's employees and should correspond to the regular number of production hours per week.

MFI-LOM-KCS	MFI-LOM-PCS	MFI-LOM-RCS	MFI-LOM-SCS
---	---	**Country Economic Indicator** (projected GDP per month)	**Sales Revenue** (projected sales per month)

Table 63: CPV Engineering from LOM-RCS and -SCS

As an initial start, this CPV should be generated for the previous three years in order to identify correlations between the GDP and the MFC's sales amounts. In the event the sales amount's amplitude correlates a time-delay with that of the GDP, the MFC should be able to forecast a potential revenue risk in the event of a decreasing GDP, or a potential revenue opportunity with the rise of an increasing GDP.

8.3 Further MFI-BPM Recommendations

The suggested BPM model for the MFI is a starting point for the adaption of BPM strategies in MFCs. It offers the framework as well as an initial set of MFI related performance indicators and business measures, derived from this research project. However this model has to be fitted to each MFC's business objectives as the competitive advantages and therefore the MFC's strategy cannot be assumed to be uniform in all MFCs. Those performance indicators and business measures which form the CPVs will therefore differ between the various MFCs. In other words, the performance indicators and business measures within the four LOM are convertible corresponding to the MFC's business objectives. Neely et al. (2000) offer a practitioner oriented approach for this adaption process derived from action research in British manufacturing and service organizations, including SMEs:

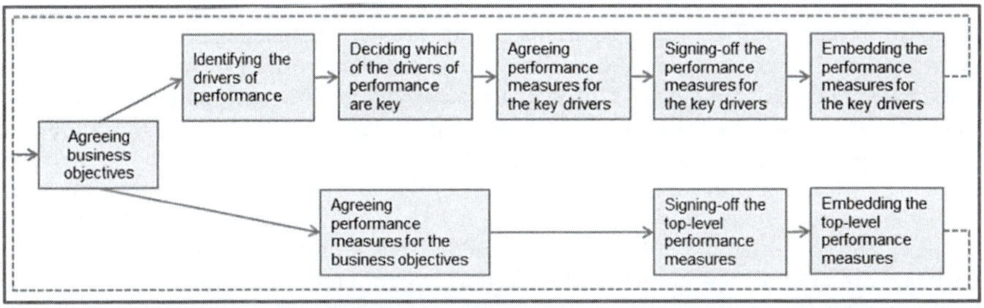

Figure 63: Performance Measures Identification and Selection Process (adopted in modified form from Neely et al., 2000, p. 1139)

This performance measures identification and selection process starts with an agreement on the business objectives involving the organization's managers. Measures are then differentiated between those related to the business objectives and those related to the drivers of performance. These performance measures correspond to the LOM performance indicators and business measures. The selected measures are then aggregated forming a) top-level performance measures for controlling the business objectives and b) those relating to the key performance drivers. This process step corresponds to the aggregation of the LOM measures into CPVs within the MFI-BPM system. This process should be repeated periodically[70] in order to react to changing business objectives and to modify the required measures accordingly. This circumstance is reflected through the dynamic capability of the LOM framework as described within the previous chapter.

To operationalize the information services of this BPM concept in the MFI, which is mainly formed by SMEs, the idea of applying standard office software suggests itself. As shown at MeFiCom (s. chapter 6.6.4) as well as in the MFI survey (s. chapter 7.2.3), the majority of the analyzed MFCs already apply standard office software (often Microsoft Excel; Singh, 2013) to measure their BPM processes. The following benefits can therefore be assumed for MFI SMEs:

- Presuming that the employees working with BPM are already familiar with the handling of such standard office software, training could therefore be limited to the operationalization of the MFI-BPM system. In addition, the risk of employee reluctance caused by an additional software tool which has to be mastered, could be reduced.
- Acquisition costs for a sophisticated software package could be avoided, as the software already exists; a benefit especially for smaller SMEs.
- Standard office software can be connected via interfaces like Open Database Connectivity (ODBC) to other software, like ERP, as further sources for performance data.
- Standard office software can supply input to other presentations or network sharing software which enables the distribution of BPM results. This is a measure of organizational interactive intergroup communication (s. chapter 2.3.3). The application of employee portals (Urbach et al., 2011) should be considered in the BPM system operationalization as an alternative way for the provision of BPM information throughout the hierarchical organization levels.

However standard office software has a limited range of functionality that becomes problematic in larger companies, especially when it comes to the IT processing of huge amounts, variety and complexity of data. To process and analyze such datasets a company can apply *Big Data Analytics* strategies (George et al., 2014). Ebner and Smolnik (2015) propose a practitioner oriented scoring model to choose an appropriate strategy out of the four main streams of big data analytics. This scoring model refers to the evaluation of twenty measures, linked to seven organizational decision factors and grouped in three temporal levels. Adapting this scoring model could assist larger MFCs in the operationalization of their customized MFI-BPM system and thereby help to improve the information services out of which the BPM system retrieves the organizational performance data.

The subtitle of this dissertation thesis reads: *An Approach Towards a Sustainable and Integrated Management System for Contract Processing Businesses*. The sustainability of the designed MFI-BPM system shall be apparent from the combination of the BPM design process being focused on the industry's requirements (s. chapter 2.5.3) with the empiric research process reflecting the organizations' requirements (s. chapters 6 and 7). But

[70] The time interval for such examinations depends on the organization's business. Neely et al. (2000, p. 1129) recommend an annual review. But as they have to regularly adapt their product and service portfolio to changing customer needs, MFCs should examine their BPM system every half year. This process could be combined during the reflection of the year-closing and mid-year reviews.

focusing solely on the requirements of the industry and organization bottom-up (s. figure 9) would neglect the management factor. Gleich (2011, p. 318) has shown the importance of management involvement in the design and implementation process of a BPM system. Neely et al. (2000) adds that this managerial commitment and involvement is not only important within the design and implementation process, but moreover in the active usage of the BPM system becoming an integrated element of the management control process. In order to achieve such a BPM sustainability it needs more than a deliberated BPM system implemented in sophisticated information services as outlined by Neely and Bourne (2000, p. 6):

> "Even when they have been through the process of designing a good measurement system and then implemented it successfully, far too often nobody manages with the measurement data. [...] There are a growing number of businesses that have put in place superb infrastructures to support their performance reporting systems. Then one observes that managers do not analyse the performance data. [...] The whole process of measuring performance is completely wasted unless action is taken on the performance data that are produced."

Regarding sustainability, Neely and Bourne's statement shows the importance of the active involvement and commitment of managers as recipients of such BPM data used in MCS. Therefore I echo Simons' judgement that "Actions Speak Louder Than Words" (1995, p. 55) to *Actions Speak Louder Than Words and Data.*

9 Conclusion

In the last chapter of my dissertation thesis I will present the common managerial and academic contributions of my research, discuss the limits within which I conducted the research project, and propose a research agenda for the future which reflects the research limitations as well as the research findings.

9.1 Common Managerial & Academic Contributions

> "Intergroup cooperation leads, as might be expected, to less discrimination than intergroup competition." (Tajfel, 1982, p. 28)

As I followed Tajfel's concept of "Intergroup Cooperation" (Tajfel, 1982; Tajfel and Turner, 1986; Ybema et al., 2009, p. 196) and as I adapted a Mode 2 Knowledge Production strategy for my qualitative research process (s. chapter 6.1.1), I do not differentiate between theoretical and managerial contributions, since my research project represents a common contribution for both. This involves creating a theoretical concept of feasibility (s. chapter 2.5.3) using generic BPM framework elements built-up on industry class characteristics and practical requirements (Drucker in Bonini et al., 1964, p. 289), to form an individual BPM model adaptable to companies within their industry class as external organizational environment. This chalice-shaped bottom-up approach, empirically validated for the MFI in this dissertation thesis, narrows the gap between theory and practice in MCS and BPM research (s. chapter 2.5.2).

Following the idea of common intergroup cooperation, the focal contribution of my work lies in the development of a cultural portrait, i.e. the cause and effect model of the MFI. This CEM holistically incorporates the following different sources of MFI information:

- MFI literature (s. chapters 3.1, 3.2, 3.3, and 4)
- International nomenclatures of economic activities (s. chapter 3.4)
- International statistical MFI data (s. chapter 3.5)
- Explorative qualitative empiric research within a major German MFC (s. chapter 6)
- Quantitative triangulation within the MFI in Germany and UK (s. chapter 7)

As a cultural portrait, the CEM is the base of the approach towards a sustainable BPM model for the MFI and is centered on the MFI characteristics following the chalice-shaped concept of BPM feasibility (s. chapter 2.5.3).

By describing in detail the research process, which follows this chalice-shaped concept of feasibility, I created a blue print for the adaption of this concept for further industry classes. This approach will contribute to the academic community by offering an alternative research process as one answer to the evolving academic MCS and BPM concerns (s. chapter 2.5.2).

As a result I combined academic rigor and managerial relevance by

1. applying a bottom-up approach (s. figure 10), as this research process is derived from a theoretical conceptualization of feasibility in order to start my structured analysis of BPM in the MFI with an analysis of the industry itself, and
2. applying a Mode 2 Knowledge Production related research methodology (s. chapter 2.5.2) by picking up Bartunek's (2011, p. 557) suggestion *"to do some empirical*

data collection" in Mode 2 research by having a joint research project on BPM between EBS and MeFiCom.

Chapter 3 includes an overview of the MFI as a widely undiscovered industry class by the academic community. This part of the research project will contribute to the academic community by offering clear structures of the MFI that can be applied as a foundation for further research projects within the MFI. Specifically, my findings stem from MFI SMEs being in the twilight zone between service and manufacturing companies. Therefore the MFI is a suitable research candidate for business schools as SMEs as well as BPM for service companies are both topics on the academic research agenda (Lange, 2009, p. 153). Further research within the MFI is therefore of mutual benefit for the academic community as well as for the MFI itself regarding the academic results. This circumstance follows the MFI community's wish as formulated by the UK Institute of Materials Finishing, in evaluating the benefits of this SIIE research project for the MFI: *"It is good for our industry, virtually unknown outside itself, to have its profile raised and be noticed as a potential research candidate for business schools etc."* (Clive Larson, personnel communication, March 27th, 2014).

The applied research framework (s. figure 7) is a first approach towards the design of a BPM meta-theory as recommended by Franco-Santos et al. (2012, p. 99, s. chapter 2.5.1). I am convinced that a BPM meta-theory allows establishing common definitions of BPM and MCS which are currently missing (s. chapters 2.3 and 2.4). The Levers of Measurement, as presented in the previous chapter, is an integrated element of the management control systems in being the foundation of this BPM meta-theory.

The Levers of Measurement, as the resulting MFI-BPM model, furthermore extends Simons' Levers of Control (1995; s. chapter 2.3.3) MCS framework within the MFI by specifying the LOC Diagnostic Control Systems. This contribution also reflects the evolving academic concern regarding MCS and BPM research, as it demonstrates that the academic LOC framework, as "action-oriented theory of control" (Simons, 1994, p. 171; Simons, 1995, p. ix), is broad enough to be adapted to the characteristics and requirements of an industry class like the MFI. This approach shows that *simplified models of the world* can be very well connected to the *real-world* (Bourne, 2008, p. 69).

9.2 Discussion & Limitations

As with any other academic study, this dissertation is not free of limitations.

Although I rigorously searched different electronic databases as well as relevant journals (s. chapter 4.1) I might have missed some relevant work within my literature review. Such work could for example have been published in journals outside the scope of my reviews or perhaps have been published in non-German or non-English-language journals.

It is the nature of ethnographic research (s. chapter 6.1.1) that the researcher becomes part of the research area which gets analyzed. Therefore I cannot fully neglect some degree of researcher bias. However I took measures in order to reduce this type of bias by choosing methodologies like the group discussion method according to Bohnsack (Bohnsack et al., 2010; s. chapter 6.4).

I applied the theoretical conceptualization of feasibility solely to one industry class. In further research the concept should be applied to other industry classes to enhance its feasibility.

I acknowledge that the suggested bottom-up approach may not result in significant time saving compared to a top-down BPM system implementation. Nevertheless, I believe that by building the BPM requirements bottom-up will help practitioners to implement *their* BPM system by reducing the risk of information overload.

Although my research focuses on *metal finishing companies of larger scale* (s. chapter 2.5.4), that does not mean that my research results are irrelevant for micro and similar small sized enterprises, as these can indeed be positively influenced by improvements of performance measuring models, such as those proven by EPA's sector strategy program (EPA, 2004). In fact the results of the MFI survey (s. chapter 7.2) showed that there are no significant differences between larger and smaller MFCs in the survey sample related to the perception of BPM.

Some of my research findings may appear idealistic – nevertheless that does not mean that they are impractical. For example:

As shown in the previous chapters innovativeness plays an important role for the prosperity of MFCs. But it is understandable that only a small number of SMEs can afford the cost for their own R&D department. But SMEs can integrate themselves in research communities like the EFDS and so participate by the community research.

The social science literature regarding the MFI is scarce and partly antiquated. However the concepts and evaluations taken from older MFI literature, which was incorporated into my dissertation, hold predominantly true even currently, as shown with Brummet and Arnett's (1967, p. 8) evaluations regarding MFI's key competitive problems with *"lack of experienced and competent"* employees as major challenge (compare to people as success factor in chapter 6.6.4 and LOM KCS in chapter 8.2.1).

It is not trivial to compare the statistical data of different countries relating to different cultures existing within these countries (s. chapter 7.2.2). The statistical data presented in chapter 3.5 can therefore only be compared restrictively with each other. However this limitation is not so critical as it was not my objective to compare country statistics with each other but to analyze industry structures within each nation.

One could argue that the analysis of statistical data solely for one focal year is just a snapshot, and that a time series analysis would be more appropriate. That is correct, but it was not the goal of my dissertation to perform in-depth statistical analyses, but to find significant peaks in the data which were then scrutinized in-depth within my empiric research. The MFI is closely connected with many other manufacturing industries, thus changes within the overall manufacturing class would immediately reflect into the MFI (related for example to the demand of temporary or leased employees).

Nevertheless it must be noted that MeFiCom is specialized in CVD and PVD metal finishing processes, being part of the vacuum vaporization share of surface finishing processes. As shown in figure 12, this share counts for 5% of all surface finishing processes in Germany in 2005. But compared to the MFI survey results, I could not find significant BPM differences between the MeFiCom and other MFCs taken from the survey sample, except the degree of decentralization and organizational ambidexterity (s. chapter 7.2).

The applied system thinking approach in the qualitatively established cause and effect model (s. chapter 6.6) is limited to applying the stocks and flows notation for the graphical interpretation of the CAQDAS code system. It does not include a simulation component (Sterman, 2000, p. 37) of the established MFI mental model. It would be of interest for future research projects to analyze how a system dynamic simulation approach, like creating an explanatory MFI model (Bossel, 2007, p. 62), is applicable for BPM research.

The rather small number of completely filled survey questionnaires within the quantitative research process restricted a multivariate data analysis. This may be owed to the length and complexity of the survey questionnaire and what restrained the respondent to complete the questionnaire. However, I based my decision to design the survey questionnaire as attached in appendices A.15 and A.16 with the goal that this survey should be a triangulation of the qualitative findings. Picking only some of the qualitative findings for the quantitative triangulation would mean a shortcoming of the research process. By reflecting on the results, it should be discussed in future research whether quantitative methods, especially surveys, are the advisable method of choice for information retrieval or data triangulation in an SME-formed industry class, or if more qualitative based methods including the direct contact with participants should be preferred.

9.3 Research Agenda for the Future

Concerning the sustainability of the MFI-BPM system, the active usage of the system as well as its dynamic capability (s. chapter 8.2) should be continuously monitored after implementation. A potential research question in this context might be: how sustainable is the implemented MFI-BPM system concerning the changes made in comparison to the initial version of the system?

The explorative propositions derived from the qualitative research (s. chapter 7.1.2) should be analyzed in more detail. Organizational ambidexterity for example shows some degree of significance of being a general MFC characteristic. Further research should determine if organizational ambidexterity is moreover a prerequisite for the economic success of MFCs.

A tendency in academic literature can be identified as utilizing the term *Performance Management* instead of BPM or MCS (s. chapters 2.3 and 2.4). Future research should find an answer as to why a third term is introduced, instead of creating clear and academically accepted definitions for already existing research fields. Is it because the term *Management* better reflects the practitioner audience of managers than *Measurement*? Or is *Performance Management* a kind of missing link between both research fields? Including the BPM meta-theory approach as presented in figure 7 and extending it if necessary by including further elements like contingency or goal-setting theories, would certainly assist in finding common definitions of MCS and BPM and their interplay which are currently missing (s. chapters 2.3 and 2.4).

Other MFI-like contract processing industry branches exist, offering products and services in order to enhance and finish raw products. Extant examples, for instance:

- Textile finishing and dyeing industry
- Food processing industry including

- Potato Processing[71]
- Grape-Vine Grafting

It would be of interest to investigate if those and other contract processing industries follow the same mechanisms the MFI does and in how far the findings of this research project contribute to these industries, referring to the ability to adapt the MFI-BPM model to these industry classes.

Regarding the information services, a focused research project including big data analysis strategies would expand the aforementioned recommendations related to BPM in the MFI, especially for larger MFCs. This approach would follow the note of Nudurupati et al. (2011; s. chapter 2.5.2) whose criticism was that Management Information Systems are rarely reflected in BPM systems.

[71] In German: *Kartoffelveredelungsindustrie*. The futures for this agricultural industry branch are traded for example on the European Derivatives Exchange: European Processing Potato Futures (FEPP) – Product ISIN: DE000A0Z3068.

References – Literature

Anthony, R. N. (1965): Planning and Control Systems: A Framework for Analysis. *Graduate School of Business Administration, Harvard University*, Boston, MA.

Anthony, R. N. (1989): Reminiscences About Management Accounting. In: *Journal of Management Accounting Research*, 1, pp. 1-20.

Anthony, R. N.; Govindarajan, V.; Hartmann, F. G. H.; Kraus, K. and Nilsson, G. (2014): Management Control Systems. First European Edition. Berkshire: McGraw-Hill Education.

Arend, R. J.; Patel, P. C. and Park, H. D. (2014): Explaining post-IPO Venture Performance Through a Knowledge-based View Typology. In: *Strategic Management Journal*, 35 (3), pp. 376-397.

Armstrong, C. E. and Shimizu, K. (2007): A Review of Approaches to Empirical Research on the Resource-Based View of the Firm. In: *Journal of Management*, 33 (6), pp. 959-986.

Armstrong, J. S. and Collopy, F. (1996): Competitor Orientation: Effects of Objectives and Information on Managerial Decisions and Profitability. In: *Journal of Marketing Research*, 33 (2), pp. 188-199.

Armstrong, J. S. and Overton, T. S. (1977): Estimating Nonresponse Bias in Mail Surveys. In: *Journal of Marketing Research,* 14 (3), pp. 396-402.

Artley, R. J. and Matthews, A. (2010): The UK Surface Industry 2010. Surface Engineering Association (SEA). Birmingham, UK.

Atif, A.; Richards, D. and Bilgin, A. (2012): Estimating Non-Response Bias in a Web-based Survey of Technology Acceptance: A Case Study of Unit Guide Information Systems. Conference Proceeding. 23rd Australasian Conference on Information Systems, 03.-05. Dec. 2012, Geelong.

Australia (2006): Australian and New Zealand Standard Industrial Classification (ANZSIC) 2006. ABS Catalogue No. 1292.0. Australian Bureau of Statistics and Statistics New Zealand.

Australia (2013): Australian and New Zealand Standard Industrial Classification (ANZSIC) 2006 (Revision 2.0). ABS Catalogue No. 1292.0. Australian Bureau of Statistics and Statistics New Zealand.

Barbour, R. S. (2014): Quality of Data Analysis. In: Flick, U. (2014): The SAGE Handbook of Qualitative Data Analysis. Los Angeles: SAGE Publications.

Barney, J. B. (1991): Firm Resources and Sustained Competitive Advantage. In: *Journal of Management*, 17 (1), pp. 99-120.

Barney, J. B. (2001): Resource-based theories of competitive advantage: A ten-year retrospective on the resource-based view. In: *Journal of Management*, 27 (6), pp. 643-650.

Barney, J. B.; Ketchen, Jr., D. J. and Wright, M. (2011): The Future of Resource-Based Theory: Revitalization or Decline? In: *Journal of Management*, 37 (5), pp. 1299-1315.

Barratt, M. and Oke, A. (2007): Antecedents of Supply Chain Visibility in Retail Supply Chains: A resource-based theory perspective. In: *Journal of Operations Management*, 25 (6), pp. 1217–1233.

Bartunek, J. M. (2011): Viewpoint - What Has Happened to Mode 2? In: *British Journal of Management*, 22 (3), pp. 555-558.

Baruch, Y. and Holtom, B. C. (2008): Survey Response Rate Levels and Trends in Organizational Research. In: *Human Relations*, 61 (8), pp. 1139-1160.

Bawden, D. and Robinson, L. (2009): The Dark Side of Information: Overload, anxiety and other paradoxes and pathologies. In: *Journal of Information Science*, 35 (2), pp. 180-191.

Bell, T. (2000): Towards a Universal Surface Engineering Road Map. In: *Surface Engineering*, 16 (2), pp. 89-90.

Bergold, J. and Flick, U. (1987): Ein-Sichten: Zugänge zur Sicht des Subjekts mittels qualitativer Forschung. Tübingen: dgvt-Verlag.

Berman, E. M.; West, J. P. and Wang, X. (1999): Using Performance Measurement in Human Resource Management: A Survey of U.S. Counties. In: *Review of Public Personnel Administration*, 19 (2), pp. 5-17.

Berry, A. J., Coad, A. F., Harris, E. P., Otley, D. T. and Stringer, C. (2009): Emerging Themes in Management Control: A Review of Recent Literature. In: *The British Accounting Review*, 41 (2009), pp. 2-20.

Bingham, C. B. and Davis, J. P. (2012): Learning Sequences: Their Existence, Effect, and Evolution. In: *Academy of Management Journal*, 55 (3), pp. 611-641.

Birnberg, J. G. (2011): Robert N. Anthony: A Pioneering Thinker in Management Accounting. In: *Accounting Horizons*, 25 (3), pp. 593-602.

Bohnsack, R.; Marotzki, W. and Meuser, M. (2006): Hauptbegriffe Qualitativer Sozialforschung. 2nd Edition. Opladen & Farmington Hills: Budrich.

Bohnsack, R. and Przyborski, A. (2009): Gruppendiskussionsverfahren und Focus Groups. In: Buber, R. and Holzmüller, H. (2009): Qualitative Marktforschung. Konzepte, Methoden, Analysen. 2nd Edition. Wiesbaden: Gabler, GWV Fachverlage GmbH.

Bohnsack, R., Przyborski, A. and Schäfer, B. (2010): Das Gruppendiskussionsverfahren in der Forschungspraxis. 2nd Edition. Opladen: Budrich.

Bonett, D. G. and Wright, T. A. (2015): Cronbach's Alpha Reliability: Interval Estimation, Hypothesis Testing, and Sample Size Planning. In: *Journal of Organizational Behavior*, 36 (1), pp. 3-15.

Bonini, C. P., Jaedicke, R. K. and Wagner, H. M. (1964): Management Controls. New Directions in Basic Research. New York: McGraw-Hill.

Booth, A; Papaioannou, D. and Sutton, A. (2012): Systematic Approaches to a Successful Literature Review. 1st Edition. London: SAGE Publication Ltd.

Bossel, H. (2007): Systems and Models. Complexity, Dynamics, Evolution, Sustainability. Norderstedt: Books on Demand.

Bourne, M. (2008): Performance Measurement: Learning from the Past and Projecting the Future. In: *Measuring Business Excellence*, 12 (4), pp. 67-72.

Bourne, M., Neely, A., Mills, J. and Platts, K. (2003): Implementing Performance Measurement Systems: A Literature Review. In: *International Journal Business Performance Management*, 5 (1), pp. 1-24.

Bourne, M.; Neely, A.; Platts, K. and Mills, J. (2002): The Success and Failure of Performance Measurement Initiatives - Perceptions of Participating Managers. In: *International Journal of Operations & Production Management*, 22 (11), pp. 1288-1310.

Box, J. F. (1987): Guinness, Gosset, Fisher, and Small Samples. In: *Statistical Science*, 2 (1), pp. 45-52.

Boyer, K. K.; Leong, G. K.; Ward, P. T. and Krajewski, L. J. (1997): Unlocking the Potential of Advanced Manufacturing Technologies. In: *Journal of Operations Management*, 15 (4), pp. 331-347.

Brace, A. (1995): The Development of an Expert System for the Identification Anodic Coating Process Defects As a Contribution to the Dissemination of Anodizing Technology. Dissertation Thesis. University of Aston, Birmingham.

Bradburn, N. M.; Sudman, S. and Wansink, B. (2004): Asking Questions. The Definitive Guide to Questionnaire Design - For Market Research, Political Polls, and Social and Health Questionnaires. Revised 1st Edition. San Francisco, CA: Jossey-Bass.

Braunsberger, K.; Wybenga, H. and Gates, R. (2007): A Comparison of Reliability Between Telephone and Web-Based Surveys. In: *Journal of Business Research*, 60 (7), pp. 758-764.

Brislin, R. (1970): Back-Translation for Cross-Cultural Research. In: *Journal of Cross-Cultural Psychology*, 1 (3), pp. 185-216.

Broadbent, J. and Laughlin, R. (2009): Performance Management Systems: A Conceptual Model. In: *Management Accounting Research*, 20 (4), pp. 283-295.

Brown, M. G. (1996): Keeping Score: Using the Right Metrics to Drive World-Class Performance. New York: Quality Resources.

Brummet, R. L. and Arnett H. E. (1967): Business and Economic Evaluation of the Metal Finishing Industry. In: *Michigan Business Reports No. 52*. Bureau of Business Research, Graduate School of Business Administration, University of Michigan.

Buber, R. and Holzmüller, H. (2009): Qualitative Marktforschung. Konzepte, Methoden, Analysen. 2nd Edition. Wiesbaden: Gabler, GWV Fachverlage GmbH.

Canada (2012): North American Industry Classification System (NAICS) Canada. Catalogue no. 12-501-X. Information Management Division, Statistics Canada, Ottawa, Ontario, Canada.

Charter, R. A. (2003): Study Samples are too Small to Produce Sufficiently Precise Reliability Coefficients. In: *The Journal of General Psychology*, 130 (2), pp. 117-129.

Charter, R. A. (2008): Statistical Approaches to Achieving Sufficiently High Test Score Reliabilities for Research Purposes. In: *The Journal of General Psychology*, 135 (3), pp. 241-251.

Chen, I. J.; and Paulraj, A. (2004): Towards a Theory of Supply Chain Management: The Constructs and Measurements. In: *Journal of Operations Management*, 22 (2), pp. 119-150.

Coase, R. H. (1937): The Nature of the Firm. In: *Economica*, New Series, 4 (16), pp. 386-405.

Coase, R. H. (1960): The Problem of Social Cost. In: *Journal of Law and Economics*, 3, pp. 1-44

Coase, R. H. (1998): The New Institutional Economics. In: *American Economic Review*, 88 (2), pp. 72-74.

Cortina, J. M. (1993): What Is Coefficient Alpha? An Examination of Theory and Applications. In: *Journal of Applied Psychology*, 78 (1), pp. 98-104.

Creswell, J. W. (2013): Qualitative Inquiry & Research Design. Choosing Among Five Approaches. 3rd Edition. Los Angeles: SAGE Publications.

Creswell, J. W. and Plano Clark, V. L. (2011): Designing and Conducting Mixed Methods Research. 2nd Edition. Thousand Oaks, CA: SAGE Publications.

Cronbach, L. J. (1951): Coefficient Alpha and the Internal Structure of Tests. In: *Psychometrika*, 16 (3), pp. 297-334.

Cugueró-Escofet, N. and Rosanas, J. M. (2013): The Just Design and Use of Management Control Systems as Requirements for Goal Congruence. In: *Management Accounting Research*, 24 (1), pp. 23-40.

Dale, E. (1957): Du Pont: Pioneer in Systematic Management. In: *Administrative Science Quarterly*, 2 (1), pp. 25-59.

Davies-Labeck, B. (2005): Fachwörterbuch Handel und Vertrieb. Dictionary of Retailing and Supply Chain. 1st Edition. Frankfurt a. M.: Deutscher Fachverlag GmbH.

De Salas, K. and Huxley, C. (2013): Enhancing Visualisation to Communicate and Execute Strategy – Strategy-to-Process Maps. In: *Journal of Strategy and Management*, 7 (2), pp. 109-126.

Denzin, N. K. and Lincoln, Y. S. (2011): The SAGE Handbook of Qualitative Research. 4th Edition. Thousand Oaks: SAGE Publications.

Dietrich, M. and Wald, A. (2014): The Challenge of the Adaption and Implementation of Management Control Systems: A Bottom-Up Approach to Improve Feasibility. In: *British Academy of Management (BAM) – Conference Proceedings*. Belfast, UK, September 2014 - Best Development Paper Award Winner in Performance Management.

Dillman, D. A.; Smyth, J. D. and Christian, L. M. (2009): Internet, Mail, and Mixed-Mode Surveys. The Tailored Design Method. 3rd Edition. Hoboken, N. J.: Wiley & Sons.

DOE (1995): Metal Manufacturing, Refining and Finishing Works: Electroplating and Other Metal Finishing Works. Industry Profile. Department of the Environment, UK.

Dresing, T. and Pehl, T. (2013): Praxisbuch Interview, Transkription & Analyse. Anleitungen und Regelsysteme für qualitativ Forschende. 5th Edition. Marburg.

Drucker, P. F. (1964): Controls, Control and Management. In: Bonini, C. P., Jaedicke, R. K. and Wagner, H. M. (1964): Management Controls. New Directions in Basic Research. New York: McGraw-Hill, pp. 286 - 296.

Duncan, R. B. (1972): Characteristics of Organizational Environments and Perceived Environmental Uncertainty. In: *Administrative Science Quarterly*, 17 (3), pp. 313-327.

Dyer, J. and Singh, H. (1998): The Relational View: Cooperative Strategy and Sources of Interorganizational Competitive Advantage. In: *Academy of Management Review*, 23 (4), pp. 660-679.

Ebner, K. and Smolnik, S. (2015): Alles eine Frage der Daten? Eine Analyse verschiedener Big-Data-Technologiestrategien und Vorschlag eines Entscheidungsmodells. In: *Controlling - Zeitschrift für Erfolgsorientierte Unternehmenssteuerung*, 27 (4/5), pp. 249-255.

EC (2003): European Commission Recommendation of 6 May 2003 concerning the Definition of Micro, Small and Medium-sized Enterprises (2003/361/EC). Official Journal of the European Union, L 124/36.

EC (2008): NACE Rev. 2: Statistical Classification of Economic Activities in the European Community. Luxembourg: Office for Official Publications of the European Communities.

Eccles, R. G. (1991): The Performance Measurement Manifesto. In: *Harvard Business Review*, 69 (1), pp. 131-137.

Eden, C. (1992): On the Nature of Cognitive Maps. In: *Journal of Management Studies*, 29 (3), pp. 261-265.

Edvinsson, L. (1997): Developing Intellectual Capital at Skandia. In: *Long Range Planning*, 30 (3), pp. 366-373.

EFS (2014): Enterprise Feedback Suite – EFS Survey. Manual 10.4 Version 1.0. Questback.

Eisenhardt, K. M. (1989a): Agency Theory: An Assessment and Review, In: *Academy of Management Review*, 14 (1), pp. 57-74.

Eisenhardt, K. M. (1989b): Building Theories from Case Study Research. In: *Academy of Management Review*, 14 (4), pp. 532-550.

EPA (1982): Economic Analysis of Proposed Effluent Standards and Limitations for the Metal Finishing Industry. United States Environmental Protection Agency. Office of Water Regulations and Standards. EPA-440/2-82-004.

EPA (1984): Guidance Manual for Electroplating and Metal Finishing Pretreatment Standards. United States Environmental Protection Agency. Office of Water Regulations and Standards. EPA-440/1-84-091g.

EPA (1992): Guides to Pollution Prevention: The Metal Finishing Industry. United States Environmental Protection Agency. Office of Research and Development. EPA/625/R-92/011.

EPA (2004): Sector Strategies: Performance Report. United States Environmental Protection Agency. National Center for Environmental Innovation (1807T). EPA 100-R-04-002.

EPA (2006): Sector Strategies: Performance Report. United States Environmental Protection Agency. National Center for Environmental Innovation (1807T). EPA 100-R-06-002.

Epstein, M. J. and Manzoni, J.-F. (1997): The Balanced Scorecard and Tableau de Bord: Translating Strategy into Action. In: *Management Accounting*, 79 (2), pp. 28-36.

EUROSTAT (2009): European Business. Facts and Figures. 2009 Edition. Luxembourg: Office for Official Publications of the European Communities.

Even, W. E. and Macpherson, D. A. (2012): Is Bigger Still Better? The Decline of the Wage Premium at Large Firms. In: *Southern Economic Journal*, 78 (4), pp. 1181-1201.

Feigh, K.; Chua, Z.; Garg, C.; Jacobsen, A.; O'Hara, J.; Rogers, W. and Shutko, J. (2011): Current State of Human Factors in Systems Design. In: *Proceedings of the Human Factors and Ergonomics Society Annual Meeting*, 55 (1), pp. 267-271.

Ferreira, A. and Otley, D. (2009): The Design and Use of Performance Management Systems: An Extended Framework for Analysis. In: *Management Accounting Research*, 20 (4), pp. 263-282.

Fetterman, D. M. (2010): Ethnography – Step-by-Step. 3rd Edition. Thousand Oaks, CA: SAGE Publications.

Fitzgerald, L.; Johnston, R.; Brignall, S.; Silvestro, R. and Voss, C. (1991): Performance Measurement in Service Businesses. London: CIMA.

Flannery, B. L. and May, D. R. (2000): Environmental Ethical Decision Making in the U. S. Metal Finishing Industry. In: *Academy of Management Journal,* 43 (4), p. 642-662.

Flick, U. (1987): Methodenangemessene Gütekriterien in der qualitativ-interpretativen Forschung. In: Bergold, J. and Flick, U. (1987): Ein-Sichten: Zugänge zur Sicht des Subjekts mittels qualitativer Forschung. Tübingen: dgvt-Verlag.

Flick, U. (2011): Triangulation: Eine Einführung. 3rd Edition. Wiesbaden: Springer Fachmedien, VS Verlag.

Flick, U. (2014): The SAGE Handbook of Qualitative Data Analysis. Los Angeles: SAGE Publications.

Foss, K. and Foss, N. J. (2005): Resources and Transaction Costs: How Property Rights Economics Furthers the Resource-Based View. In: *Strategic Management Journal*, 26 (6), pp. 541-553.

Foss, N. J.; Lyngsie, J. and Zahra, S. A. (2013): The Role of External Knowledge Sources and Organizational Design in the Process of Opportunity Exploitation. In: *Strategic Management Journal*, 34 (12), pp. 1453-1471.

Franco-Santos, M. and Bourne, M. (2005): An Examination of the Literature Relating to Issues Affecting How Companies Manage Through Measures. In: *Production Planning and Control*, 16 (2), pp. 114-124.

Franco-Santos, M.; Kennerley, M.; Micheli, P.; Martinez, V.; Mason, S.; Marr, B.; Gray, D. and Neely, A. (2007): Towards a definition of a business performance measurement system. In: *International Journal of Operations & Production Management*, 27 (8), pp. 784-801.

Franco-Santos, M.; Lucianetti, L. and Bourne, M. (2012): Contemporary Performance Measurement Systems: A Review of their Consequences and a Framework for Research. In: *Management Accounting Research*, 23 (2), pp. 79-119.

Garengo, P. and Biazzo, S. (2012): Unveiling Strategy in SMEs Through Balanced Scorecard Implementation: A Circular Methodology. In: *Total Quality Management*, 23 (1), pp. 79–102.

Gawron, V. J.; Cipriano, L.; Fleishman, E.; Hegge, F.; Lehman, E.; Meister, D. and Reising, J. (1990): Guide for Human Performance Measurements. In: *Proceedings of the Human Factors and Ergonomics Society Annual Meeting*, 34 (17), pp. 1238-1240.

Geertz, C., Luchesi, B. and Bindemann, R. (1987): Dichte Beschreibung. Beiträge zum Verstehen kultureller Systeme. Frankfurt a.M.: Suhrkamp.

George, G.; Haas, M. R. and Pentland, A. (2014): From the Editors: Big Data and Management. In: *Academy of Management Journal*, 57 (2), pp. 321-326.

Giannopoulos, G., Holt, A., Khansalar, E. and Cleanthous, S. (2013): The Use of the Balanced Scorecard in Small Companies. In: *International Journal of Business and Management*, 8 (14), pp. 1-22.

Gibbons, M.; Limoges, C.; Nowotny, H.; Schwartzman, S.; Scott, P. and Trow, M. (1994): The New Production of Knowledge: the Dynamics of Science and Research in Contemporary Societies. London: SAGE.

Gibbs, G. R. (2014): Using Software in Qualitative Analysis. In: Flick, U. (2014): The SAGE Handbook of Qualitative Data Analysis. Los Angeles: SAGE Publications.

Gläser, J. and Laudel, G. (2010): Experteninterviews und Qualitative Inhaltsanalyse. 4th Edition. Wiesbaden: VS Verlag, Springer Fachmedien.

Gleich, R. (2011): Performance Measurement: Konzepte, Fallstudien und Grundschema für die Praxis. 2nd Edition. München: Verlag Franz Vahlen.

Gleich, R. (2012): Balanced Scorecard: Best-Practice-Lösungen für die Unternehmens-steuerung. Freiburg: Haufe-Lexware.

Gleich, R.; Motwani, J. and Wald, A. (2008): Process Benchmarking: A new Tool to Improve the Performance of Overhead Areas. In: *Benchmarking: An International Journal,* 15 (3), pp. 242-256.

Gond, J.; Grubnic, S.; Herzig, C. and Moon, J. (2012): Configuring Management Control Systems: Theorizing the integration of strategy and sustainability. In: *Management Accounting Research*, 23 (3), pp. 205-223.

Grant, R. M. (1996): Toward a Knowledge-Based Theory of the Firm. In: *Strategic Management Journal*, 17 (Winter Special Issue), pp. 109-122.

Gray, J. V.; Tomlin, B. and Roth, A. V. (2009): Outsourcing to a Powerful Contract Manufacturer: The Effect of Learning-by-Doing. In: *Production and Operations Management*, 18 (5), pp. 487-505.

Greene, J. C. (2007): Mixed Methods in Social Inquiry. 1st Edition. San Francisco, CA: Jossey-Bass.

Grosswiele, L., Röglinger, M. and Friedl, B. (2013): A decision framework for the consolidation of performance measurement systems. In: *Decision Support Systems*, 54 (2), pp. 1016–1029.

Hair, J. F.; Black, W. C.; Babin, B. J. and Anderson, R. E. (2010): Multivariate Data Analysis. A Global Perspective. 7th Edition. Upper Saddle River: Pearson Education.

Hall, M. (2008): The effect of comprehensive performance measurement systems on role clarity, psychological empowerment and managerial performance. In: *Accounting Organizations and Society*, 33 (2–3), pp. 141-163.

Hambrick, D. C. and Mason, P. A. (1984): Upper Echelons: The Organization as a Reflection of its Top Managers. In: *American Management Review*, 9 (2), pp. 193-206.

Hammersley, M. (1998): Reading Ethnographic Research: A Critical Guide. 2nd Edition. London: Routledge.

Hammersley, M. and Atkinson, P. (1995): Ethnography. Principles in Practice. 2nd Edition. London: Taylor & Francis Ltd.

Hassini, E.; Surti, C. and Searcy, C. (2012): A Literature Review and a Case Study of Sustainable Supply Chains with a Focus on Metrics. In: *International Journal of Production Economics*, 140 (1), pp. 69-82.

Helfferich, C. (2011): Die Qualität Qualitativer Daten – Manual für die Durchführung Qualitativer Interviews. 4th Edition. Wiesbaden: VS Verlag, Springer Fachmedien.

Hofmann, E. (2014): Interorganizational Operations Management: Von der Strategie bis zur finanziellen Steuerung der Performance in Supply Chains. *Habilitation Thesis University St. Gallen, Switzerland.* Springer Gabler Fachmedien, Wiesbaden.

Hofstede, G.; Hofstede, G. J. and Minkov, M. (2010): Cultures and Organizations: Software of the Mind. 3rd Edition. Boston: McGraw-Hill Education.

Hult, G. T. M.; Ketchen Jr., D. J.; Griffith, D. A.; Chabowski, B. R.; Hamman, M. K.; Johnson Dykes, B.; Pollitte, W. A. and Cavusgil, S. T. (2008): An Assessment of the Measurement of Performance in International Business Research. In: *Journal of International Business Studies*, 39 (6), pp. 1064-1080.

Hypko, P.; Tielebein, M. and Gleich, R. (2010): Benefits and Uncertainties of Performance-Based Contracting in Manufacturing Industries: An Agency Theory Perspective. In: *Journal of Service Management*, 21 (4), pp. 460-489.

Idson, T. L. and Oi, W. Y. (1999): Workers Are More Productive in Large Firms. In: *American Economic Review*, 89 (2), pp. 104-108.

Industry Canada (2000): Industry Snapshot: Metal Finishing in Canada. D.B. Caldwell Research Associates Inc. Cat. No. C45-2/3-2000-6.

Jansen, J. J. P.; Tempelaar, M. P.; van den Bosch, F. A. J. and Volberda, H. W. (2009): Structural Differentiation and Ambidexterity: The Mediating Role of Integration Mechanisms. In: *Organization Science*, 20 (4), pp. 797-811.

Jansen, J. J. P.; van den Bosch, F. A. J. and Volberda, H. W. (2006): Exploratory Innovation, Exploitative Innovation, and Performance: Effects of Organizational Antecedents and Environmental Moderators. In: *Management Science*, 52 (11), pp. 1661-1674.

Jensen, M. C. and Meckling, W. H. (1976): Theory of the Firm: Managerial Behavior, Agency Costs and Ownership Structure. In: *Journal of Financial Economics*, 3, pp. 305-360.

Jesson, J. K.; Matheson, L. and Lacey, F. M. (2011): Doing your Literature: Traditional and Systematic Techniques. 1st Edition. London: SAGE Publications Ltd.

Kaplan, R. S. and Norton, D. P. (1992): The Balanced Scorecard - Measures that drive Performance. Reprint in: Harvard Business Review on Measuring Corporate Performance. 1998, 1st Edition. Boston: Harvard Business School Press.

Kaplan, R. S. and Norton, D. P. (1996): The Balanced Scorecard: Translating Strategy into Action. Boston: Harvard Business Review Press.

Kaplan, R. S. and Norton, D. P. (2003): Strategy Maps. In: *Strategic Finance*, 85 (9), pp. 27-35.

Kaplan, R. S. and Norton, D. P. (2004): How Strategy Maps Frame an Organization's Objectives. In: *Financial Executive*, 20 (2), pp. 40-45.

Keegan, D. P.; Eiler, R. G. and Jones, C. R. (1989): Are your Performance Measures Obsolete? In: *Management Accounting*, 70 (12), pp. 45-50.

Kim, J. and Mahoney, J. T. (2005): Property Rights Theory, Transaction Costs Theory, and Agency Theory: An Organizational Economics Approach to Strategic Management. In: *Managerial and Decision Economics*, 26 (4), pp. 223–242.

König, U. (2010): Trends and Innovation in German Surface Technology. Central Association of Surface Treatment Professionals (ZVO). In: *EURAILmag Business & Technology*, 21, pp. 66-70.

Koufteros, X.; Verghese, A. J. and Lucianetti, L. (2014): The Effect of Performance Measurement Systems on Firm Performance: A Cross-Sectional and a Longitudinal Study. In: *Journal of Operations Management*, 32 (6), pp. 313-336.

Kowal, S. and O'Connell, D. C. (2014): Transcription as a Crucial Step of Data Analysis. In: Flick, U. (2014): The SAGE Handbook of Qualitative Data Analysis. Los Angeles: SAGE Publications.

Krueger, R. A. and Casey, M. A. (2009): Focus groups. A Practical Guide for Applied Research. 4th Edition. Los Angeles: SAGE Publications.

Kuckartz, U. (2014): Qualitative Inhaltsanalyse. Methoden, Praxis, Computer-unterstützung. 2nd Edition. Weinheim: Beltz Juventa.

Kunc, M. (2008): Using systems thinking to enhance strategy maps. In: *Management Decision*, 46 (5), pp. 761-778.

Lange, I. C. (2009): Leistungsmessung Industrieller Dienstleistungen: Prozess- und Leistungstransparenz als Basis für das Management von Produkt-Service-Systemen. Dissertation. DISS. ETH Nr. 18757. ETH Zürich.

Langley, A. (1999): Strategies for theorizing from process data. In: *Academy of Management Review*, 24 (4), pp. 691-710.

Lapan, S. D.; Quartaroli, M. T. and Riemer, F. J. (2012): Qualitative Research. An Introduction to Methods and Designs. 1. Edition. San Francisco: Jossey-Bass.

Larson, C. (2012a): Global Comparisons of Metal Finishing Sectors: Part 1, Variations in Economic and Structural Factors. In: *Transactions of the Institute of Metal Finishing*, 90 (4), pp. 179-187.

Larson, C. (2012b): Global Comparisons of Metal Finishing Sectors: Part 2, Some Technology and Operational Variations. In: *Transactions of the Institute of Metal Finishing*, 90 (5), pp. 232-236.

Larson, C. (2012c): Global Comparisons of Metal Finishing Sectors: Part 3, Estimate of World Output. In: *Transactions of the Institute of Metal Finishing*, 90 (6), pp. 285-287.

Larson, C. (2014): Editorial. In: *Transactions of the Institute of Materials Finishing*, 92 (4), p. 177.

Larson, C. (2015): After the Recession – How is the Metal Finishing Sector Recovering? In: *Transactions of the Institute of Materials Finishing*, 93 (2), pp. 62-65.

Larson, T. J. and Brown, H. J. (1997): Designing Metrics That Fit: Rethinking Corporate Environmental Performance Measurement Systems. In: *Environmental Quality Management*, 6 (3), pp. 81-88.

Laursen, K. and Salter, A. (2006): Open for Innovation: The Role of Openness in Explaining Innovation Performance among U.K. Manufacturing Firms. In: *Strategic Management Journal*, 27 (2), pp. 131-150.

Lavie, D. (2006): The Competitive Advantage of Interconnected Firms: An Extension of the Resource-Based View. In: *Academy of Management Review*, 31 (3), pp. 638-658.

Lebas, M. J. (1995): Performance Measurement and Performance Management. In: *International Journal of Production Economics*, 41 (1), pp. 23-35.

Lee, H.; Kim, M. S. and Kim, K. K. (2014): Interorganizational Information Systems Visibility and Supply Chain Performance. In: *International Journal of Information Management*, 34 (2), pp. 285-295.

Lee, T. J. (1977): An Economic Appraisal of the Metal Finishing Industry. University of Adelaide. Bachelor Thesis.

Lightfoot, H., Baines, T. and Smart, P. (2013): The Servitization of Manufacturing: A Systematic Literature Review of Interdependent Trends. In: *International Journal of Operations & Production Management*, 33 (11/12), pp. 1408-1434.

Lincoln, Y. S. and Guba, E. G. (1985): Naturalistic Inquiry. Beverly Hills: SAGE Publications.

Lincoln, Y. S. and Guba, E. G. (2013): The Constructivist Credo. Walnut Creek: Left Coast Press Inc.

Lindell, M. K. and Whitney, D. J. (2001): Accounting for Common Method Variance in Cross-Selectional Research Designs. In: *Journal of Applied Psychology*, 86 (1), pp. 114-121.

Lugosi, P. (2006): Between Overt and Covert Research: Concealment and Disclosure in an Ethnographic Study of Commercial Hospitality. In: *Qualitative Inquiry*, 12 (3), pp. 541-561.

Lynch, R. L. and Cross, K. F. (1991): Measure Up! Yardsticks for Continuous Improvement. Cambridge, MA: Basil Blackwell Inc.

Malmi, T. and Brown, D. A. (2008): Management Control Systems as a Package - Opportunities, Challenges and Research Directions. In: *Management Accounting Research*, 19 (4), pp. 287-300.

Manfreda, K. L.; Bosnjak, M.; Berzelak, J.; Haas, I. and Vehovar, V. (2008): Web Surveys Versus Other Survey Modes: A Meta-Analysis Comparing Response Rates. In: *International Journal of Market Research*, 50 (1), pp. 79-104.

March, J. G. (1991): Exploration and exploitation in organizational learning. In: *Organization Science*, 2 (1), pp. 71-87.

Martinez-Noya, A.; Garcia-Canal, E. and Guillen, M. F. (2013): R&D Outsourcing and the Effectiveness of Intangible Investments: Is Proprietary Core Knowledge Walking out of the Door? In: *Journal of Management Studies*, 50 (1), pp. 67–91.

Matthews, A.; Artley, R. J.; Holiday P. and Stevenson, P. (1992): The UK Engineering Coatings Industry in 2005 – A Report Produced for the Department of Trade and Industry. The Research Center in Surface Engineering. University of Hull, UK.

Matthews, A.; Artley, R. J. and Holiday P. (1998): 2005 revisited - The UK surface engineering industry to 2010. An Update of the 'UK Engineering Coatings Industry in 2005' Report. National Surface Engineering Centre (NASURF), UK.

Maxwell, J. A. (2013): Qualitative Research Design – An Interactive Approach. 3rd Edition. Thousand Oaks: SAGE Publications.

Maxwell, J. A. and Chmiel, M. (2014): Notes Toward a Theory of Qualitative Data Analysis. In: Flick, U. (2014): The SAGE Handbook of Qualitative Data Analysis. Los Angeles: SAGE Publications.

Mayring, P. (2010): Qualitative Inhaltsanalyse: Grundlagen und Techniken. 11th Edition. Weinheim: Beltz.

McGahan, A. M. and Porter, M. E. (1997): How much does Industry matter, really? In: *Strategic Management Journal*, 18, Supplement 1, pp. 15-30.

Melnyk, S. A.; Umit Bititci, U.; Platts, K.; Tobias, J. and Andersen, B. (2014): Is Performance Measurement and Management fit for the Future? In: *Management Accounting Research*, 25 (2), pp. 173-186.

Merchant, K. A. and Van der Stede, W. A. (2006): Field-Based Research in Accounting: Accomplishments and Prospects. In: *Behavioral Research in Accounting*, 18 (1), pp. 117-134.

Mesquita, L. F.; Anand, J. and Brush, T. H. (2008): Comparing the resource-based and relational views: knowledge transfer and spillover in vertical alliances. In: *Strategic Management Journal*, 29 (9), pp. 913-941.

Micheli, P. and Mari, L. (2014): The Theory and Practice of Performance Measurement. In: *Management Accounting Research*, 25 (2), pp. 147-156.

Miles, M. B.; Huberman, A. M. and Saldaña, J. (2014): Qualitative Data Analysis – A Methods Sourcebook. 3rd Edition. Thousand Oaks, CA: SAGE Publications.

Mintzberg, H. (1987): Crafting Strategy. In: *Harvard Business Review*, 65 (4), pp. 66-75.

Mouritsen, J., Larsen, H. T. and Bukh, P. N. (2001): Valuing the future: intellectual capital supplements at Skandia. In: *Accounting, Auditing & Accountability Journal*, 14 (4), pp. 399-422.

Naderer, G. and Balzer, E. (2011): Qualitative Marktforschung in Theorie und Praxis Grundlagen – Methoden – Anwendungen. 2nd Edition. Wiesbaden: Gabler Verlag, Springer Fachmedien.

Nanni Jr., A. J.; Dixon, J. R. and Vollmann, T. E. (1992): Integrated Performance Measurement: Management Accounting to Support the New Manufacturing Realities. In: *Journal of Management Accounting Research*, 4 (1), pp. 1-19.

Neely, A. (2005): The Evolution of Performance Measurement Research: Developments in the last decade and a research agenda for the next. In: *International Journal of Operations & Production Management*, 25(12), pp. 1264–1277.

Neely, A. (2011): Business Performance Measurement. Unifying Theory and Integrating Practice. 2nd Edition. *Cambridge: Cambridge University Press.*

Neely, A.; Adams, C. and Crowe, P. (2001): The Performance Prism in Practice. In: *Measuring Business Excellence*, 5 (2), pp. 6-13.

Neely, A. and Bourne, M. (2000): Why Measurement Initiatives Fail. In: *Measuring Business Excellence*, 4 (4), pp. 3-6.

Neely, A.; Gregory, M. and Platts, K. (2005): Performance Measurement System Design - A Literature Review and Research Agenda. In: *International Journal of Operations & Production Management*, 25 (12), pp. 1228-1263.

Neely, A.; Mills, J.; Platts, K.; Gregory, M. and Richards, H. (1994): Realizing Strategy through Measurement. In: *International Journal of Operations & Production Management*, 14 (3), pp. 140-152.

Neely, A.; Mills, J.; Platts, K.; Richards, H.; Gregory, M.; Bourne, M. and Kennerley, M. (2000): Performance Measurement Systems Design: Developing and Testing a Process-Based Approach. In: *International Journal of Operations & Production Management*, 20 (10), pp. 1119-1145.

Nixon, W. A. J. and Burns, J. (2005): Management Control in the 21st Century. In: *Management Accounting Research*, 16 (3), pp. 260-268.

North, D. C. (1991): Institutions. In: *Journal of Economic Perspectives*, 5 (1), pp. 97-112.

Nudurupati, S. S.; Bititci, U. S.; Kumar, V. and Chan, F. T. S. (2011): State of the Art Literature Review on Performance Measurement. In: *Computers & Industrial Engineering*, 60 (2), pp. 279–290.

Otley, D. (1999): Performance Management: A Framework for Management Control Systems Research. In: *Management Accounting Research*, 10 (4), pp. 363-382.

Otley, D. (2011): Accounting Performance Measurement: A Review of its Purpose and Practices. In: Neely, A. (2011): Business Performance Measurement. Unifying Theory and Integrating Practice. 2nd Edition. Cambridge: Cambridge University Press.

Otley, D. T.; Broadbent, J. M. and Berry, A. J. (1995): Research in Management Control: An Overview of its Development. In: *British Journal of Management*, 6, pp. S31-S34.

Pagell, M. and Krause, D. R. (2004): Re-Exploring the Relationship between Flexibility and the External Environment. In: *Journal of Operations Management*, 21 (6), pp. 629-649.

Patton, M. Q. (2002): Qualitative Research and Evaluation Methods. 3rd Edition. Thousand Oaks, CA: SAGE Publications.

Penrose, E. (1955): Research on the Business Firm Limits to the Growth and Size of Firms. In: *American Economic Review*, 45 (2), pp. 531-544.

Peteraf, M. A. and Barney, J. B. (2003): Unraveling the Resource-Based Tangle. In: *Managerial and Decision Economics*, 24 (4), pp. 309-324.

Petrash, G. (1996): Dow's Journey to a Knowledge Value Management Culture. In: *European Management Journal*, 14 (8), pp. 365-373.

Pettigrew, A. M. (1990): Longitudinal Field Research on Change: Theory and Practice. In: *Organization Science*, 1 (3), pp. 267-292.

Pettigrew, A. M. (1992): The Character and Significance of Strategy Process Research. In: *Strategic Management Journal*, 13 (Winter Special Issue), pp. 5-16.

Pezet, A. (2009): The History of the French Tableau de Bord (1885-1975): Evidence from the Archives. In: *Accounting Business and Financial History*, 19 (2), pp.103-125.

Porter, M. E. (1979): How competitive forces shape strategy. In: *Harvard Business Review*, 57 (2), pp. 137-145.

Porter, M. E. (1980): Industry Structure and Competitive Strategy: Keys to Profitability. In: *Financial Analysts Journal*, 36 (4), pp. 30-41.

Porter, M. E. (1981): The contributions of industrial organizations to strategic management. In: *Academy of Management Review*, 6 (4), pp. 609-620.

Porter, M. E. (2008): The Five Competitive Forces That Shape Strategy. In: *Harvard Business Review,* 86 (1), pp. 78-93.

Porter, M. E. and Kramer, M. R. (2011): Creating Shared Value. In: *Harvard Business Review*, 89 (1/2), pp. 62-77.

Porter Liebeskind, J. (1996): Knowledge, Strategy, and the Theory of the Firm. In: *Strategic Management Journal*, 17 (Winter Special Issue), pp. 93-107.

Powell, T. C. (1995): Total Quality Management as Competitive Advantage: A Review and Empirical Study. In: *Strategic Management Journal*, 16 (1), pp. 15-37.

Pratt, M. G. (2000): The Good, the Bad, and the Ambivalent: Managing Identification among Amway Distributors. In: *Administrative Science Quarterly*, 45 (3), pp. 456-493.

Raisch, S. and Birkinshaw, J. (2008): Organizational Ambidexterity: Antecedents, Outcomes, and Moderators. In: *Journal of Management*, 34 (3), pp. 375-409.

REACH (2006): Regulation concerning the Registration, Evaluation, Authorisation and Restriction of Chemicals (REACH), establishing a European Chemicals Agency. Regulation (EC) No. 1907/2006 of the European Parliament and of the Council. Official Journal of the European Union. L 396.

Riemer, F. J. (2012): Ethnographic Research. In: Lapan, S. D.; Quartaroli, M. T. and Riemer, F. J. (2012): Qualitative Research. An Introduction to Methods and Designs. 1. Edition. San Francisco: Jossey-Bass.

Riempp, G. (2004): Integrierte Wissensmanagementsysteme – Architektur und praktische Anwendung. Berlin, Heidelberg: Springer.

Roulston, K. (2014): Analysing Interviews. In: Flick, U. (2014): The SAGE Handbook of Qualitative Data Analysis. Los Angeles: SAGE Publications.

Saldaña, J. (2010): The Coding Manual for Qualitative Researchers. 1st Edition. Thousand Oaks: SAGE Publications.

Sarkis, J. (2006): The Adoption of Environmental and Risk Management Practices: Relationships to Environmental Performance. In: *Annals of Operations Research*, 145 (1), pp. 367-381.

Schaffer, B. S. and Riordan, C. M. (2003): A Review of Cross-Cultural Methodologies for Organizational Research: A Best-Practices Approach. In: *Organizational Research Methods*, 6 (2), pp. 169-215.

Schmalensee, R. (1985): Do Markets Differ Much? In: *American Economic Review*, 75 (3), pp. 341-351.

Schreier, M. (2014a): Varianten qualitativer Inhaltsanalyse: Ein Wegweiser im Dickicht der Begrifflichkeiten. In: *Forum: Qualitative Social Research*, 15 (1).

Schreier, M. (2014b): Qualitative Content Analysis. In: Flick, U. (2014): The SAGE Handbook of Qualitative Data Analysis. Los Angeles: SAGE Publications.

Schwaiger, M. and Meyer, A. (2009): Theorien und Methoden der Betriebswirtschaft. Handbuch für Wissenschaftler und Studierende. München: Verlag Franz Vahlen.

SCOR (2010): Supply Chain Operations Reference (SCOR®) Model Overview – Version 10.0. White Paper. Supply Chain Council. Cypress, USA.

Scott, C.; Lundgren, H. and Thompson, P. (2011): Guide to Supply Chain Management. Berlin: Springer Verlag.

Selviaridis, K. and Norrman, A. (2014): Performance-Based Contracting in Service Supply Chains: A Service Provider Risk Perspective. In: *Supply Chain Management: An International Journal*, 19 (2), pp. 153-172.

Silsand, L. and Ellingsen, G. (2014): Generification by Translation: Designing Generic Systems in Context of the Local. In: *Journal of the Association for Information Systems*, 15, Special Issue, pp. 177-196.

Simons, R. (1994): How new Top Managers use Control Systems as Levers of Strategic Renewal. In: *Strategic Management Journal*, 15 (3), pp. 169-189.

Simons, R. (1995): Levers of Control. How Managers use Innovative Control Systems to Drive Strategic Renewal. Boston: Harvard Business School Press.

Simons, R. (2005): Levers of Organization Design: How Managers Use Accountability Systems for Greater Performance and Commitment. Boston: Harvard Business School Press.

Simons, R. (2014): Performance Measurement and Control Systems for Implementing Strategy. 1st New International Edition. Essex: Pearson Education Limited.

Simsek, Z.; Heavey, C.; Veiga, J. F. and Souder, D. (2009): A Typology for Aligning Organizational Ambidexterity's Conceptualizations, Antecedents, and Outcomes. In: *Journal of Management Studies*, 46 (5), pp. 864-894.

Singh, S. (2013): Supply Chain Forecasting & Planning: Move On from Microsoft Excel? In: *Foresight: The International Journal of Applied Forecasting*, 31, pp. 6-13.

Speckbacher, G.; Bischof, J. and Pfeiffer, T. (2003): A Descriptive Analysis on the Implementation of Balanced Scorecards in German-speaking Countries. In: *Management Accounting Research*, 14 (4), pp. 361-388.

Starkey, K. and Madan, P. (2001): Bridging the Relevance Gap: Aligning Stakeholders in the Future of Management Research. In: *British Journal of Management*, 12, Special Issue, pp. S3-S26.

Sterman, J. D. (2000): Business Dynamics. Systems Thinking and Modeling for a Complex World. Boston: Irwin/McGraw-Hill.

Stölzle, W. and Hofmann, E. (2007): Process-based View of Strategy (PBV): Impulse für die Konzeption eines strategischen Logistikmanagements. In: Hausladen, Iris (2007): Management am Puls der Zeit - Strategien, Konzepte und Methoden. TCW Verlag, München, pp. 1047-1070.

Strauß, E. and Zecher, C. (2013): Management Control Systems: A Review. In: *Journal of Management Control*, 23 (4), pp. 233-268.

Student (1908): The Probable Error of a Mean. In: *Biometrika*, 6 (1), pp. 1-25.

Szmigin, I. T. D. (1993): Managing Quality in Business-to-business Services. In: *European Journal of Marketing*, 27 (1), pp. 2–21.

Tajfel, H. (1982): Social Psychology of Intergroup Relations. In: *Annual Review of Psychology*, 33 (1), pp. 1-39.

Tajfel, H. and Turner, J. C. (1986): The Social Identity Theory of Intergroup Behavior. In: S. Worchel & W. G. Austin: Psychology of Intergroup Relations. 2nd Edition. Pp. 7-24. Chicago: Nelson-Hall

Teece, D. J.; Pisano, G. and Shuen A. (1997): Dynamic Capabilities and Strategic Management. In: *Strategic Management Journal*, 18 (7), pp. 509-533.

Tessier, S. and Otley, D. (2012): A Conceptual Development of Simons' Levers of Control Framework. In: *Management Accounting Research*, 23 (3), pp. 171-185.

Thiery, L. and Wojczykowski, K. (2013): Impact of REACH Regulation on the Global Finishing Market. In: *Products Finishing,* 77 (5), p. 12.

Thornberg, R. and Charmaz, K. (2014): Grounded Theory and Theoretical Coding. In: Flick, U. (2014): The SAGE Handbook of Qualitative Data Analysis. Los Angeles: SAGE Publications.

Tranfield, D. and Starkey, K. (1998): The Nature, Social Organization and Promotion of Management Research: Towards Policy. In: *British Journal of Management*, 9 (4), pp. 341-353.

Trompenaars, F. and Hampden-Turner, C. (2012): Riding the Waves of Culture: Understanding Diversity in Global Business. 3rd Edition. Boston: Nicholas Brealey Publishing.

Trumpp, C.; Endrikat, J.; Zopf, C. and Guenther, E. (2015): Definition, Conceptualization, and Measurement of Corporate Environmental Performance: A Critical Examination of a Multidimensional Construct. In: *Journal of Business Ethics*, 126 (2), pp. 185-204.

UN (2008): International Standard Industrial Classification of All Economic Activities, Revision 4. United Nations. Department of Economic and Social Affairs, Statistical papers Series M No. 4/Rev.4.

Urbach, N.; Smolnik, S. and Riempp, G. (2011): Determining the Improvement Potentials of Employee Portals using a Performance-Based Analysis. In: *Business Process Management Journal*, 17 (5), pp. 829-845.

Van De Ven, A. H. (1992): Suggestions for Studying Strategy Process: A Research Note. In: *Strategic Management Journal*, 13, pp. 169-188.

Van Voorhis, K. R. (1981): The DuPont Model Revisited: A Simplified Application to Small Business. In: *Journal of Small Business Management*, 19 (2), pp. 45-51.

Walker, K. and Wan, F. (2012): The Harm of Symbolic Actions and Green-Washing: Corporate Actions and Communications on Environmental Performance and Their Financial Implications. In: *Journal of Business Ethics*, 109 (2), pp. 227-242.

Watkins, J. P. (1998): Towards a Reconsideration of Social Evolution: Symbiosis and its Implications for Economics. In: Journal of Economic Issues, 31 (1), pp. 87-105.

Watts, T. and McNair-Connolly, C. J. (2012): New performance measurement and management control systems. In: *Journal of Applied Accounting Research*, 13 (3), pp. 226-241.

Webb, A. R. (2004): Managers' Commitment to the Goals Contained in a Strategic Performance Measurement System. In: *Contemporary Accounting Research*, 21 (4), pp. 925-958.

Widener, S. K. (2007): An Empirical Analysis of the Levers of Control Framework. In: *Accounting, Organizations and Society*, 32 (7-8), pp. 757–788.

Williamson, O. E. (2000): The New Institutional Economics: Taking Stock, Looking Ahead. In: *Journal of Economic Literature*, 38 (3), pp. 595-614.

Wongrassamee, S.; Gardiner, P. D. and Simmons, J. E. L. (2003): Performance Measurement Tools: the Balanced Scorecard and the EFQM Excellence Model. In: *Measuring Business Excellence*, 7 (1), pp. 14-29.

Ybema, S., Yanow, D., Wels, H. and Kamsteeg, F. (2009): Organizational Ethnography. Studying the Complexities of Everyday Life. London: SAGE Publications.

Yin, R. K. (2009): Case Study Research: Design and Methods. Applied social research methods series, Vol. 5. 4th Edition. Los Angeles: SAGE Publications.

Zhu, X. and Zolkiewski, J. (2015): Exploring Service Failure in a Business-to-Business Context. In: Journal of Services Marketing, 29 (5).

References – Internet

ASIG (2015): German Occupational Safety Act – Arbeitssicherheitsgesetz (ASiG). Bundesministerium der Justiz und für Verbraucherschutz, Germany. Retrieved June 28th, 2015, from
http://www.gesetze-im-internet.de/bundesrecht/asig/gesamt.pdf

BDSG (2015): German Federal Data Protection Act – Bundesdatenschutzgesetz (BDSG). Bundesministerium der Justiz und für Verbraucherschutz, Germany. Retrieved April 12th, 2015, from
http://www.gesetze-im-internet.de/englisch_bdsg/index.html

DESTATIS (2015a): GENESIS Online Database. Employment Size Classes. Table 42271-0005. Retrieved March 12th, 2015, from
https://www-genesis.destatis.de/genesis/online/data;jsessionid=AF155B7CA150F92CA8D2B1AEF22DF787.tomcat_GO_1_1?operation=abruftabelleAbrufen&selectionname=42271-0005&levelindex=1&levelid=1426164615056&index=1

DESTATIS (2015b): German Statistical Business Register. Methodologies. Retrieved March 12th, 2015, from
http://www.statistikportal.de/statistik-portal/de_entMethDef.asp

EFDS (2014): Member list of the European Society of Thin Films EFDS. Retrieved August 15th, 2014, from
http://www.efds.org/mitgliederueberblick.html?order_by=sortname

EUROSTAT (2015a): Structural business statistics (SBS). Overview. Eurostat. European Commission. Retrieved March 11th, 2015, from
http://ec.europa.eu/eurostat/web/structural-business-statistics/overview

EUROSTAT (2015b): Structural business statistics (SBS). Reference Metadata in Euro SDMX Metadata Structure (ESMS). Eurostat. European Commission. Retrieved March 12th, 2015, from
http://ec.europa.eu/eurostat/cache/metadata/en/sbs_esms.htm

EUROSTAT (2015c): Structural Business Statistics (SBS). National Methodologies. RAMON - Reference And Management Of Nomenclatures. Eurostat. European Commission. Retrieved March 12th, 2015, from
http://ec.europa.eu/eurostat/ramon/nomenclatures/index.cfm?TargetUrl=LST_NOM_DTL&StrNom=NAT_SBS&StrLanguageCode=EN&IntPcKey=25915925&StrLayoutCode=HIERARCHIC

EUROSTAT (2015d): Structural business statistics (SBS). Data Explorer. Eurostat. European Commission. Retrieved March 12th, 2015, from
http://appsso.eurostat.ec.europa.eu/nui/show.do?dataset=sbs_na_ind_r2&lang=en

F4 (2015): f4 Transcription Support Software. Retrieved April 19th, 2015, from
https://www.audiotranskription.de/english/f4.htm

G20 (2013): Gruppe der Zwanzig. Die Bundesregierung der Bundesrepublik Deutschland. Retrieved September 19th, 2013, from

http://www.bundesregierung.de/Content/DE/StatischeSeiten/Breg/G8G20/G20-uebersicht.html

Google (2015): Google Scholar Alert (German Version). Retrieved March 20th, 2015, from **http://scholar.google.de/scholar_alerts?view_op=create_alert_options**

IMF (2013): World Economic Outlook Database, April 2013. International Monetary Fund - Data and Statistics. Retrieved September 19th, 2013, from **http://www.imf.org/external/pubs/ft/weo/2013/01/weodata/weoselgr.aspx**

IMF (2014): Editorial – Transactions of the IMF, 92 (4), pp. 177-177. Retrieved May 17th, 2015, from **http://www.maneyonline.com/doi/full/10.1179/0020296714Z.000000000193**

IOM3 (2000): Foresight in Surface Engineering. Surface Engineering Committee of The Institute of Materials, Minerals and Mining. Retrieved February 14th, 2015 from **www.iom3.org/fileproxy/2144**

ISO (2015a): ISO 9001:2008 online platform. International Organization for Standardization. Retrieved April 4th, 2015 from **https://www.iso.org/obp/ui/#iso:std:iso:9001:ed-4:v2:en**

ISO (2015b): ISO 14000 online platform. International Organization for Standardization. Retrieved June 26th, 2015 from **https://www.iso.org/obp/ui/#iso:pub:PUB200002:en**

MFI CEM (2015): Qualitatively established MFI Cause and Effect Model. BPM in the MFI Research Project. File size: 0.3 MB. EBS-SIIE, Mike Dietrich, Oestrich-Winkel. Retrieved May 3rd, 2015 from **http://home.arcor.de/mikedietrich79/BPM-MFI/20140528_BPM-MFI_QDA-CEM_EBS-SIIE_Mike-Dietrich.pdf**

OANDA (2015): Foreign Exchange Rates – Currency Converter. Retrieved May 25th, 2015 from **http://www.oanda.com/currency/converter/**

SAP (2015): SAP Business Warehouse product website. Retrieved April 4th, 2015 from **http://www.sap.com/pc/tech/data-warehousing/software/netweaver-business-warehouse/index.html**

SEA (2014): Member list of the Surface Engineering Association SEA. Retrieved August 15th, 2014, from **http://www.sea.org.uk/supplier-search/index.htm**

UNSD (2013): National Classifications. United Nations Statistics Division. United Nations Department of Economic and Social Affairs. Retrieved September 19th, 2013, from **http://unstats.un.org/unsd/cr/ctryreg/default.asp?Lg=1**

USCB (2014a): Microsoft Excel table with a translation of NAICS 2012 into ISIC rev. 4 classes. United States Census Bureau. Retrieved January18th, 2014, from **http://www.census.gov/eos/www/naics/concordances/2012_NAICS_to_ISIC_4.xls**

USCB (2014b): PDF with NAICS 2012 definition file. United States Census Bureau. Retrieved January18th, 2014, from

http://www.census.gov/eos/www/naics/2012NAICS/2012_Definition_File.pdf

USCB (2015a): Annual Survey of Manufactures. Overview. United States Census Bureau. Retrieved March, 13th, 2015, from **http://factfinder.census.gov/faces/affhelp/jsf/pages/metadata.xhtml?lang=en&type=program&id=program.en.ASM#**

USCB (2015b): Annual Survey of Manufactures. Methodology. United States Census Bureau. Retrieved March, 13th, 2015, from **http://factfinder.census.gov/faces/affhelp/jsf/pages/metadata.xhtml?lang=en&type=survey&id=survey.en.ASM_ASM#**

USCB (2015c): County Business Patterns. Overview. United States Census Bureau. Retrieved March, 13th, 2015, from **http://www.census.gov/econ/cbp/index.html**

USCB (2015d): County Business Patterns. Methodology. United States Census Bureau. Retrieved March, 13th, 2015, from **http://www.census.gov/econ/cbp/methodology.htm**

USCB (2015e): Fact Finder – Data Explorer. United States Census Bureau. Retrieved March, 13th, 2015, from **http://factfinder.census.gov/faces/nav/jsf/pages/searchresults.xhtml?refresh=t**

VDMA (2014): Industrial Sector Guide, German Engineering Federation – Surface Technology VDMA. Retrieved August 15th, 2014, from **http://ot.vdma.org/article/-/articleview/1422461**

VwVws (2005): Verwaltungsvorschrift wassergefährdende Stoffe 2005, Umwelt Bundesamt. Retrieved April 4th, 2015, from **http://www.umweltbundesamt.de/dokument/verwaltungsvorschrift-wassergefaehrdende-stoffe-0**

ZVO (2014): Member List of the Zentralverband Oberflächentechnik ZVO. Retrieved August 15th, 2014, from **http://www.zvo.org/no_cache/einkaufsfuehrer-online.html**

References – Applied Software

dr. dresing & pehl GmbH - **f4** – Release: 5.20.0:
Transcription support software.

IBM – **SPSS Statistics Version 20** – Release 20.0.0:
Statistics software for the quantitative analysis of the MFI survey.

Microsoft – **Excel 2010** – Release 14.0.6112.5000:
Spreadsheet analysis software for MFI data analysis, table and figure creation.

Microsoft – **PowerPoint 2010** – Release 14.0.6112.5000:
Presentation software applied in the group discussion.

Microsoft – **Visio Premium 2010** – Release 14.0.6112.5000:
Diagramming and vector graphics software for the generation of the MFI cause and effect model.

Microsoft – **Word 2010** – Release 14.0.6112.5000:
Writing software for dissertation, interview guide, etc.

Questback – **EFS** – Release 10.4:
EFS Survey & EFS Reporting+ - Survey software.

VERBI gmbh – **MAXQDA 11 Plus** – Release: 11.0.9b:
CAQDAS software.

Appendices

A.1 Classification of the MFI within the G20

MFI Classification of the G20 Countries and the UN

G20 Country Data		G20 MFI Classification of Economic Activities										
G20 Countries + UN [1)]	IMF WEO NGDPD 2012 [Billion US-$] [2)]	Classification System	Class. System Abbrev.	Current Release	Section	Division	Group / Class	Group / Class Description	Class / Subclass	Class / Subclass Description	Data Source
United States	15,685	North American Industry Classification System US	NAICS US	2012	31-33	332	33281	Coating, engraving, heat treating, and allied activities	**332811**	**Metal heat treating**	MFI-Class-01
									332812	**Metal coating, engraving (except jewelry and silverware), and allied services to manufacturers**	
									332813	**Electroplating, plating, polishing, anodizing, and coloring**	
China	8,227	Industrial Classification for National Economic Activities	ICNEA	rev. 9, 2011	C	25	254	Metalworking activities	2541	Forging of metals	MFI-Class-02
									2542	**Powder metallurgy**	
									2543	**Heat treatment of metals**	
									2544	**Treatment of metal surface**	
									2549	Other metalworking activities	
Japan	5,964	Japan Standard Industrial Classification	JSIC	rev. 12, 2007	E	24	246	Metal coating, engraving and heat treating, except enameled ironware	**2461**	**Coating metal products**	MFI-Class-03
									2462	**Galvanized and other hot-dip coated metal products, except coated steel**	
									2463	**Engraving on metal**	
									2464	**Electroplated metal products, except coated steel**	
									2465	**Heat treated metal**	
									2469	**Miscellaneous treatment of metal surface**	
Germany	3,401	Classification of Economic Activities	WZ	2008	C	25	25.6	Treatment and coating of metals; machining	**25.61**	**Treatment and coating of metals**	MFI-Class-04
									25.62	Machining	
France	2,609	French Industry Classification	NAF	rev. 2	C	25	25.6	Treatment and coating of metals; machining	**25.61Z**	**Treatment and coating of metals**	MFI-Class-05
									25.62A	Cutting-off	
									25.62B	Industrial mechanical engineering	
United Kingdom	2,441	UK Standard Industrial Classification of Economic Activities	UKSIC	2007	C	25	25.6	Treatment and coating of metals; machining	**25.61**	**Treatment and coating of metals**	MFI-Class-06
									25.62	Machining	
Brazil	2,396	National Economic Activity Classification	CNAE	V2.1	C	25	2539-0	Services machining, welding, metal treatment and coating	2539-0/01	Services machining, turning and welding	MFI-Class-07
									2539-0/02	**Services treatment and coating metals**	

MFI Classification of the G20 Countries and the UN

G20 Country Data		G20 MFI Classification of Economic Activities									
G20 Countries + UN [1)]	IMF WEO NGDPD 2012 [Billion US-$] [2)]	Classification System	Class. System Abbrev.	Current Release	Section	Division	Group / Class	Group / Class Description	Class / Subclass	Class / Subclass Description	Data Source
Russia	2,022	Russian Classification of Economic Activities	NACE	2007	D	28	28.5	Treatment and coating of metals; processing hardware using general mechanical engineering	**28.51**	**Treatment and coating of metals**	MFI-Class-08
									28.52	Processing of metal products using the basic technological processes of mechanical engineering	
Italy	2,014	Classification of Economic Activity	ATECO	2007	C	25	25.6	Treatment and coating of metals; general mechanical engineering	**25.61**	**Treatment and coating of metals**	MFI-Class-09
									25.62	General mechanical engineering	
India	1,825	National Industrial Classification	NIC	2008	---	25	2592	Machining; treatment and coating of metals	**25920**	**Machining; treatment and coating of metals**	MFI-Class-10
Canada	1,819	North American Industry Classification System Canada	NAICS Canada	2012	31-33	332	33281	Coating, Engraving, Heat Treating, and Allied Activities	**332810**	**Coating, engraving, cold and heat treating and allied activities**	MFI-Class-11
Australia	1,542	Australian and New Zealand Standard Industrial Classification	ANZIC	rev. 2, 2006	C	C22	C229	Other fabricated metal product manufacturing	**C2293**	**Metal coating and finishing**	MFI-Class-12
Mexico	1,177	North American Industry Classification System México	NAICS / SCIAN México	2007	31-33	332	33281	Metal coating and finishings	**33281**	**Metal coating and finishings**	MFI-Class-13
South Korea	1,156	Korean Standard Industrial Classification	KSIC	rev. 9, 2008	C	25	2592	Treatment and coating of metals	**25921**	**Heat treatment of metals**	MFI-Class-14
									25922	**Plating and anodizing of metals**	
									25923	**Coating and similar treatment of metals**	
									25924	Engraving, cutting and similar processing of metals or other materials	
									25929	Other treatment and coating of metals n. e. c.	
Indonesia	878	Indonesian Standard Classification of Economic Activities	KBLI	2005	C	25	2592	Service industry working for various special metals and articles of metal	**25920**	**Service industry working for various special metals and articles of metal**	MFI-Class-15
Turkey	794	National Classification of Economics Activities Turkey	NACE	rev.2	C	25	25.6	Treatment and coating of metals; machining	**25.61**	**Treatment and coating of metals**	MFI-Class-16
									25.62	Machining	

MFI Classification of the G20 Countries and the UN

G20 Country Data		G20 MFI Classification of Economic Activities									
G20 Countries + UN [1)]	IMF WEO NGDPD 2012 [Billion US-$] [2)]	Classification System	Class. System Abbrev.	Current Release	Section	Division	Group / Class	Group / Class Description	Class / Subclass	Class / Subclass Description	Data Source
Saudi Arabia	727	International Standard Industrial Classification of All Economic Activities	ISIC	rev. 4	C	25	259	Manufacture of other fabricated metal products; metalworking service activities	**2592**	**Treatment and coating of metals; machining**	MFI-Class-17
Argentina	475	National Classification of Economics Activities 2010 of Argentina	NClaEA	2010	C	25	2592	Treatment and coating of metals and metals in general works	**25920**	**Treatment and coating of metals and metals in general works**	MFI-Class-18
South Africa	384	Standard Industrial Classification of all Economic Activities	SIC	7th Edition	C	25	2592	Treatment and coating of metals; machining	**25921**	**Treating and coating of metals**	MFI-Class-19
									25922	General mechanical engineering on a fee or contract basis	
European Union	16,584	General Industrial Classification of Economic Activities within the European Communities	NACE	rev. 2	C	25	25.6	Treatment and coating of metals; machining	**25.61**	**Treatment and coating of metals**	MFI-Class-20
									25.62	Machining	
United Nations	---	International Standard Industrial Classification of All Economic Activities	ISIC	rev. 4	C	25	259	Manufacture of other fabricated metal products; metalworking service activities	**2592**	**Treatment and coating of metals; machining**	MFI-Class-21

NGDPD* Total G20 (excl. EU/UN)	55,535	NGDPD* Total World	71,707	NGDPD G20 to World [%]	77.4%

*in [Billion US-$]

Data Sources for G20 MFI Classification of Economic Activities:

1): G20 - Gruppe der Zwanzig. Die Bundesregierung der Bundesrepublik Deutschland. Retrieved 19.09.2013, from **http://www.bundesregierung.de/Content/DE/StatischeSeiten/Breg/G8G20/G20-uebersicht.html**

MFI-Class-01: United States Census Bureau, North American Industry Classification System. Retrieved 19.09.2013, from **http://www.census.gov/eos/www/naics/**

MFI-Class-02: Directorate-General of Budget, Accounting and Statistics, R. O. C. Retrieved 19.09.2013, from **http://www.dgbas.gov.tw/public/Attachment/17514272871.xls**

MFI-Class-03: Statistics Bureau, Japan. Retrieved 19.09.2013, from **http://www.stat.go.jp/english/index/seido/sangyo/san07-3a.htm#e**

MFI-Class-04: DESTATIS, Germany. Retrieved 19.09.2013, from **https://www.destatis.de/DE/Methoden/Klassifikationen/GueterWirtschaftklassifikationen/Content75/KlassifikationWZ08.html**

MFI-Class-05: National Institute of Statistics and Economic Studies, France. Retrieved 19.09.2013, from **http://www.insee.fr/fr/methodes/default.asp?page=nomenclatures/naf2008/naf2008.htm**

MFI-Class-06: Office for National Statistics, United Kingdom. Retrieved 19.09.2013, from **http://www.ons.gov.uk/ons/guide-method/classifications/current-standard-classifications/standard-industrial-classification/index.html**

MFI-Class-07: Comissão Nacional de Classificação, Brazil. Retrieved 19.09.2013, from **http://www.cnae.ibge.gov.br/**

MFI-Class-08: Federal State Statistics Service, Russian Federation. Retrieved 19.09.2013, from **http://www.gks.ru/metod/classifiers.html**

MFI-Class-09: Italian National Institute of Statistics. Retrieved 19.09.2013, from **http://www3.istat.it/strumenti/definizioni/ateco/**

MFI-Class-10: Ministry of Statistics and Programme Implementation, India. Retrieved 19.09.2013, from **http://mospi.nic.in/Mospi_New/site/inner.aspx?status=2&menu_id=129**

MFI-Class-11: Statistics Canada, North American Industry Classification System. Retrieved 19.09.2013, from **http://www.statcan.gc.ca/concepts/industry-industrie-eng.htm**

2): IMF – International Monetary Fund (2013): World Economic Outlook Database, April 2013. International Monetary Fund - Data and Statistics. Retrieved 19.09.2013, from **http://www.imf.org/external/pubs/ft/weo/2013/01/weodata/weoselgr.aspx**

MFI-Class-12: Australian and New Zealand Standard Industrial Classification (ANZSIC) 2006 (Revision 2.0). ABS Catalogue No. 1292.0. Australian Bureau of Statistics and Statistics New Zealand, 2013.

MFI-Class-13: Instituto Nacional de Estadística y Geografía (INEGI), Mexico. Retrieved 19.09.2013, from **http://www.inegi.org.mx/sistemas/scian/default.aspx**

MFI-Class-14: Statistics Korea. Retrieved 19.09.2013, from **http://kostat.go.kr/kssc/main/MainAction.do?method=sub&catgrp=ekssc&catid1=ekssc01**

MFI-Class-15: Indonesia Investment Coordinating Board. Retrieved 19.09.2013, from **http://www3.bkpm.go.id/file_uploaded/Klasifikasi%20baku%20Lapangan%20Usaha%20Indonesia%20(KBLI)%202009-1.pdf**

MFI-Class-16: Turkish Statistical Institute. Retrieved 19.09.2013, from **http://tuikapp.tuik.gov.tr/DIESS/SozlukDetayiGetirAction.do?surumId=358&duzey=0&ustKod=yok**

MFI-Class-17: Central Department Of Statistics & Information, Kingdom of Saudi Arabia - Email contact with Mr. Mohammed Aldafas, 21.09.2013, 11:21am.

MFI-Class-18: Dirección General de Estadística y Cencsos, Argentina. Retrieved 19.09.2013, from **http://www.estadistica.chubut.gov.ar/archivos/publicaciones/CLANAE%202010.pdf**

MFI-Class-19: Standard Industrial Classification of all Economic Activities (SIC) Seventh Edition, Report No. 09-90-02, Statistics South Africa, 2012.

MFI-Class-20: NACE Rev. 2: Statistical Classification of Economic Activites in the European Community. Luxembourg: Office for Official Publications of the European Communities, 2008.

MFI-Class-21: International Standard Industrial Classification of All Economic Activities, Revision 4. United Nations. Department of Economic and Social Affairs, Statistical papers Series M No. 4/Rev.4, 2008.

A.2 Eurostat – Detailed MFI Enterprise Statistics in 2012

Last update: 05.01.2015 17:11:13
Source of data: Eurostat
NACE rev2 Code: **C-25.61** - Treatment and coating of metals
Internet source: http://appsso.eurostat.ec.europa.eu/nui/submitViewTableAction.do // Retrieved 11.03.2015

Extracted on: 11.03.15
Data Table: sbs_na_ind_r2
Focus Year: 2012 (latest data available)

Year: 2012	Turnover or gross premiums written [MEUR]	Number of enterprises	Number of persons employed	*Number of employees*	*Number of unpaid persons employed*	Number of employees in full time equivalent units	Number of persons employed per enterprise	Turnover per person employed [TEUR]	Production value [MEUR]	Total purchases of goods and services [MEUR]
Germany	8,299.1	3,250	71,918	69,920	1,998	65,152	22	115.4	8,245.8	4,709.2
Italy	4,439.8	3,662	35,607	30,802	4,805	25,659	10	124.7	4,500.6	2,831.6
France	3,284.4	1,961	24,173	23,834	339	21,562	:	:	3,238.2	1,895.0
United Kingdom	2,029.5	1,266	20,182	19,865	317	19,158	16	100.6	1,885.1	951.7
Spain	1,847.7	1,685	15,351	14,663	688	14,090	9	120.4	1,889.8	1,216.3
Poland	1,680.7	3,089	24,493	20,941	3,552	19,880	8	68.6	1,519.7	1,265.1
Netherlands	1,382.1	861	7,713	7,233	480	6,438	9	179.2	1,371.9	873.7
Belgium	818.1	451	4,517	4,142	375	3,520	10	181.1	792.8	508.0
Austria	779.9	239	5,117	4,985	132	4,760	21	152.4	772.4	477.1
Sweden	655.3	792	5,571	4,957	614	4,530	7	117.6	646.6	339.2
Denmark	440.0	286	3,924	3,779	145	2,812	14	112.1	446.3	233.8
Portugal	356.4	375	4,575	4,481	94	4,461	12	77.9	315.2	242.2
Norway	351.8	327	2,034	1,938	96	1,875	6	173.0	350.9	192.6
Slovakia	325.4	2,252	5,897	3,848	2,049	3,830	3	55.2	313.7	212.1
Finland	266.0	330	2,396	2,247	149	1,954	7	111.0	265.5	151.5
Greece	219.9	481	1,910	1,600	310	1,569	4	115.1	212.3	169.5
Hungary	209.3	408	4,362	4,225	137	4,059	11	48.0	173.1	126.9
Romania	201.0	164	3,474	3,457	17	3,426	21	57.9	201.1	153.9
Croatia	115.4	246	2,682	2,539	143	1,993	11	43.0	100.1	74.2
Slovenia	66.2	185	998	909	89	:	5	66.4	64.5	39.4
Bulgaria	53.6	195	1,841	1,718	123	1,671	9	29.1	52.0	37.9
Ireland	41.3	21	264	263	1	257	13	156.3	37.4	25.8
Lithuania	31.1	68	800	784	16	731	12	38.9	30.8	20.7
Latvia	23.7	77	493	481	12	378	6	48.1	23.3	15.4
Bosnia and Herzegovina	15.1	39	568	518	50	519	15	26.7	14.1	9.1
Former Yugoslav Republic of Macedonia	3.3	40	:	218	:	218	:	:	3.0	2.4
Cyprus	1.9	5	31	31	0	31	6	61.2	2.0	1.2

No data available for Malta, Estonia, Luxembourg, Czech Republic, Iceland, Liechtenstein, Switzerland and Turkey.

: No SBS data available.

Last update: 05.01.2015 17:11:13
Source of data: Eurostat
NACE rev2 Code: **C-25.61** - Treatment and coating of metals
Internet source: http://appsso.eurostat.ec.europa.eu/nui/submitViewTableAction.do // Retrieved 11.03.2015

Extracted on: 11.03.15
Data Table: sbs_na_ind_r2
Focus Year: 2012 (latest data available)

Year: 2012	Personnel costs [MEUR]	*Wages and Salaries [MEUR]*	*Social security costs [MEUR]*	Average personnel costs (pers. costs per emp.) [TEUR]	Share of personnel costs in turnover [%]	Share of pers. costs in total purchases of goods & services [%]	Purchases of energy products (in value) [MEUR]	Share of energy costs in total purchases of goods & services [%]	Gross investment in tangible goods [MEUR]	*Gross investment in machinery and equipment [MEUR]*	Investment rate (Gross invest / turnover) [%]
Germany	**2,619.4**	2,180.6	438.8	**37.5**	**31.6**	**55.6**	**506.3**	**11**	**427.3**	366.2	**5**
Italy	**1,123.5**	797.9	325.6	**36.5**	**25.3**	**39.7**	**15.3**	**1**	**198.1**	137.3	**4**
France	**1,039.4**	735.1	304.2	:	:	:	**103.9**	**5**	**170.4**	:	**5**
United Kingdom	**595.0**	526.9	68.2	**30.0**	**29.3**	**62.5**	**108.1**	**11**	**103.7**	70.1	**5**
Spain	**499.0**	390.0	109.0	**34.0**	**27.0**	**41.0**	**112.9**	**9**	**120.3**	96.5	**7**
Poland	**216.6**	177.7	38.9	**10.3**	**12.9**	**17.1**	:	:	**78.2**	54.3	**5**
Netherlands	**338.5**	271.9	66.6	**46.8**	**24.5**	**38.7**	**46.5**	**5**	**50.3**	42.9	**4**
Belgium	**193.4**	135.4	58.0	**46.7**	**23.6**	**38.1**	**45.8**	**9**	**37.7**	26.8	**5**
Austria	**226.8**	179.7	47.1	**45.5**	**29.1**	**47.5**	**33.7**	**7**	**53.8**	38.9	**7**
Sweden	**243.1**	174.9	68.2	**49.0**	**37.1**	**71.7**	**15.0**	**4**	**29.9**	25.5	**5**
Denmark	**171.0**	154.7	16.3	**45.3**	**38.9**	**73.1**	**17.3**	**7**	**13.8**	10.7	**3**
Portugal	**76.3**	59.2	17.1	**17.0**	**21.4**	**31.5**	**14.4**	**6**	**22.0**	11.2	**6**
Norway	**124.4**	104.5	19.9	**64.2**	**35.4**	**64.6**	**8.2**	**4**	**11.1**	7.1	**3**
Slovakia	**51.6**	39.1	12.5	**13.4**	**15.9**	**24.3**	**12.9**	**6**	**13.2**	8.6	**4**
Finland	**89.3**	69.7	19.7	**39.8**	**33.6**	**59.0**	**4.7**	**3**	**13.1**	9.9	**5**
Greece	**38.6**	30.1	8.5	**24.1**	**17.6**	**22.8**	**12.2**	**7**	**5.0**	2.9	**2**
Hungary	**44.4**	35.7	8.8	**10.5**	**21.2**	**35.0**	**9.4**	**7**	**22.8**	16.2	**11**
Romania	**23.7**	18.7	5.0	**6.9**	**11.8**	**15.4**	**7.7**	**5**	**16.4**	14.9	**8**
Croatia	**30.6**	26.1	4.5	**12.1**	**26.5**	**41.3**	**3.6**	**5**	**3.5**	1.9	**3**
Slovenia	**17.4**	15.3	2.2	**19.2**	**26.3**	**44.2**	**5.8**	**15**	**3.0**	2.3	**5**
Bulgaria	**6.9**	5.9	0.9	**4.0**	**12.9**	**18.2**	**1.9**	**5**	**4.2**	2.3	**8**
Ireland	**10.4**	9.1	1.3	**39.7**	**25.2**	**40.4**	**1.4**	**5**	**0.2**	0.0	**0**
Lithuania	**7.6**	5.9	1.8	**9.8**	**24.4**	**37.0**	**1.3**	**6**	**2.3**	2.0	**7**
Latvia	**3.4**	2.8	0.7	**7.1**	**14.3**	**22.2**	**1.4**	**9**	**1.8**	1.7	**8**
Bosnia and Herzegovina	**4.0**	3.8	0.2	**7.8**	**26.5**	**44.1**	**1.0**	**11**	**1.9**	1.3	**13**
Former Yugoslav Republic of Macedonia	**0.7**	0.6	0.2	**3.4**	**21.2**	**31.3**	:	:	:	:	:
Cyprus	**0.6**	0.5	0.1	**20.9**	**31.6**	**55.1**	**0.1**	**8**	**0.0**	0.0	**0**

No data available for Malta, Estonia, Luxembourg, Czech Republic, Iceland, Liechtenstein, Switzerland and Turkey.

: No SBS data available.

A.3 US Census – Detailed MFI Enterprise Statistics in 2013

Last update: 27.02.2015
Source of data: United States Census Bureau
NAICS 2012 Code: **33281** - Coating, Engraving, Heat Treating, and Allied Activities
Internet source: http://factfinder.census.gov/faces/nav/jsf/pages/searchresults.xhtml?refresh=t // Retrieved 13.03.2015

Extracted on: 13.03.15
Data Table: AM1331GS101
Focus Year: 2013 (latest data available)

Industry Class in Year 2013	**Total value of shipments and receipts for services - TVSS [MUSD]**	Number of employees	**Annual payroll [MUSD]**	*Production workers annual wages [MUSD]*	*Total fringe benefits [MUSD]*	*Other personnel costs [MUSD]*	Share of annual payroll in TVSS [%]	**Total cost of materials [MUSD]**	*Cost of materials, parts, containers, packaging, etc. used [MUSD]*	*Cost of resales [MUSD]*	*Cost of purchased fuels consumed [MUSD]*	*Cost of purchased electricity [MUSD]*
NAICS 31-33 - Manufacturing	**5,846,768**	11,082,237	**602,878**	340,635	186,178	76,065	**10.3**	**3,456,983**	3,135,559	177,520	37,581	51,132
NAICS 33281 - Coating, Engraving, ...	**27,133**	122,480	**5,465**	3,645	1,387	433	**20.1**	**10,771**	9,466	148	317	489
Share of 33281 in 31-33 [%]	***0.5***	*1*	***0.9***	*1.1*	*0.7*	*0.6*		***0.3***	*0.3*	*0.1*	*0.8*	*1.0*

Industry Class in Year 2013	*Cost of contract work [MUSD]*	Share of total cost of Materials in TVSS [%]	Share of electricity cost in total cost of materials [%]	**Total rental payments or lease payments [MUSD]**	*Thereof for buildings and other structures [MUSD]*	*Thereof for machinery and equipment rentals [MUSD]*	Share of total rental or lease payments in TVSS [%]	**Total other operating expenses [MUSD]**	*Temp. staff & leased employee expenses [MUSD]*	*Expensed computer hardware & other equip. [MUSD]*	*Expensed purchases of software [MUSD]*	*Data processing & oth. purchased comp. services [MUSD]*
NAICS 31-33 - Manufacturing	55,192	**59.1**	**1.5**	**35,231**	24,598	10,633	**0.6**	**372,132**	32,000	3,789	3,001	5,879
NAICS 33281 - Coating, Engraving, ...	351	**39.7**	**4.5**	**445**	345	100	**1.6**	**2,628**	370	18	11	19
Share of 33281 in 31-33 [%]	*0.6*			***1.3***	*1.4*	*0.9*		***0.7***	*1.2*	*0.5*	*0.4*	*0.3*

Industry Class in Year 2013	*Communication services [MUSD]*	*Repair & mainten. services of buildings and/or machinery [MUSD]*	*Refuse removal (including hazardous waste) services [MUSD]*	*Advertising and promotional services [MUSD]*	*Purchased professional and technical services [MUSD]*	*Taxes and license fees [MUSD]*	*All other operating expenses [MUSD]*	Share of total oth. op. exp. in TVSS [%]	Share of temp. & leased emp. exp. in total oth. op. exp. [%]	Share of refuse removal services in total oth. op. exp. [%]	Share of advertising & promotional services in total oth. op. exp. [%]
NAICS 31-33 - Manufacturing	4,781	46,218	13,736	13,093	30,112	17,813	201,710	**6.4**	**8.6**	**3.7**	**3.5**
NAICS 33281 - Coating, Engraving, ...	39	378	162	29	161	132	1,310	**9.7**	**14.1**	**6.1**	**1.1**
Share of 33281 in 31-33 [%]	*0.8*	*0.8*	*1.2*	*0.2*	*0.5*	*0.7*	*0.6*				

A.4 Literature Review within Scientific Databases

Database Research						Database Research Output				
						Overall				
Search ID	Database	Only Acad. Papers Select.	Search over "all fields" / "all content"	Search String (copy&paste as inserted in Database input field)	Date of Online Search	Time Span of Hits - From	Time Span of Hits - Till	No. of Hits	thereof relevant for BPM-MFI	Comments
LSB001	EBSCO Business Source Complete	YES	YES	"contract processing"	23.05.2013	1996	2011	**6**	0	No BPM-MFI relevant article found. Mostly articles related to tax and investments.
LSB002		YES	YES	"surface finishing" "business performance"	23.05.2013	---	---	0	0	---
LSB003		YES	YES	"finishing industry" "performance measurement"	23.05.2013	---	---	0	0	---
LSB004		YES	YES	"finishing industry" "business performance"	23.05.2013	---	---	0	0	---
LSB005		YES	YES	"finishing industry" "performance management"	23.05.2013	---	---	0	0	---
LSB006		YES	YES	"metal finishing" "performance management"	23.05.2013	---	---	0	0	---
LSB007		YES	YES	"metal finishing" "performance measurement"	23.05.2013	---	---	0	0	---
LSB008		YES	YES	"metal finishing" "performance"	23.05.2013	1998	1999	**2**	0	No BPM-MFI relevant article found. Both articles related to environment and pollution.
LSB009		YES	YES	"metal finishing" "business performance"	23.05.2013	---	---	0	0	---
LSB010		YES	YES	"metal finishing" "business"	23.05.2013	1997	2012	**2**	0	No relevant articles found. One court decision and one upcoming event.
LSB011	Google Scholar	n. a.	n. a.	"contract processing" "business performance"	09.06.2013	n. a.	n. a.	**20**	0	No BPM-MFI relevant article found. Most entries are related to IT information systems.
LSB012		n. a.	n. a.	"surface finishing" "business performance"	09.06.2013	n. a.	n. a.	**46**	0	No BPM-MFI relevant article found. Most entries are related to technical processes and environment.
LSB013		n. a.	n. a.	"finishing industry" "business performance"	09.06.2013	n. a.	n. a.	**49**	0	Most sources are focused on environmental performance and related to the US EPA Sector Strategy program 1995-2005 including one article regarding an MFI overview from 1996.
LSB014		n. a.	n. a.	"finishing industry" "performance measurement"	09.06.2013	n. a.	n. a.	**28**	0	Mainly focused on textile finishing and environment. Two articles concerning innovation performance in the textile finishing sector in Turkey.
LSB015		n. a.	n. a.	"metal finishing" "performance measurement"	09.06.2013	n. a.	n. a.	**71**	0	No BPM-MFI relevant article found. Mostly environmental driven (some reference the AMJ-EPA article by Flannery and May (2000)).
LSB016		n. a.	n. a.	"metal finishing" "performance management"	09.06.2013	n. a.	n. a.	**89**	0	No BPM-MFI relevant article found. Mostly environmental driven (some reference the AMJ-EPA article by Flannery and May (2000)).
LSB017		n. a.	n. a.	"finishing industry" "performance management"	09.06.2013	n. a.	n. a.	**59**	0	No BPM-MFI relevant article found. Mostly environmental driven (some reference the AMJ-EPA article by Flannery and May (2000)).
LSB018		n. a.	n. a.	"metal finishing" "business performance"	09.06.2013	n. a.	n. a.	**95**	0	No BPM-MFI relevant article found. Mostly environmental driven, focused on EPA and the AMJ-article by Flannery and May (2000).
LSB019		n. a.	n. a.	"finishing" "Leistungsmessung"	09.06.2013	n. a.	n. a.	**49**	0	No BPM-MFI relevant article found.
LSB020		n. a.	n. a.	"Veredlung" "Leistungsmessung"	09.06.2013	n. a.	n. a.	**25**	0	No relevant article found. Mainly "Veredlung" in the sense of improving a certain condition.
LSB021		n. a.	n. a.	"Veredelung" "Leistungsmessung"	09.06.2013	n. a.	n. a.	**84**	0	No relevant article found. Mainly "Veredlung" in the sense of improving a certain condition.
LSB022		n. a.	n. a.	"Veredelung" "performance management"	09.06.2013	n. a.	n. a.	**34**	0	No relevant article found. Mainly "Veredlung" in the sense of improving datasets or forecasts (Business Interlligence e. g.).
LSB023		n. a.	n. a.	"Veredelung" "performance measurement"	09.06.2013	n. a.	n. a.	**29**	0	No relevant article found. Mainly "Veredlung" in the sense of improving datasets or forecasts (Business Interlligence e. g.).
LSB024		n. a.	n. a.	"Veredelung" "business performance"	09.06.2013	n. a.	n. a.	**10**	0	No relevant article found. Mainly "Veredlung" in the sense of improving datasets or forecasts (Business Interlligence e. g.).
LSB025		n. a.	n. a.	"Veredlung" "business performance"	09.06.2013	n. a.	n. a.	**1**	0	Not relevant. Bachelor Thesis / Case Study "Rasselstein", Germany.
LSB026		n. a.	n. a.	"Veredlung" "performance management"	09.06.2013	n. a.	n. a.	**2**	0	No BPM-MFI relevant article found.
LSB027		n. a.	n. a.	"Veredlung" "performance measurement"	09.06.2013	n. a.	n. a.	**2**	0	No BPM-MFI relevant article found.

n. a. = not applicable

Database Research						Database Research Output – Overall				
Search ID	Database	Only Acad. Papers Select.	Search over "all fields" / "all content"	Search String (copy&paste as inserted in Database input field)	Date of Online Search	Time Span of Hits - From	Time Span of Hits - Till	No. of Hits	thereof relevant for BPM-MFI	Comments
LSB028	Elsevier - SciVerse Science Direct	n. a.	YES	"contract processing"	09.06.2013	1985	2013	36	0	No BPM-MFI relevant article found. Mainly technical, IT and process related articles.
LSB029		n. a.	YES	"surface finishing" "business performance"	09.06.2013	1995	2013	7	0	No BPM-MFI relevant article found. Mainly related to enviroment and energy saving.
LSB030		n. a.	YES	"metal finishing" "performance measurement"	09.06.2013	1992	2013	10	0	No BPM-MFI relevant article found. Articley mainly related to environmental performance.
LSB031		n. a.	YES	"metal finishing" "performance management"	09.06.2013	2001	2012	10	0	No BPM-MFI relevant article found.
LSB032		n. a.	YES	"finishing industry" "business performance"	09.06.2013	2006	2012	13	0	No BPM-MFI relevant article found. Nearly all hits focused on environmental themes.
LSB033		n. a.	YES	"metal finishing" "business performance"	09.06.2013	1996	2013	20	0	No BPM-MFI relevant article found. Hits mostly focused on environmental themes.
LSB034		n. a.	YES	"finishing industry" "performance measurement"	09.06.2013	1995	2012	3	0	No BPM-MFI relevant article found.
LSB035		n. a.	YES	"finishing industry" "performance management"	09.06.2013	2006	2012	6	0	No BPM-MFI relevant article found. Hits focused on environmental or ethical themes.
LSB036	Emerald Management Xtra	n. a.	YES	"metal finishing" "performance measurement"	09.06.2013	n. a.	n. a.	6	0	No relevant article found (EPA, "Metal Finishing"-Journal as references).
LSB037		n. a.	YES	"contract processing"	09.06.2013	n. a.	n. a.	15	0	No BPM-MFI relevant article found.
LSB038		n. a.	YES	"surface finishing" "business performance"	09.06.2013	n. a.	n. a.	1	0	No BPM-MFI relevant article found.
LSB039		n. a.	YES	"metal finishing" "performance management"	09.06.2013	n. a.	n. a.	1	0	No BPM-MFI relevant article found (one work study to an equal pay act from 1972).
LSB040		n. a.	YES	"metal finishing" "business performance"	09.06.2013	n. a.	n. a.	7	0	No BPM-MFI relevant article found.
LSB041		n. a.	YES	"finishing industry" "performance measurement"	09.06.2013	n. a.	n. a.	2	0	Articles on Turkish textile (not metal) finishing industry - same article as in LSB014.
LSB042		n. a.	YES	"finishing industry" "performance management"	09.06.2013	n. a.	n. a.	0	0	---
LSB043		n. a.	YES	"finishing industry" "business performance"	09.06.2013	n. a.	n. a.	3	0	No BPM-MFI relevant article found.
LSB044	JSTOR Business I, II, III	n. a.	n. a.	"contract processing"	09.06.2013	n. a.	n. a.	8	0	No BPM-MFI relevant article found. Hits related to public management topics (China, USA).
LSB045		n. a.	n. a.	"surface finishing" "business performance"	09.06.2013	n. a.	n. a.	2	0	No BPM-MFI relevant article found. Not related to MFI.
LSB046		n. a.	n. a.	"metal finishing" "performance measurement"	09.06.2013	n. a.	n. a.	3	0	No BPM-MFI relevant article found.
LSB047		n. a.	n. a.	"metal finishing" "performance management"	09.06.2013	n. a.	n. a.	0	0	---
LSB048		n. a.	n. a.	"metal finishing" "business performance"	09.06.2013	n. a.	n. a.	2	0	No BPM-MFI relevant article found.
LSB049		n. a.	n. a.	"finishing industry" "performance measurement"	09.06.2013	n. a.	n. a.	2	0	No BPM-MFI relevant article found.
LSB050		n. a.	n. a.	"finishing industry" "performance management"	09.06.2013	n. a.	n. a.	0	0	---
LSB051		n. a.	n. a.	"finishing industry" "business performance"	09.06.2013	n. a.	n. a.	2	0	No BPM-MFI relevant article found.
LSB052	SAGE Publications	n. a.	n. a.	"contract processing"	09.06.2013	n. a.	n. a.	5	0	No BPM-MFI relevant article found.
LSB053		n. a.	n. a.	"surface finishing" "business performance"	09.06.2013	n. a.	n. a.	1	0	No BPM-MFI relevant article found. Article refers to SixSigma methodology within the Automotive industry.
LSB054		n. a.	n. a.	"metal finishing" "performance measurement"	09.06.2013	n. a.	n. a.	1	0	No BPM-MFI relevant article found. Article references AMJ-EPA article (Flannery and May, 2000).
LSB055		n. a.	n. a.	"metal finishing" "performance management"	09.06.2013	n. a.	n. a.	2	0	No BPM-MFI relevant article found.
LSB056		n. a.	n. a.	"metal finishing" "business performance"	09.06.2013	n. a.	n. a.	0	0	---
LSB057		n. a.	n. a.	"finishing industry" "performance measurement"	09.06.2013	n. a.	n. a.	1	0	No BPM-MFI relevant article found. Article references AMJ-EPA article (Flannery and May, 2000).
LSB058		n. a.	n. a.	"finishing industry" "performance management"	09.06.2013	n. a.	n. a.	2	0	No BPM-MFI relevant article found.
LSB059		n. a.	n. a.	"finishing industry" "business performance"	09.06.2013	n. a.	n. a.	0	0	---

n. a. = not applicable

Database Research						Database Research Output – Overall				
Search ID	Database	Only Acad. Papers Select.	Search over "all fields" / "all content"	Search String (copy&paste as inserted in Database input field)	Date of Online Search	Time Span of Hits - From	Time Span of Hits - Till	No. of Hits	thereof relevant for BPM-MFI	Comments
LSB060	WAO - World Affairs online	n. a.	YES	"contract processing"	09.06.2013	n. a.	n. a.	0	0	---
LSB061		n. a.	YES	"surface finishing" "business performance"	09.06.2013	n. a.	n. a.	0	0	---
LSB062		n. a.	YES	"metal finishing" "performance measurement"	09.06.2013	n. a.	n. a.	0	0	---
LSB063		n. a.	YES	"metal finishing" "performance management"	09.06.2013	n. a.	n. a.	0	0	---
LSB064		n. a.	YES	"metal finishing" "business performance"	09.06.2013	n. a.	n. a.	0	0	---
LSB065		n. a.	YES	"finishing industry" "performance measurement"	09.06.2013	n. a.	n. a.	0	0	---
LSB066		n. a.	YES	"finishing industry" "performance management"	09.06.2013	n. a.	n. a.	0	0	---
LSB067		n. a.	YES	"finishing industry" "business performance"	09.06.2013	n. a.	n. a.	0	0	---
LSB068	WISO (German)	n. a.	YES	"contract processing"	09.06.2013	n. a.	n. a.	**11**	0	No BPM-MFI relevant article found.
LSB069		n. a.	YES	"surface finishing" "business performance"	09.06.2013	n. a.	n. a.	0	0	No BPM-MFI relevant article found.
LSB070		n. a.	YES	"Veredelung" "performance management"	09.06.2013	n. a.	n. a.	**7**	0	No BPM-MFI relevant article found.
LSB071		n. a.	YES	"Veredelung" "performance measurement"	09.06.2013	n. a.	n. a.	**5**	0	No relevant article found - Article focused on "Personalentwicklung".
LSB072		n. a.	YES	"Veredelung" "business performance"	09.06.2013	n. a.	n. a.	**5**	0	No relevant article found - Article focused on "Personalentwicklung".
LSB073		n. a.	YES	"Veredlung" "business performance"	09.06.2013	n. a.	n. a.	**1**	0	No relevant article found - Article focused on "Data Warehouse".
LSB074		n. a.	YES	"Veredlung" "performance management"	09.06.2013	n. a.	n. a.	**2**	0	No BPM-MFI relevant article found.
LSB075		n. a.	YES	"Veredlung" "performance measurement"	09.06.2013	n. a.	n. a.	**2**	0	No BPM-MFI relevant article found.
LSB076		n. a.	YES	"Veredlung" "Leistungsmessung"	09.06.2013	n. a.	n. a.	0	0	No BPM-MFI relevant article found.
LSB077		n. a.	YES	"Veredelung" "Leistungsmessung"	09.06.2013	n. a.	n. a.	**7**	0	No BPM-MFI relevant article found. Articles mainly related to Human Resources.
LSB078		n. a.	YES	"finishing" "Leistungsmessung"	09.06.2013	n. a.	n. a.	**5**	0	No BPM-MFI relevant article found.
LSB079		n. a.	YES	"metal finishing" "performance measurement"	09.06.2013	n. a.	n. a.	0	0	---
LSB080		n. a.	YES	"metal finishing" "performance management"	09.06.2013	n. a.	n. a.	0	0	---
LSB081		n. a.	YES	"metal finishing" "business performance"	09.06.2013	n. a.	n. a.	0	0	---
LSB082		n. a.	YES	"finishing industry" "performance measurement"	09.06.2013	n. a.	n. a.	**1**	0	Dictionary Entry (Davies-Labeck, 2015).
LSB083		n. a.	YES	"finishing industry" "performance management"	09.06.2013	n. a.	n. a.	0	0	---
LSB084		n. a.	YES	"finishing industry" "business performance"	09.06.2013	n. a.	n. a.	**1**	0	Dictionary Entry (Davies-Labeck, 2015).

n. a. = Not applicable.

Scientific Database	For further information please refer to:
EBSCO Business Source Complete	http://www.ebscohost.com/academic/business-source-complete
Google Scholar	http://scholar.google.com/
Elsevier - SciVerse ScienceDirect	http://www.sciencedirect.com/
Emerald Management Xtra	http://www.emeraldinsight.com/
JSTOR Business I, II, III	http://www.jstor.org/
SAGE Publications	http://www.sagepub.com/home.nav
WAO - World Affairs online	http://www.fiv-iblk.de/database/database_information.htm
WISO (German)	https://www.wiso-net.de/

A.5 Semi-Structured Expert Interviews – Interview Guide

Leitfaden Experteninterview

Teilnehmer: [redacted] (Experte)
Funktion: [redacted]

[redacted] (Interviewer)
Funktion: [redacted] | *Doktorand*
EBS Universität für Wirtschaft und Recht / EBS Business School

Datum: 05. Februar 2014

Uhrzeit: ______ - ______

Ort: [redacted]

______

[redacted]

Zur Person:

Bitte stelle Dich kurz vor: was ist Deine Funktion bei [redacted], seit wie vielen Jahren arbeitest Du in der Metallveredlungsindustrie – wie viele davon bei [redacted]?

Aus dem Forschungsprojekt abgeleitete Themenkomplexe:

1. Themenkomplex: Metallveredlungs-Branche

1.1 Aus Deiner Sicht: was sind die Branchenbesonderheiten der Metallveredlungsindustrie[1], insbesondere im Vergleich zur produzierenden Industrie wie etwa dem Maschinenbau?

[1] Englisch: "Metal Finishing Industry" - MFI

1.2 Wenn Du an die vergangenen zehn Jahren (2003-2013) denkst: wie hat sich die MFI entwickelt bzw. verändert?

2. Themenkomplex: [redacted] als Unternehmen der Metallveredlungs-Branche

2.1 Mit Blick auf die eben von Dir genannten Branchenbesonderheiten: Was macht [redacted] zu einem typischen Vertreter der MFI?

2.2 Welchen Einfluss haben [redacted]'s Konzern-Core Values[3] auf die Ausrichtung von [redacted]?:

- *„Integrity - Wir bauen auf Vertrauen*
- *Team Spirit - Mir ist das Team wichtig, und ich bin dem Team wichtig.*
- *Excellence - Alles was wir tun, tun wir hervorragend.*
- *Innovation - Ideen kommen von jedem und von überall aus der Organisation"*

- Werden die Core Values im Alltag gelebt?

[redacted]

2.3 Wo siedelst Du die Ausrichtung von [redacted] als Organisation an?:

- o als eine Organisation welche ihre bestehenden Unternehmensprozesse ständig optimiert („Exploitation“) oder
- o als Organisation welche sich ständig innovativ weiterentwickelt („Exploration“)
- o oder eine Mischung von beidem?

- Was sind die Gründe für diese Art der Ausrichtung bei [redacted]?

2.4 Nach welchen Regeln oder Vorgaben wird bei [redacted] die betriebliche Leistung von Prozessen oder Funktionsbereichen gemessen[4]?

- Wer stellt diese Regeln oder Vorgaben auf?

- Wie oft ändern sich diese Regeln und wie werden diese Änderungen kommuniziert?

[4] Englisch: „Business Performance Measurement“ - BPM

3. Themenkomplex: Business Performance Measurement in [redacted]'s Funktionsbereichen

3.1 Welche internen oder externen Gruppen, Prozesse bzw. Aspekte, haben bei der betrieblichen Leistungsmessung in Deinem Funktionsbereich die größte Relevanz?

3.2 Welche Leistungskennzahlen bzw. Key Performance Indicators (KPIs) haben bei der Leistungsmessung in Deinem Bereich die größte Bedeutung? Wie sind diese aufgebaut?

- Welche Kennzahlen nutzt Du um *zukünftige* Entwicklungen abzuleiten?

- Welche *externen Indikatoren*[5] nutzt Du als Informationsquelle zur Leistungsmessung?

3.3 Welche – bspw. IT-gestützten – Instrumente bzw. Werkzeuge benutzt Du zur betrieblichen Leistungsmessung innerhalb Deines Bereiches?

[5] Externe Indikatoren wir bspw. PM-Index oder Produktionsindex im verarb. Gewerbe, etc.

3.4 Wenn Du an die vergangenen drei Jahre (2011-2013) denkst: haben sich diese Instrumente grundlegend verändert bzw. wurden bestehende durch neue Instrumente ersetzt?

3.5 Von welchen Adressaten außerhalb Deines Funktionsbereiches ist Dir bekannt, dass sie die Ergebnisse Deiner Leistungsmessung nutzen? Für welche Zwecke?

- Von welchen anderen Funktionsbereichen nutzt Du die Ergebnisse betrieblicher Leistungsmessung? Für welche Zwecke?

3.6 Versetzte Dich in Deine Vorgesetzten: was sind die bedeutendsten strategischen Unsicherheiten bei [redacted]? Auf welche Weise beeinflussen diese strategischen Unsicherheiten die Gestaltung der Unternehmensstrategie?

- Auf welche Weise trittst Du mit Deinen Mitarbeitern in Dialog um auf diese strategischen Unsicherheiten reagieren zu können? Anhand welcher Kennzahlen legt ihr dabei Vorgehensweisen fest?

3.7 Betrachte bitte den aktuellen Status der betrieblichen Leistungsmessung bei [redacted]:

- Worin bestehen, bezogen auf das gesamte Unternehmen, die bedeutendsten Herrausforderungen oder Verbesserungspotentiale von BPM?

- W orin bestehen, bezogen auf Deinen Funktionsbereich, die bedeutendsten Herrausforderungen oder Verbesserungspotentiale bzgl. BPM?

4. Abschließende Frage:

4.1 Welches Thema oder welchen Aspekt siehst Du im Zusammenhang mit dem Forschungsprojekt „BPM in the MFI" ferner als interessant an?

VIELEN DANK F ÜR DEINE UNTERSTÜTZUNG!

© Mike Dietrich, 2013

A.6 Semi-Structured Expert Interviews – Invitation Letter

Informationsblatt Experteninterview

Wir informieren Sie hiermit über das Forschungsprojekt „Business Performance Measurement in the Metal Finishing Industry“, für welches wir Sie gerne interviewen möchten, und über unser Vorgehen. Der Datenschutz verlangt Ihre ausdrückliche und informierte Einwilligung, dass wir das Interview aufzeichnen und auswerten.

Die verantwortliche akademische Leitung des Projektes liegt bei Prof. Dr. Ronald Gleich, Executive Director des Strascheg Institute for Innovation and Entrepreneurship (SIIE) der EBS Universität für Wirtschaft und Recht. Durchgeführt wird das Forschungsprojekt von Mike Dietrich, Doctoral Candidate des SIIE.

Das Forschungsprojekt wird in Kooperation mit [redacted] ausgeführt und ist in zwei aufeinander aufbauende Forschungsgebiete geteilt, welche die folgenden Forschungsfragen behandeln:

<u>Forschungsgebiet 1</u> *<u>(wird Themenkomplex 1 im Interview entsprechen):</u>*

Metallveredlungsindustrie („Metal Finishing Industry – MFI“)

(Gewichtung im Forschungsprojekt: ~20%)

Was sind die Branchenbesonderheiten der Metallveredlungsindustrie im Vergleich zur produzierenden Industrie?

<u>Forschungsgebiet 2</u> *<u>(wird Themenkomplexen 2 und 3 im Interview entsprechen):</u>*

Leistungsmessung („Business Performance Measurement – BPM“) in Organisationen der Metallveredlungsindustrie

Gewichtung im Forschungsprojekt: ~80%

Auf welche Art und Weise messen Unternehmen der Metallveredlungsindustrie ihre betriebliche Leistung im Rahmen ihrer Management Control Systems?

Die Durchführung der Studie geschieht auf der Grundlage der Bestimmungen des Bundesdatenschutzgesetzes. Der Interviewer unterliegt der Schweigepflicht und ist auf das Datengeheimnis verpflichtet. Die Arbeit dient allein wissenschaftlichen Zwecken. In Ergänzung gilt die zwischen [redacted] und dem SIIE geschlossene Geheimhaltungsvereinbarung, die diesem Informationsblatt als Kopie beiliegt.

Wir sichern Ihnen folgendes Verfahren zu, damit Ihre Angaben nicht mit Ihrer Person in Verbindung gebracht werden können:

- Wir gehen sorgfältig mit dem Erzählten um: Wir nehmen das Gespräch digital auf; diese digitale Aufzeichnung wird abgetippt (Transkript) und anschließend gelöscht.
- Wir anonymisieren, d. h. wir verändern alle Personen-, Orts-, und sonstigen eindeutig Ihrer Person zuordenbaren Informationen. „Management-Team-Mitglied“ stellt die niedrigste Stufe dar, die wir in der Zuordnung der gewonnen Inhalte verwenden; eine Nennung der von Ihnen getätigten Aussagen in Verbindung mit Ihrer direkten Position innerhalb der Organi-

sation findet somit nicht statt. Des Weiteren wird die Organisation [redacted]" pseudonymisiert durch „Metal Finishing Company – MeFiCom"

- Ihre persönlichen Daten wie Name, Telefonnummer, E-Mail-Adresse, die im Rahmen des Forschungsprojektes erfasst wurden, werden am Ende des Projektes aus unseren Unterlagen gelöscht, so dass lediglich das anonymisierte Transkript existiert. Die von Ihnen unterschriebene Erklärung zur Einwilligung in die Auswertung wird in einem gesonderten Ordner an einer gesicherten und nur dem Doctoral Candidate zugänglichen Stelle aufbewahrt. Sie dient lediglich dazu, bei einer Überprüfung durch den Datenschutzbeauftragten nachweisen zu können, dass Sie mit der Auswertung einverstanden sind. Sie kann mit Ihrem Interview nicht mehr in Verbindung gebracht werden.

- Die Abschrift wird nicht veröffentlicht und ist nur universitätsintern für die Auswertung zugänglich. In Veröffentlichungen können einzelne Sätze des Interviews als Zitate eingeführt werden, selbstverständlich ohne dass erkennbar ist, von welcher Person sie stammen.

Die Datenschutzbestimmungen verlangen auch, dass wir Sie noch einmal ausdrücklich darauf hinweisen, dass aus einer Nichtteilnahme keine Nachteile entstehen. Sie können Antworten auch bei einzelnen Fragen verweigern. Auch die Einwilligung ist freiwillig und kann jederzeit von Ihnen widerrufen und die Löschung des Interviews von Ihnen verlangt werden.

Wir bedanken uns für Ihre Bereitschaft uns Auskunft zu geben und hoffen, dass unsere wissenschaftliche Arbeit dazu dienen wird sowohl einen theoretischen als aber auch einen praxisrelevanten Beitrag zur Forschung auf dem Gebiet der Management Control Systems mit Fokus auf Business Performance Measurement zu leisten.

Oestrich-Winkel, den 27. Januar 2014

Prof. Dr. Ronald Gleich
Executive Director SIIE

Mike Dietrich
Doctoral Candidate SIIE

EBS Universität für Wirtschaft und Recht
EBS Business School

Department of Innovation Management & Entrepreneurship (IME)
Strascheg Institute for Innovation and Entrepreneurship (SIIE)

EBS Campus Rheingau
Burgstraße 5
65375 Oestrich-Winkel

Vorlage zu diesem Informationsblatt angepasst und übernommen von:
Helfferich, Cornelia (2011): Die Qualität qualitativer Daten. 4. Aufl. Wiesbaden: VS, Verl. für Sozialwissenschaften.

A.7 Semi-Structured Expert Interviews – Informed Consent

Forschungsprojekt: Business Performance Measurement in the Metal Finishing Industry – BPM in the MFI

Einwilligungserklärung Experteninterview

Ich bin über das Vorgehen bei der Auswertung der persönlichen, „freien" Interviews mit einem Handzettel informiert worden (u. a.: die Abschrift gelangt nicht an die Öffentlichkeit, Anonymisierung bei der Abschrift, Löschung der Aufzeichnung, Aufbewahrung dieser Einwilligungserklärung nur im Zusammenhang mit dem Nachweis des Datenschutzes und nicht zusammenführbar mit dem Interview).

Ich bin damit einverstanden, dass einzelne Sätze des Interviews, die nicht mit meiner Person in Verbindung gebracht werden können, als Material für wissenschaftliche Zwecke und für die Weiterentwicklung der Forschung genutzt werden können.

Unter diesen Bedingungen erkläre ich mich bereit, das Interview zu geben und bin damit einverstanden, dass es digital aufgezeichnet, abgetippt, anonymisiert und ausgewertet wird.

Bingen, den ___. Februar 2014

____

Unterschrift

Vorname Nachname

Vorlage zu dieser Einwilligungserklärung angepasst und übernommen von:
Helfferich, Cornelia (2011): Die Qualität qualitativer Daten. 4. Aufl. Wiesbaden: VS, Verl. für Sozialwissenschaften.

EBS Universität für Wirtschaft und Recht

SIIE.

A.8 Group Discussion Presentation

SIIE. Strategic Institute for Innovation & Entrepreneurship

EBS Business School

0. Kommunikationsregeln

1) Jeder anwesende Experte ist gleichberechtigt.
2) Keine feste Zuweisung von Redezeit.
3) Es soll nacheinander und nicht parallel gesprochen werden.
4) Die Positionen und Meinungen aus den Einzelinterviews dürfen wiederholt werden, um diese mit und in der Gruppe zu diskutieren.

Ziel:
Offene Diskussion zwischen allen beteiligten Experten.

1 Mike Dietrich - 27.02.2014 - BPM in the MFI - Gruppendiskussion

SIIE. Strategic Institute for Innovation & Entrepreneurship

EBS Business School

1. Offene Diskussion zu Business Performance Measurement - Leitfragen

Was sind die Gründe für das Fehlen einer
➢ **Betrieblichen Leistungsmessung**
(Business Performance Measurement)
bei [redacted],
➢ **die für alle Funktionsbereiche**
➢ **verlässliche Daten**
➢ **als Basis für zukünftige Aktionen liefert?**

Aus welchen **Elementen** sollte eine solche **integrierte, „ganzheitliche“ Betriebliche Leistungsmessung** bestehen?

2 Mike Dietrich - 27.02.2014 - BPM in the MFI - Gruppendiskussion

A.9 CAQDAS – Document System

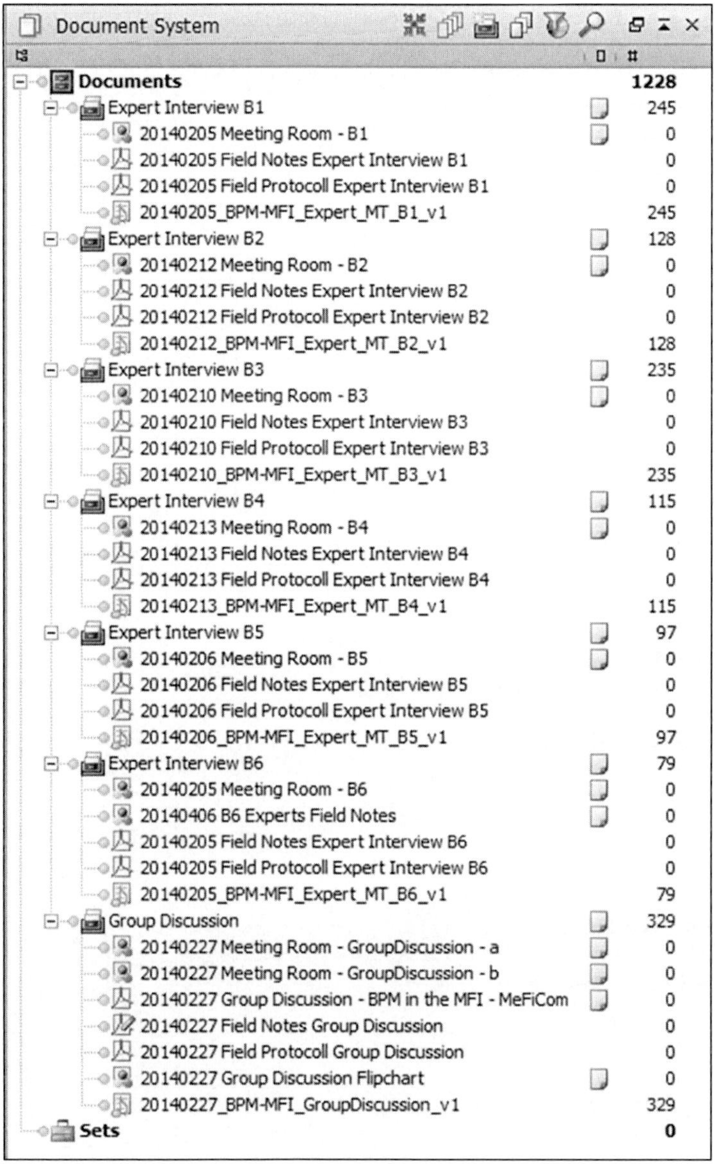

Note: The figures to the right of the MAXQDA Document System represent the number of coded transcript passages in each document (1,070 transcript passages, of which 158 are additionally coded as illustrative examples).

A.10 CAQDAS – Code System

This appendix includes MAXQDA screenshots of the complete generated hierarchical code system. Each category/code splits up into sub-categories/subcodes until the lowest level of itemization is reached. The figures to the right of the screenshots represent the number of coded segments being assigned to each code.

Metal Finishing Industry:

- Metal Finishing Industry — M 0
  - Differences To Other Industries — 0
    - Innovativeness — 6
    - Profitability — 6
    - Specialization — 5
    - Business Model — 4
    - Environment — 4
    - Customizable Services And Products — 3
    - Research And Development — 2
    - Delivery Times — 1
  - Industrial Service Character — 0
    - Time — 0
      - Fast Moving Industry — 5
      - Delivery Times — 1
    - Service Orientation — 5
    - Flexibility — 3
    - Quality — 2
    - Customer Accessibility And Proximity — 1
  - Environment — 0
    - Positive - Resource Conservation — 3
    - Negative - Energy Consumption — 2
    - Negative - Hazardous Consumables — 1
  - SME — 6
  - Structural Change In Industry — 0
    - Insourcing — 7
    - Servitization — 6
    - Globalization — 2
    - Increasing Competition — 2
    - Environment — 1
  - Cost Drivers — 1
    - Fixed Costs — 2
      - Personnel Cost — 3

Metal Finishing Company – Belief Systems:

- Metal Finishing Company — M — 0
  - Belief Systems — 0
    - Corporate Strategy — 21
      - Exploration — 5
        - Innovation — 8
          - Product Innovation — 13
          - Service Innovation — 7
        - Inorganic Growth - Acquisitions — 4
      - Exploitation — 3
        - Market Share — 9
          - Mature Market — 11
          - Emerging Market — 4
          - Competition — 3
      - Organizational Ambidexterity — 11
      - Institutional Change — 12
    - Core Values — 7
      - Real Life Practice — 11
      - Team Spirit — 5
      - Excellence — 5
      - Innovation — 4
      - Integrity — 4

Metal Finishing Company – Boundary Systems:

- Boundary Systems — 2
  - Business Model — 13
    - Decentralized Processing — 8
      - Customer Accessibility And Proximity — 8
    - Holistic Service Provider — 13
    - System Manufacturer — 5
    - International Network — 4
    - Business Model Innovation — 9
  - Legislation — 3
    - Environment — 9
      - REACH-Agreement — 3
    - Labor — 2
    - Finance — 2
  - Audits — 4
  - Contracting — 3
  - Challenge - Market Fluctuation — 2

Metal Finishing Company – Diagnostic Control Systems - I:

- Diagnostic Control Systems — 0
  - Set Standards — 2
    - Standards And Rules — 19
    - Decision And Policy Makers — 8
  - Business Performance Measurement - Measure Output — 4
    - Internal Controls — 4
      - Critical Performance Variables - Real KPI — 6
        - Employed KPIs — 0
          - Sales — 2
          - Production — 0
            - Prod - Coating Meter — 1
            - Prod - GSR Per Manhour — 3
            - Prod - Utilization Rate — 1
          - Management — 0
            - Management - LTA — 4
            - Management - Daily Sales vs. Oder Backlog — 3
          - Finance — 0
            - Fin - Pers. Exp. To GSR Ratio — 1
            - Fin - Days Sales Outstanding — 1
          - HR — 0
            - HR - Employee Opinion Survey — 1
            - HR - Time To Fill — 2
      - Business Measures - Pseudo-KPI — 0
        - Employed Business Measures — 1
          - Gross Sales Revenue - GSR — 23
          - Cross Profit — 23
            - Watchlist — 7
          - Operating Result — 2
          - EBIT — 12
          - Order Backlog — 4
          - HR Measures — 4
          - Customer Complaints Count — 1
      - Cost Controls — 4
        - Pricing — 13
        - Cost Drivers — 4
          - Fixed Costs — 7
            - Personnel Cost — 11
            - Depreciation — 3
          - Variable Costs — 1
            - Energy Cost — 4
            - Labor Leasing — 5
            - Research And Development — 2
            - Logistics — 1

Metal Finishing Company – Diagnostic Control Systems - II:

- Challenges 0
  - Future Focus 11
    - Budget Processes 4
    - Forecast Processes 15
  - KPI-Engineering 6
    - KPI Evolution And Maturing 5
      - Sustainability 9
      - Interdependency 2
  - Standardization And Harmonization 2
    - Global Comparability 20
  - Order Backlog 3
- External Indicators 5
  - Purchase Manager Index 3
  - GDP 3
  - VDMA KPIs 1
  - IFO Index 1
- Information Supply 0
  - Technical Solutions And Tools 5
    - System Diversification And Variability 9
    - ERP System 10
    - Data Warehouse 8
    - HR Performance Measurement 10
    - CRM System 6
    - Standardized Software 0
      - Network Folder 1
      - Microsoft Excel 3
        - Benchmarking 4
    - EMails 2
  - Challenges 1
    - Information Systems Design 0
      - Development Of New Systems 6
        - Customer Profitability Analysis 5
        - Cost Of Sales Analysis 1
        - Simulation System 7
      - Streamlining Of Existing Systems 5
        - System Consolidation And Combination 11
    - Data Focus 0
      - Data Reliability And Validity 8
      - Currentness Of Data 4
    - Information Overload 9
      - Information Imbalance 5
      - Information Retrieval 9
      - Information Delivery 1
    - Edge on Information 4

Metal Finishing Company – Diagnostic Control Systems - III:

- Success Factors 4
  - People Focus 0
    - Challenges 1
      - Employee Engagement 14
        - Compensation 5
          - Incentives Goal Achievement 6
          - Competitiveness 3
          - Performance-Based Strategy 3
        - Passive Headhunting 1
      - Scarce Resource 0
        - Headcount - Internal Availability 5
        - Recruitment - Market Availability 6
      - Confidence vs. Control 5
      - Demographic Change 4
    - Characteristics 0
      - Experience 12
        - Gut Feeling 6
      - Expertise 6
  - Time Focus 0
    - Challenges 0
      - Delivery Times - Delivery Reliability 19
      - Time-To-Market 4
  - Quality Focus 6
    - Increasing Quality Requirements 3
    - Zero-Defect Quality 2
  - Service Focus 0
    - Customer Significance 2
      - Customer Driven Business 5
      - Customer Complaints 7
      - Data Reliability And Validity 5
    - Servitization 13
    - Empathy - Problem Solver & Care Taker 12
  - Production Focus 4
    - Challenges 1
      - Production Planning 18
      - Decentralized Processing 2
      - Production Equipment 4
  - Market Leader 1
    - Pioneer 4
    - Barometric-Firm Leadership 11
    - Role Model 3
    - Competitiveness 3
  - Flexibility vs. Bureaucracy 11
  - Financial Focus 4

Metal Finishing Company – Interactive Control Systems:

- Interactive Control Systems — 1
  - Interactive Intergroup Communication — 7
    - Business Performance Measurement Results — 28
    - Strategic Uncertainties Communication — 12
      - Business Measure - KPI — 3
    - Strategy — 12
    - Provoke New Initiatives And Strategies — 5
    - Office Grapevine — 2
  - Strategic Uncertainties — 3
    - Customer Portfolio Diversification — 1
      - Automotive Industry Dependency — 7
    - Inorganic Growth - Acquisitions — 2
      - Merging Companies — 5
    - Forecast Processes — 6
    - Boundary System — 0
      - Global Economy — 1
        - Emerging Market — 2
      - Political Risks - Global View — 1
      - Legislation — 1
    - Commutability — 4
    - Insourcing — 3

A.11 MFI Survey – Invitation Email – English

To the Management or Head of Finance of your Organization

Dear Sir or Madam,

Please find attached the invitation to a survey related to the Dissertation Research Project "Business Performance Measurement in the Metal Finishing Industry" by Mr Mike Dietrich, Doctoral Candidate at the European Business School, Germany.

We kindly ask you to spend about **15 to 20 minutes** of your precious time to complete this survey as it is a fundamental part of the whole dissertation research project. The goal of the research is to analyse the Metal Finishing Industry (MFI) in order to build a foundation for a Business Performance Measurement model that is suitable for MFI organizations. Your company was chosen to take part in the survey as you are either member of the *Surface Engineering Association (SEA)*, *Institute of Materials Finishing (IMF)* and/or the *European Society of Thin Films (EFDS)*.

Please follow the link to start the survey:

http://ww2.unipark.de/uc/dietrich_bpm_mfi/

(In case the survey does not start automatically, please copy the link and paste it into the address bar of your browser.)

We are convinced that you, as an industry insider, can add valuable insights to our research. We are interested in your personal evaluation and experience in the industry - there are no "right or wrong" answers. You have the opportunity to leave your contact information in order to receive a short report on the survey results as a reward for your assistance.

Please answer all questions by the date of **September 30th, 2014**.

Your evaluations and data will be processed and treated anonymously and strictly confidently. The data and information are solely processed for this research project and are not passed on to a third party.

For further information regarding the research project please refer to the attached documents and the latest editorial of the *Transactions of the Institute of Materials Finishing*.
Please contact Mr Mike Dietrich for further information regarding the survey:

Email: mike.dietrich@students.ebs.edu Phone: [redacted]

Thank you very much for your research inputs!

Sincerely yours

Prof. Dr. Ronald Gleich Mike Dietrich, Dipl. Wirtsch. Ing. (FH), GradIMF

The Research Project is supported by:

[redacted] SEA imf EFDS

A.12 MFI Survey – Invitation Email – German

An die Geschäftsführung oder den Leiter der Finanzabteilung Ihres Unternehmens

Sehr geehrte Damen und Herren,

im Rahmen des Dissertations-Forschungsprojektes "Business Performance Measurement in the Metal Finishing Industry" von Herrn Mike Dietrich, Doktorand an der European Business School, Oestrich-Winkel, erhalten Sie diese Einladung zur Teilnahme an einer Umfrage unter Unternehmen der Metallveredlungsindustrie.

Wir bitten Sie an der Umfrage teilzunehmen (Dauer **ca. 15 - 20 Minuten**), da diese ein fundamentaler Bestandteil des gesamten Dissertations-Forschungsprojektes ist. Ziel der Forschung ist es die Metallveredlungsindustrie zu analysieren, um damit die Basis für ein betriebliches Leistungsmessungs-Modell zu bilden, welches passend für Metallveredlungs-Unternehmen ist. Ihr Unternehmen wurde zur Teilnahme an der Umfrage ausgewählt, da es entweder Mitglied im *Verband Deutscher Maschinen- und Anlagenbau (VDMA)* oder in der *Europäischen Forschungsgesellschaft Dünne Schichten e. V. (EFDS)* ist.

Die Umfrage finden Sie unter folgendem Link:

http://ww2.unipark.de/uc/dietrich_bpm_mfi/

(Falls sich die Umfrage nicht durch Anklicken des Links automatisch öffnen sollte, so kopieren Sie bitte die Internetadresse und fügen Sie diese in die Adresszeile Ihres Browsers ein.)

Wir sind überzeugt, dass Sie als Branchenkenner wichtige Hinweise für unsere Forschung liefern können. Wir sind an Ihrer persönlichen Einschätzung und Industrieerfahrung interessiert - es gibt somit keine "richtigen" oder "falschen" Antworten. Als Dankeschön für Ihre Unterstützung haben Sie die Möglichkeit Ihre Kontaktdaten zu hinterlassen, um einen Bericht über die Umfrageergebnisse zu erhalten.

Bitte beantworten Sie alle Fragen bis zum **30. September 2014**.

Alle Angaben werden selbstverständlich streng vertraulich und anonymisiert behandelt. Die Daten werden ausschließlich zu Forschungszwecken verwendet und nicht an Dritte weitergegeben.

Weitere Informationen zu dem Forschungsprojekt, sowie Empfehlungsschreiben des *Verbandes Deutscher Maschinen- und Anlagenbau (VDMA)* sowie des *Institutes of Materials Finishing (IMF)* können Sie den angefügten Dokumenten entnehmen.

Für weitere Auskünfte zu der Studie steht Ihnen Mike Dietrich gerne zur Verfügung:

E-Mail: mike.dietrich@students.ebs.edu Telefon: [redacted]

Herzlichen Dank für Ihre Unterstützung des Forschungsprojektes!

Mit freundlichen Grüßen

Prof. Dr. Ronald Gleich Mike Dietrich, Dipl. Wirtsch. Ing. (FH), GradIMF

Das Forschungsprojekt wird unterstützt durch:

[redacted]

A.13 MFI Survey – Invitation Letter – English

EBS Business School

EBS Universität
für Wirtschaft und Recht
Gustav-Stresemann-Ring 3
D-65189 Wiesbaden
www.ebs.edu

EFMD EQUIS ACCREDITED

SIIE.
Strascheg Institute for Innovation & Entrepreneurship

Prof. Dr. Ronald Gleich
Strascheg Institute for Innovation and Entrepreneurship (SIIE)
Rheingaustraße 1
Besucheradresse: Burgstraße 5
65375 Oestrich-Winkel
Telefon +49 611 7102 1331
Fax +49 611 7102 10 1331
ronald.gleich@ebs.edu
www.ebs.edu/siie

EBS Universität | SIIE · Gustav-Stresemann-Ring 3 · D-65189 Wiesbaden

To the Management or Head of Finance
of your Organization

August 31th, 2014

RESEARCH PROJECT: "Business Performance Measurement in the Metal Finishing Industry"

Dear Sir or Madam,

"Business Performance Measurement in the Metal Finishing Industry" is the dissertation research project of Mr Mike Dietrich, who is a doctoral candidate at the Strascheg Institute for Innovation & Entrepreneurship (SIIE) of the European Business School, Germany. In addition to the present project, Mr Dietrich has been working as a financial analyst in the Metal Finishing Industry since several years. The goal of the research is to analyse the Metal Finishing Industry (MFI) in order to build a foundation for a Business Performance Measurement model that is suitable for MFI organizations. Your company was chosen to take part in the survey as you are either member of the SEA, IMF and/or the EFDS.

We kindly ask you to spend about **15 to 20 minutes** of your precious time to complete this survey as it is a fundamental part of the whole dissertation research project. We are convinced that you, as an industry insider, can add valuable insights to our research. We are interested in your personal evaluation and experience in the industry - there are no "right or wrong" answers. You have the opportunity to leave your contact information in order to receive a short report on the survey results as a reward for your assistance.

Please answer all questions by the date of **September 30th, 2014.**

Your evaluations and data will be processed and treated anonymously and strictly confidently. The data and information are solely processed for this research project and are not passed on to a third party. Please contact **Mr Mike Dietrich** for further information regarding the survey:

E-mail: mike.dietrich@students.ebs.edu Phone: [redacted]

Thank you very much for your research inputs!

Sincerely yours

Prof. Dr. Ronald Gleich

Mike Dietrich, Dipl. Wirtsch. Ing. (FH), GradIMF

Supported by: [redacted] SEA imf EFDS

EBS Universität für Wirtschaft und Recht gemeinnützige GmbH, Wiesbaden | Amtsgericht Wiesbaden HRB 19951 | Umsatzsteuer-ID DE 113 891 213
Aufsichtsrat: Robert Koehler, Vorsitzender
Geschäftsführung: Professor Dr. Rolf Wolff, Olaf Bernd Güllich, Professor Dr. Richard Raatzsch

DZ Bank AG Frankfurt	Deutsche Bank AG Wiesbaden	Nassauische Sparkasse	Rheingauer Volksbank eG	Deutsche Postbank AG Frankfurt
Kto. 28 644 (500 604 00)	Kto. 4 052 510 (510 700 21)	Kto. 454 013 220 (510 500 15)	Kto. 20 24 24 26 (510 915 00)	Kto. 12 25 609 (500 100 60)
IBAN:	IBAN:	IBAN:	IBAN:	IBAN:
DE22 500 604 00 00000 28644	DE77 510 700 21 0405251000	DE06 510 500 15 0454013220	DE95 510 915 00 00 20242426	DE47 500 100 60 000 1225609
BIC: GENODEFF	BIC: DEUTDEFF510	BIC: NASSDE55	BIC: GENODE51RGG	BIC:PBNKDEFF

A.14 MFI Survey – Invitation Letter – German

EBS Business School

EBS Universität
für Wirtschaft und Recht
Gustav-Stresemann-Ring 3
65189 Wiesbaden
www.ebs.edu

EFMD EQUIS ACCREDITED

SIIE.
Strascheg Institute for Innovation & Entrepreneurship

Prof. Dr. Ronald Gleich
Strascheg Institute for Innovation and Entrepreneurship (SIIE)
Rheingaustraße 1
Besucheradresse: Burgstraße 5
65375 Oestrich-Winkel
Telefon +49 611 7102 1331
Fax +49 611 7102 10 1331
ronald.gleich@ebs.edu
www.ebs.edu/siie

EBS Universität | SIIE · Gustav-Stresemann-Ring 3 · 65189 Wiesbaden

An die Geschäftsführung oder
den Leiter der Finanzabteilung
Ihres Unternehmens

31. August 2014

FORSCHUNGSPROJEKT: „Business Performance Measurement in the Metal Finishing Industry"

Sehr geehrte Damen und Herren,

"Business Performance Measurement in the Metal Finishing Industry" ist das Dissertations-Forschungsprojekt von Herrn Mike Dietrich, Doktorand am Strascheg Institute for Innovation & Entrepreneurship (SIIE) der European Business School, Oestrich-Winkel. Parallel arbeitet Herr Dietrich seit einigen Jahren als Financial Analyst in der Metallveredlungsindustrie (Metal Finishing Industry - MFI). Ziel der Forschung ist es die Metallveredlungsindustrie zu analysieren, um damit die Basis für ein betriebliches Leistungsmessungs-Modell zu bilden, welches passend für Metallveredlungs-Unternehmen ist. Ihr Unternehmen wurde zur Teilnahme an der Umfrage ausgewählt, da es entweder Mitglied im VDMA oder in der EFDS ist.

Wir bitten Sie an der Umfrage teilzunehmen (Dauer ca. **15 - 20 Min.**), da diese ein fundamentaler Bestandteil des gesamten Dissertations-Forschungsprojektes ist. Wir sind überzeugt, dass Sie als Branchenkenner wichtige Hinweise für unsere Forschung liefern können. Wir sind an Ihrer persönlichen Einschätzung und Industrieerfahrung interessiert - es gibt somit keine "richtigen" oder "falschen" Antworten. Als Dankeschön für Ihre Unterstützung haben Sie die Möglichkeit Ihre Kontaktdaten zu hinterlassen, um einen Bericht über die Umfrageergebnisse zu erhalten.

Bitte beantworten Sie alle Fragen bis zum **30. September 2014**.

Alle Angaben werden selbstverständlich streng vertraulich und anonymisiert behandelt. Die Daten werden ausschließlich zu Forschungszwecken verwendet und nicht an Dritte weitergegeben.

Mike Dietrich gibt Ihnen gerne weitere Auskünfte zu der Studie:

E-Mail: mike.dietrich@students.ebs.edu Telefon: [redacted]

Herzlichen Dank für Ihre Unterstützung des Forschungsprojektes!

Mit freundlichen Grüßen

Prof. Dr. Ronald Gleich

Mike Dietrich, Dipl. Wirtsch. Ing. (FH), GradIMF

Unterstützt durch: [redacted] SEA imf EFDS

EBS Universität für Wirtschaft und Recht gemeinnützige GmbH, Wiesbaden | Amtsgericht Wiesbaden HRB 19951 | Umsatzsteuer-ID DE 113 891 213
Aufsichtsrat: Robert Koehler, Vorsitzender
Geschäftsführung: Professor Dr. Rolf Wolff, Olaf Bernd Güllich, Professor Dr. Richard Raatzsch

DZ Bank AG Frankfurt	Deutsche Bank AG Wiesbaden	Nassauische Sparkasse	Rheingauer Volksbank eG	Deutsche Postbank AG Frankfurt
Kto. 28 644 (500 604 00)	Kto. 4 052 510 (510 700 21)	Kto. 454 013 220 (510 500 15)	Kto. 20 24 24 26 (510 915 00)	Kto. 12 25 609 (500 100 60)
IBAN:	IBAN:	IBAN:	IBAN:	IBAN:
DE22 500 604 00 00000 28644	DE77 510 700 21 0405251000	DE06 510 500 15 0454013220	DE95 510 915 00 00 20242426	DE47 500 100 60 000 1225609
BIC: GENODEFF	BIC: DEUTDEFF510	BIC: NASSDE55	BIC: GENODE51RGG	BIC:PBNKDEFF

A.15 MFI Survey – Questionnaire – English

Print version

Display options

Info: Here you can change the display options. If you select a language which doesn't have own text elements, the text elements of the default language will be displayed.

Edit display options:
- ☒ Print options
- ☐ Show internal links
- ☒ Show filters
- ☐ Show triggers
- ☒ Show plausibility checks
- ☐ Enable randomization of answer options
- ☐ Show pretest comments
- ☐ Show to-dos

Language: English / Deutsch

Save settings

Information on survey Business Performance Measurement in the Metal Finishing Industry

Survey no.	170722
Author	Mike Dietrich
Coauthors	
Start	2014-08-24 00:00:00
End	2014-11-30 00:00:00

Questionnaire

1 Start

Survey within the Research Project

„Business Performance Measurement in the Metal Finishing Industry"

Bitte wählen Sie Ihre Sprache aus: // Please choose you language:

- Deutsch
- English

2 Intro_01

Survey within the Research Project

„Business Performance Measurement in the Metal Finishing Industry"

"Business Performance Measurement in the Metal Finishing Industry" is the dissertation research project of Mr Mike Dietrich, who is a doctoral candidate at the *Strascheg Institute for Innovation & Entrepreneurship (SIIE)* of the *European Business School, Germany.* In addition to the present project, Mr Dietrich has been working as a financial analyst in the Metal Finishing Industry since several years. The goal of the research is to analyse the Metal Finishing Industry (MFI) in order to build a foundation for a Business Performance Measurement model that is suitable for MFI organizations.

We kindly ask you to complete this survey as it is a fundamental part of the whole dissertation research project. We are convinced that you, as an industry insider, can add valuable insights to our research. We are interested in your personal evaluation and experience in the industry - there are no "right or wrong" answers.

<u>The survey is structured into three parts - each of them will be introduced by a brief description to explain their relevance in the context of our research project:</u>
1. Statistical Data
2. Overall Characteristics of your Organization

3. Business Performance Measurement

Please answer all questions by the date of September 30th, 2014.

You have the opportunity to leave your contact information at the end of the survey in order to receive a short report on the survey results as a reward for your assistance.

Your evaluations and data will be processed and treated anonymously and strictly confidently. The data and information is solely for the research project and will not be passed on to a third party.

Please contact **Mr Mike Dietrich** for further information regarding the survey: E-mail: **mike.dietrich@students.ebs.edu**
Phone: [redacted]

The Research Project is supported by:

[redacted]	**Surface Engineering Association, UK**
European Society of Thin Films	**Institute of Materials Finishing, UK**

3 Intro_02_TransitionToB1

Survey within the Research Project

„Business Performance Measurement in the Metal Finishing Industry"

Survey Part 1

The first survey part is focussed on statistical data regarding your person and your organization.

"Organization" in the context of our research relates to the legal entity in which you are working.
Example: if your organization is part of an international group of companies, we kindly ask you to focus on your local company rather than on the parent company.

In the context of our research this data is required in order to situate your organization into a suitable group of different organizational clusters.
This organizational clustering is necessary for the purpose of analysis to determine, for instance, if there are any differences for example between small or medium sized metal finishing organizations.

4 Page_01

1. Statistical Data

"Organization" in the context of our research relates to the legal entity in which you are working.
Example: if your organization is part of an international group of companies, we kindly ask you to focus on your local company rather than on the parent company.

What is your function in your organization?

- Chief Executive Officer (CEO)
- General Manager
- Chief Financial Officer (CFO)
- Chief Information Officer (CIO)
- Head of Production
- Head of Sales
- Head of HR
- Other (please indicate):

In which department are you working?

- Management
- Production
- Sales
- Purchase
- Finance - Accounting
- Finance - Controlling
- Engineering and Design
- Human Resources (HR)
- Information Technology (IT)
- Research and Development (R&D)
- Other (please indicate):

How many years have you been working within the metal finishing industry?

- 0-5 years
- 6-10 years
- 11-15 years
- 16-20 years
- 21-25 years
- 26-30 years
- 31-35 years
- 36-40 years
- 41-45 years
- >45 years

5 Page_02

1. Statistical Data

"Organization" in the context of our research relates to the legal entity in which you are working.
Example: if your organization is part of an international group of companies, we kindly ask you to focus on your local company rather than on the parent company.

In which area do you classify your organization regarding the majority of your operations?

- Metal finishing company
- Manufacturing company with metal finishing department
- Supplier for metal finishing companies
- Metal finishing system manufacturer
- Other (please indicate):

Which metal finishing / surface engineering processes are applied by your organization? Or for which processes does your organization deliver systems / equipment?

(Multiple answers allowed)

- Paint finishing (incl. wet and powder painting)
- Chemical / Physical Vapour Deposition (CVD / PVD)
- Electro plating (incl. chrome plating)
- Galvanization (incl. hot-dip galvanization, phosphate conversion)
- Heat treatment (incl. nitriding)
- Thermal Spraying (incl. plasma spraying)
- Pre- / Post-treatment
- Other (please indicate):

How many years has your current organization been in operation?

- 0-5 years
- 6-10 years
- 11-15 years
- 16-20 years
- 21-25 years
- 26-30 years
- 31-35 years
- 36-40 years
- 41-45 years
- >45 years

6 Page_03a

1. Statistical Data

"Organization" in the context of our research relates to the legal entity in which you are working.
Example: if your organization is part of an international group of companies, we kindly ask you to focus on your local company rather than on the parent company.

Please indicate the country where your organization is located:

Germany
United Kingdom
France
Italy
Switzerland
Austria
Poland
Ireland
Other (please indicate):

Other (please indicate):

Please indicate your local currency:

EUR - EURO
CHF - Swiss Franc
GBP - Pound Sterling
USD - US Dollar
PLN - Polish Złoty
Other (please indicate):

Other (please indicate):

What was the total turnover of your organization in 2013 ?

(Scaling factor: please enter the value in 1.000 local currency - example: "1.000.000 EURO" should be entered as "1000" - without point or comma)

7 Page_03b

Plausibility checks:

Only production worker: (v_373) > In total: (v_372)

1. Statistical Data

"Organization" in the context of our research relates to the legal entity in which you are working.
Example: if your organization is part of an international group of companies, we kindly ask you to focus on your local company rather than on the parent company.

How many employees are currently working for your organization?

(Please enter the number of Fulltime-Equivalents - FTEs. *Reference date: June 30th, 2014.*)

In total:

Only production worker:

How many production sites does your organization have?

(Please enter the number of production sites on June 30th, 2014 as reference date)

In the country

where your organization is located:

In all other countries:

8 Page_04_TransitionB1B2

Survey within the Research Project

„Business Performance Measurement in the Metal Finishing Industry"

Survey Part 2

Thank you for finishing the first survey part.

The second survey part is focused on the overall characteristics of your organization; those characteristics are required to understand your organization's strategy, orientation and behaviour.

These organizational characteristics are the base, upon which will be developed a Business Performance Measurement Model, therefore we kindly ask you to answer all of the following questions.
Please answer all questions according to your perception and experience - there are no "right" or "wrong" answers.

"Organization" in the context of our research relates to the legal entity in which you are working.
Example: if your organization is part of an international group of companies, we kindly ask you to focus on your local company rather than on the parent company.

9 Page_05

2. Organization Characteristics

"Organization" in the context of our research relates to the legal entity in which you are working.
Example: if your organization is part of an international group of companies, we kindly ask you to focus on your local company rather than on the parent company.

Please evaluate the following statements regarding your organization's characteristics:

	Strongly Disagree						Strongly Agree
We anticipate and respond to customers' evolving needs and wants.	○	○	○	○	○	○	○
We emphasize the evolution of formal and informal customer complaints.	○	○	○	○	○	○	○
We follow-up with customers for quality/service feedback.	○	○	○	○	○	○	○
We interact with customers to set reliability, responsiveness, and other standards.	○	○	○	○	○	○	○
Satisfying customer needs is the central purpose of our business.	○	○	○	○	○	○	○
Customer focus is reflected in our business planning.	○	○	○	○	○	○	○
We produce products and/or offer services that satisfy or exceed customer expectations.	○	○	○	○	○	○	○

	Strongly Disagree						Strongly Agree
We always try to increase our organization's direct personal contacts with customers.	○	○	○	○	○	○	○
We are actively seeking customer inputs to determine their requirements.	○	○	○	○	○	○	○
We are using customer requirements as the basis for quality.	○	○	○	○	○	○	○
Involving customers in product/service design is crucial for our organization.	○	○	○	○	○	○	○

10 Page_06

2. Organization Characteristics

"Organization" in the context of our research relates to the legal entity in which you are working.
Example: if your organization is part of an international group of companies, we kindly ask you to focus on your local company rather than on the parent company.

Please evaluate the following statements related to your organization's characteristics:

	Strongly Disagree						Strongly Agree
Our strategy cannot be described as concentrating solely on offering products with the lowest price.	○	○	○	○	○	○	○
Our strategy is based on quality performance rather than price.	○	○	○	○	○	○	○
We place greater emphasis on innovation than price.	○	○	○	○	○	○	○
We place greater emphasis on customer service than price.	○	○	○	○	○	○	○
Our strategy places importance on delivering products/services with high performance.	○	○	○	○	○	○	○
We emphasize launching new products/services quickly.	○	○	○	○	○	○	○

Please rank the following factors to place 1, 2 and 3 according to their priority within your organization:

(Please click the factor with the highest priority for your organization first - „1" has the highest priority.)

	1	2	3
Costs/Price	○	○	○
Quality	○	○	○
Delivery Reliability	○	○	○

11 Page_07

2. Organization Characteristics

"Organization" in the context of our research relates to the legal entity in which you are working.
Example: if your organization is part of an international group of companies, we kindly ask you to focus on your local company rather than on the parent company.

Please evaluate the following statements related to possible uncertainites within your organization:

	Never Predictable						Always Predictable
Actual users of our products/services are ...	○	○	○	○	○	○	○
Competitors for the supply of raw materials are ...	○	○	○	○	○	○	○
Competitors for our customers are ...	○	○	○	○	○	○	○
Government regulations controlling our industry are ...	○	○	○	○	○	○	○
The public's political views and attitudes toward our industry are ...	○	○	○	○	○	○	○

	Strongly Disagree						Strongly Agree
Our industry is characterized by rapidly changing technology.	○	○	○	○	○	○	○
If we do not keep up with the changes in technology, it will be difficult for us to remain competitive.	○	○	○	○	○	○	○
The rate of process obsolescence is high in our industry.	○	○	○	○	○	○	○
The production technology changes frequently and sufficiently.	○	○	○	○	○	○	○

	Strongly Disagree						Strongly Agree

Our master production schedule has a high percentage of variation in demand.							
Customer demands fluctuate drastically from week to week.							
Our internal supply requirements (for example for consumables) vary drastically from week to week.							
We keep weeks of inventory of critical materials to meet changing customer demands.							
The volume and/or composition of customer demands are difficult to predict.							

12 Page_08

2. Organization Characteristics

"Organization" in the context of our research relates to the legal entity in which you are working.
Example: if your organization is part of an international group of companies, we kindly ask you to focus on your local company rather than on the parent company.

Please evaluate the following statements related to the behaviour of your organization:

	Strongly Disagree						Strongly Agree
Our organization accepts demands that go beyond existing products and services.							
We commercialize products and services that are completely new to our organization.							
We frequently utilize new opportunities in new markets.							
Our organization regularly uses new distribution channels.							

	Strongly Disagree						Strongly Agree
We frequently make small adjustments to our existing products and services.							
We improve the production efficiency for our products and/or the provision efficiency for our services.							
We reduce costs by having efficiency effects channelled through the volume increase of existing products.							
Our organization expands services for existing clients.							

13 Page_09

2. Organization Characteristics

"Organization" in the context of our research relates to the legal entity in which you are working.
Example: if your organization is part of an international group of companies, we kindly ask you to focus on your local company rather than on the parent company.

Please rate the importance of the following flexibility factors for your organization:

	Extremely Unimportant						Extremely Important
Delivery flexibility: The ability of an organization to effectively respond to changes in planned delivery dates.							
Volume flexibility: The ability of an organization to effectively increase or decrease aggregate production in							

response to customers.

Mix flexibility: The ability of an organization to effectively produce a wide variety of different products or offer a wide variety of different services.	◦	◦	◦	◦	◦	◦	◦
Changeover flexibility: The ability of an organization to effectively handle additions and subtractions to the product or service mix over time.	◦	◦	◦	◦	◦	◦	◦
Modification flexibility: The ability of an organization to effectively implement minor changes in current products or services that result from corrective actions or changing customer requirements.	◦	◦	◦	◦	◦	◦	◦

Please indicate your order backlog in working days (on average):

◦ 0 working days - we are working "just-in-time"

◦ 1 to 5 working days

◦ 6 to 10 working days

◦ 11 to 15 working days

◦ 16 to 20 working days

◦ More than 20 working days

14 Page_10_TransitionB2B3

Survey within the Research Project

„Business Performance Measurement in the Metal Finishing Industry"

Survey Part 3

Thank you very much for finishing the second survey part.

The third and last survey part is focused on "Business Performance Measurement (BPM)" within your organization. *"Business Performance Measurement"* in the context of our research can be described as the process to evaluate the effectiveness and efficiency of your organization's activities*.

"Business Measures" and / or "Key Performance Indicators (KPIs)" can be applied as factors/measures within "Business Performance Measurement" processes:

- *"Business Measures"* in the context of our research are financial as well as non-financial measures like Turnover, Gross Profit, Number of FTEs, Complaint Count, e. g.
- *"Key Performance Indicators"* are multi-dimensional measures* that for example combine a business measure with other dimensions (another business measure or other factors like time). KPI examples: Turnover per FTE, Lost-Time-Accidents per Month, e. g.

Some organizations are using "Business Performance Measurement Systems (BPM Systems)" in which the different Business Measures or KPIs are combined and analysed:

- These systems are often adapted from so called "generic" BPM System Models (existing, mainly theory driven models like the Balanced Scorecard* or the Performance Prism*). These generic BPM System Models are regularly adapted according to the organizational characteristics and requirements before the organization's BPM System is implemented.
- A BPM system can of course also directly be created in order to reflect the organization's requirements and not being based on a generic BPM System.

Traditional Performance Measurement Systems are mainly focused on financial data and indicators. Modern Business Performance Measurement Systems* cover financial as well as non-financial measures in order to analyse and control interrelations between the different measures.

**For further details on BPM, KPI and BPM Systems you may refer to Neely, A. D. (2011): Business performance measurement. Unifying theory and integrating practice. 2nd Edition. Cambridge: Cambridge University Press.*

In our Research Project we are aiming for the development of a BPM System for the Metal Finishing Industry. In this context we kindly ask you to explain to us how your organization is *measuring its business performance.*

15 Page_11

Plausibility checks:

The responsibility for setting BPM rules/guidelines in our organization is NOT defined. (v_294) == quoted
AND (**Corporate Headquarters (v_137)** == quoted
OR Executive Board (v_138) == quoted
OR Supervisory Board (v_139) == quoted
OR Owner (v_140) == quoted
OR Local Management (v_141) == quoted
OR Local Department Heads (v_142) == quoted
OR Other (please indicate): (v_143) == quoted)

3. Business Performance Measurement

"Business Performance Measurement" in the context of our research can be described as the process to evaluate the effectiveness and efficiency of your organization's activities.

Who is responsible for setting the rules/guidelines for Business Performance Measurement within your organization?
(Multiple answers allowed)

- Corporate Headquarters
- Executive Board
- Supervisory Board
- Owner
- Local Management
- Local Department Heads
- Other (please indicate):
- The responsibility for setting BPM rules/guidelines in our organization is NOT defined.

For your major product or service, please indicate your position with respect to your competitors on the following dimensions for the last two years:

	Significantly Lower						Significantly Higher
Market share growth	○	○	○	○	○	○	○
Sales growth	○	○	○	○	○	○	○
Return on investment (ROI)	○	○	○	○	○	○	○
Growth in ROI	○	○	○	○	○	○	○
Return on sales (ROS)	○	○	○	○	○	○	○

16 Page_12a

3. Business Performance Measurement

"Business Measures" and / or "Key Performance Indicators (KPIs)" can be applied as factors/measures within "Business Performance Measurement" processes:

- *"Business Measures"* in the context of our research are financial as well as non-financial measures like Turnover, GrossProfit, Number of FTEs, Complaint Count, e. g.
- *"Key Performance Indicators"* are multi-dimensional measures that for example combine a business measure with other dimensions (another business measure or other factors like time). KPI examples: Turnover per FTE, Lost-Time-Accidents per Month, e. g.

Please evaluate the importance of the following measures for your organization:

	Extremely Unimportant						Extremely Important
Unit product costs	○	○	○	○	○	○	○
Costs versus budget	○	○	○	○	○	○	○
Incoming customer parts quality	○	○	○	○	○	○	○
In process quality	○	○	○	○	○	○	○
Field failure under warranty	○	○	○	○	○	○	○
On-time delivery to customer	○	○	○	○	○	○	○
Delivery lead time	○	○	○	○	○	○	○
Product/machine change over times	○	○	○	○	○	○	○

Please indicate the approximate ratio between fixed costs (personnel cost, e. g.) and variable costs (expenses for material or consumables, e. g.) to the total costs of your organization:

- Fixed Costs 0% - Variable Costs 100%
- Fixed Costs 10% - Variable Costs 90%
- Fixed Costs 20% - Variable Costs 80%
- Fixed Costs 30% - Variable Costs 70%
- Fixed Costs 40% - Variable Costs 60%
- Fixed Costs 50% - Variable Costs 50%
- Fixed Costs 60% - Variable Costs 40%
- Fixed Costs 70% - Variable Costs 30%
- Fixed Costs 80% - Variable Costs 20%
- Fixed Costs 90% - Variable Costs 10%
- Fixed Costs 100% - Variable Costs 0%

17 Page_12b

Plausibility checks:

Our organization is NOT using external business / market indicators for Business Performance Measurement. (v_334) == quoted
AND (**GDP - Gross Domestic Product (v_330)** == quoted
OR **PMI - Purchasing Managers Index (v_331)** == quoted
OR **Others (please indicate): (v_332)** == quoted)

3. Business Performance Measurement

"Business Measures" and / or "Key Performance Indicators (KPIs)" can be applied as factors/measures within "Business Performance Measurement" processes:

- *"Business Measures"* in the context of our research are financial as well as non-financial measures like Turnover, Gross Profit, Number of FTEs, Complaint Count, e. g.
- *"Key Performance Indicators"* are multi-dimensional measures that for example combine a business measure with other dimensions (another business measure or other factors like time). KPI examples: Turnover per FTE, Lost-Time-Accidents per Month, e. g.

Please name up to five of the most important Key Performance Indicators (KPI) and / or Business Measures you are using in order to assess your organization's performance:

(Please name the KPI / Business Measure and a brief description related to the significance / information value for your organization. Each text field entry is limited to enter a maximum of 255 characters.)

	KPI / Business Measure	Description
KPI / Business Measure No. 1		
KPI / Business Measure No. 2		
KPI / Business Measure No. 3		
KPI / Business Measure No. 4		
KPI / Business Measure No. 5		

Please indicate external business / market indicators that are relevant for your organization's Business Performance Measurement:

(Multiple answers allowed)

- GDP - Gross Domestic Product
- PMI - Purchasing Managers Index
- Others (please indicate):
- Our organization is NOT using external business / market indicators for Business Performance Measurement.

18 Page_12c

Plausibility checks:

We are NOT using IT Systems / Software for measuring our organization's business performance. (v_311) == quoted
AND (**Standardized Office Software (for example Microsoft Excel or Access, OpenOffice) (v_295)** == quoted
OR **Enterprise Resource Planning (ERP) Software (v_296)** == quoted
OR **Reporting based on Data Warehouses (for example IBM Business Warehouse) (v_297)** == quoted
OR **Customer Relationship Management (CRM) Software (v_298)** == quoted
OR **Others (please indicate): (v_309)** == quoted)

3. Business Performance Measurement

Please indicate the IT systems / software you are using for measuring your organization's business performance:

(Multiple answers allowed)

- Standardized Office Software (for example Microsoft Excel or Access, OpenOffice)
- Enterprise Resource Planning (ERP) Software
- Reporting based on Data Warehouses (for example IBM Business Warehouse)
- Customer Relationship Management (CRM) Software
- Others (please indicate):
- We are NOT using IT Systems / Software for measuring our organization's business performance.

19 Page_13

3. Business Performance Measurement

Some organizations are using "Business Performance Measurement Systems (BPM Systems)" in which the different Business Measures or KPIs are combined and analysed:

- These systems are often adapted from so called "generic" BPM System Models (existing, mainly theory driven models like the Balanced Scorecard or the Performance Prism). These generic BPM System Models are regularly adapted according to the organizational characteristics and requirements before the organization's BPM System is implemented.
- A BPM system can of course also directly be created in order to reflect the organization's requirements and not being based on a generic BPM System.

Are you using a BPM-System within your organization?

- YES
- NO

20 Filter Filter_BPM_EN_YES

v_113 BPM_ExistsSystem	Are you using a BPM-System within your organization? - BPM_ExistsSystem (From page 19: Page_13)	equal 1

20.1 Page_14.1_BPM_YES

3. Business Performance Measurement

Some organizations are using "Business Performance Measurement Systems (BPM Systems)" in which the different Business Measures or KPIs are combined and analysed:

- These systems are often adapted from so called "generic" BPM System Models (existing, mainly theory driven models like the Balanced Scorecard or the Performance Prism). These generic BPM System Models are regularly adapted according to the organizational characteristics and requirements before the organization's BPM System is implemented.
- A BPM system can of course also directly be created in order to reflect the organization's requirements and not being based on a generic BPM System.

Please indicate the functions for which your BPM-System is used:

(Multiple answers allowed)

- Organizational Strategy Setting ("Plan")
- Organizational Strategy Implementation ("Do")
- Organizational Strategy Control ("Check")
- Organizational Strategy Alignment ("Act" - in case of strategy deviations)
- Personal Target Setting
- Personal Feedback Processes (including incentives, gratifications and penalties)
- Information Supply (Management Information System, e.g.)
- Communication (strategy, target achievement, e. g., are being communicated via the BPM-System)

- Operational Management of the Organization
- Other (please indicate):

On which generic system is your BPM based?

- Balanced Scorecard
- Performance Pyramid
- Performance Prism
- SCOR-Model
- EFQM-Model
- Other (please indicate):
- Our BPM-System is NOT based on a generic BPM System Model - it is our own development.

20.1.1 Filter Filter_BPM_GenSys_Yes

	v_352 BPM_GenSys	On which generic system is your BPM based? - BPM_GenSys (From page 20.1: Page_14.1_BPM_YES)	equal 1
or	v_352 BPM_GenSys	On which generic system is your BPM based? - BPM_GenSys (From page 20.1: Page_14.1_BPM_YES)	equal 2
or	v_352 BPM_GenSys	On which generic system is your BPM based? - BPM_GenSys (From page 20.1: Page_14.1_BPM_YES)	equal 3
or	v_352 BPM_GenSys	On which generic system is your BPM based? - BPM_GenSys (From page 20.1: Page_14.1_BPM_YES)	equal 4
or	v_352 BPM_GenSys	On which generic system is your BPM based? - BPM_GenSys (From page 20.1: Page_14.1_BPM_YES)	equal 5
or	v_352 BPM_GenSys	On which generic system is your BPM based? - BPM_GenSys (From page 20.1: Page_14.1_BPM_YES)	equal 6

20.1.1.1 Page_15.1_GenSys_YES

3. Business Performance Measurement

Please describe the main alterations your organization has performed in order to transform the generic BPM System Model into your organization's BPM System:

Are you satisfied with your BPM-System?

- YES
- NO

20.1.1.1.1 Filter Filter_BPM_Satisfy_NO

v_114 BPM_Yes_Satisfaction	Are you satisfied with your BPM-System? - BPM_Yes_Satisfaction (From page 20.1.1.1: Page_15.1_GenSys_YES)	equal 2

20.1.1.1.1.1 Page_16_BPM_Satisfaction_NO

3. Business Performance Measurement

Regarding your answer not being satisfied with your BPM-System - what could be improved?

20.1.2 Filter Filter_BPM_GenSys_No

v_352 BPM_GenSys On which generic system is your BPM based? - BPM_GenSys (From page 20.1: Page_14.1_BPM_YES) equal 7

20.1.2.1 Page_15.2_GenSys_No

3. Business Performance Measurement

Please indicate the main characteristics of your organization's BPM-System:

(Multiple answers allowed)

- Includes financial indicators
- Includes non-financial indicators
- The management uses the BPM-System
- The department heads are using the BPM-System
- The finance department uses the BPM-System
- Other (please indicate):

Are you satisfied with your BPM-System?

- YES
- NO

20.1.2.1.1 Filter Filter_BPM_Satisfy_NO

v_359 BPM_Yes_Satisfaction	Are you satisfied with your BPM-System? - BPM_Yes_Satisfaction (From page 20.1.2.1: Page_15.2_GenSys_No)	equal 2

20.1.2.1.1.1 Page_16_BPM_Satisfaction_NO

3. Business Performance Measurement

Regarding your answer not being satisfied with your BPM-System - what could be improved?

21 Filter Filter_BPM_EN_NO

v_113 BPM_ExistsSystem	Are you using a BPM-System within your organization? - BPM_ExistsSystem (From page 19: Page_13)	equal 2

21.1 Page_14.2_BPM_NO

3. Business Performance Measurement

What are the reasons for the absence of a BPM-system in your organization?

(Multiple answers allowed.)

- Implementation too time consuming
- Implementation too expensive
- Sustaining too time consuming
- Sustaining too expensive
- Costs disproportional to benefit
- Already implemented system is not used anymore – what are the reasons?:
- Other reasons (please indicate):

From your point of view, which main characteristics should a <u>potential</u> BPM-System for your organization include?

(Multiple answers allowed)

- Should include financial indicators
- Should include non-financial indicators
- The management should use the BPM-System

- The department heads should use the BPM-System
- The finance department should use the BPM-System
- Other (please indicate):

Are you planning to implement a BPM-system within the next six months?

- YES
- NO

22 FINAL

You have nearly finished the survey
Business Performance Measurement in the Metal Finishing Industry

On this side you have the opportunity to leave comments and/or your thoughts regarding the survey topic or the survey itself. You can also leave your email address in order to receive a short report on the results of the survey.

Here you have the opportunity to leave remarks, suggestions or additions related to the survey:

Please leave your email address in case you would like to receive a short report on the results of the survey:

You will submit your survey data by clicking the "Continue" button.

23 Endseite

Your evaluations and data will be processed and treated anonymously.

We assure you that your personal data will be strictly separated from the anonymous analysis of the questionnaire; your contact information (email address if entered) will only be used for sending the short report.

Please contact **Mr Mike Dietrich** for further information regarding the survey: E-mail: **mike.dietrich@students.ebs.edu** Phone: [redacted]

Thank you very much for supporting the dissertation research project

Business Performance Measurement in the Metal Finishing Industry

A.16 MFI Survey – Questionnaire – German

Print version

Display options

Info: Here you can change the display options. If you select a language which doesn't have own text elements, the text elements of the default language will be displayed.

Edit display options:

- Print options
- Show internal links
- Show filters
- Show triggers
- Show plausibility checks
- Enable randomization of answer options
- Show pretest comments
- Show to-dos

Language

English
Deutsch

Save settings

Information on survey Business Performance Measurement in the Metal Finishing Industry

Survey no.	170722
Author	Mike Dietrich
Coauthors	
Start	2014-08-24 00:00:00
End	2014-11-30 00:00:00

Questionnaire

1 Start

Umfrage im Rahmen des Forschungsprojektes

„Business Performance Measurement in the Metal Finishing Industry"

Bitte wählen Sie Ihre Sprache aus: // Please choose you language:

- Deutsch
- English

2 Intro_01

Umfrage im Rahmen des Forschungsprojektes

„Business Performance Measurement in the Metal Finishing Industry"

"Business Performance Measurement in the Metal Finishing Industry" ist das Dissertations-Forschungsprojekt von Herrn Mike Dietrich, Doktorand am *Strascheg Institute for Innovation & Entrepreneurship (SIIE)* der *European Business School, Oestrich-Winkel.* Parallel arbeitet Herr Dietrich seit einigen Jahren als Financial Analyst in der Metallveredlungsindustrie (Metal Finishing Industry - MFI).

Ziel der Forschung ist es die Metallveredlungsindustrie zu analysieren, um damit die Basis für ein betriebliches Leistungsmessungs-Modell zu bilden, welches passend für Metallveredlungs-Unternehmen ist. Wir bitten Sie an der Umfrage teilzunehmen, da diese ein fundamentaler Bestandteil des gesamten Dissertations-Forschungsprojektes ist.
Wir sind überzeugt, dass Sie als Branchenkenner wichtige Hinweise für unsere Forschung liefern können. Wir sind an Ihrer persönlichen Einschätzung und Industrieerfahrung interessiert - es gibt somit keine "richtigen" oder "falschen" Antworten.

Die Umfrage ist in drei Teile gegliedert - jeder Teil wird durch eine kurze Beschreibung der jeweiligen Bedeutung im Rahmen des Forschungsprojektes eingeleitet:
1. Statistische Daten
2. Generelle Charakteristika Ihres Unternehmens
3. Betriebliche Leistungsmessung - Business Performance Measurement

Bitte beantworten Sie alle Fragen bis zum 30. September 2014.

Als Dankeschön für Ihre Unterstützung haben Sie die Möglichkeit am Ende der Umfrage Ihre Kontaktinformationen zu hinterlassen, um einen Kurzbericht über die Umfrageergebnisse zu erhalten.

Ihre Angaben und Daten werden anonymisiert und strikt vertraulich verarbeitet. Die Daten und Angaben werden ausschließlich für dieses Forschungsprojekt erhoben und werden nicht an Dritte weitergegeben.

Für weitere Fragen kontaktieren Sie bitte **Herrn Mike Dietrich**: E-Mail: **mike.dietrich@students.ebs.edu** Telefon: [redacted]

Das Forschungsprojekt wird unterstützt von:

[redacted] | **Surface Engineering Association, UK**

European Society of Thin Films | Institute of Materials Finishing, UK

3 Intro_02_TransitionToB1

Umfrage im Rahmen des Forschungsprojektes

„Business Performance Measurement in the Metal Finishing Industry"

Umfrageteil 1

Der erste Umfrageteil befasst sich mit statistischen Daten bezüglich Ihrer Person und Ihres Unternehmens.

"Unternehmen" im Kontext unserer Forschung bezieht sich auf das Rechtssubjekt, für das Sie arbeiten.
Beispiel: Falls Ihr Unternehmen Teil eines internationalen Konzerns ist, so bitten wir Sie sich bei der Beantwortung der Fragen auf Ihr lokales Unternehmen, anstatt auf die Konzernmutter zu beziehen.

Wir benötigen diese Daten im Rahmen unserer Forschung, um Ihr Unternehmen in die verschiedenen Organisationsgruppen einordnen zu können.
Diese organisatorische Einordnung ist von hoher Bedeutung um beispielsweise mögliche Unterschiede zwischen kleinen und mittelständigen Unternehmen zu analysieren.

4 Page_01

1. Statistische Daten

"Unternehmen" im Kontext unserer Forschung bezieht sich auf das Rechtssubjekt, für das Sie arbeiten.
Beispiel: Falls Ihr Unternehmen Teil eines internationalen Konzerns ist, so bitten wir Sie sich bei der Beantwortung der Fragen auf Ihr lokales Unternehmen, anstatt auf die Konzernmutter zu beziehen.

Welche Position haben Sie in Ihrem Unternehmen?

- Chief Executive Officer (CEO)
- Geschäftsführer
- Chief Financial Officer (CFO)
- Chief Information Officer (CIO)
- Produktionsleiter
- Vertriebsleiter
- Personalleiter
- Andere (bitte angeben):

In welcher Abteilung arbeiten Sie?

- Geschäftsführung
- Produktion
- Vertrieb
- Einkauf
- Finanzabteilung - Buchhaltung
- Finanzabteilung - Controlling
- Konstruktionsabteilung
- Personalabteilung
- EDV/IT-Abteilung
- Forschungsabteilung
- Andere (bitte angeben):

Seit wie vielen Jahren sind Sie in der Metallveredlungsbranche tätig?

0-5 Jahre	26-30 Jahre
6-10 Jahre	31-35 Jahre
11-15 Jahre	36-40 Jahre
16-20 Jahre	41-45 Jahre
21-25 Jahre	>45 Jahre

5 Page_02

1. Statistische Daten

"Unternehmen" im Kontext unserer Forschung bezieht sich auf das Rechtssubjekt, für das Sie arbeiten.
Beispiel: Falls Ihr Unternehmen Teil eines internationalen Konzerns ist, so bitten wir Sie sich bei der Beantwortung der Fragen auf Ihr lokales Unternehmen, anstatt auf die Konzernmutter zu beziehen.

In welchen Bereich ordnen Sie Ihr Unternehmen primär ein?

- Metallveredlungsunternehmen
- Produzierendes Unternehmen mit Metallveredlungsabteilung
- Zulieferer für Metallveredlungsunternehmen
- Metallveredlungs-Systemhersteller
- Andere (bitte angeben):

Welche Metallveredlungsprozesse/-verfahren werden in Ihrem Unternehmen angewendet? Oder für welche Prozesse liefert Ihr Unternehmen Systeme / Ausrüstung?

(Mehrfachnennungen erlaubt)

- Lackierungen (inkl. Nass- und Pulverlackieren)
- Chemical / Physical Vapor Deposition (CVD, PVD)
- Elektroplattieren (inkl. Verchromen)
- Galvanisieren (inkl. Verzinken, Phosphatieren)
- Wärmebehandlung (inkl. Nitrieren/Nitridieren)
- Thermische Spritzverfahren (inkl. Plasmaspritzen)
- Vor- / Nachbehandlung
- Andere (bitte angeben):

Seit wie vielen Jahren besteht Ihr gegenwärtiges Unternehmen?

0-5 Jahre	26-30 Jahre
6-10 Jahre	31-35 Jahre
11-15 Jahre	36-40 Jahre
16-20 Jahre	41-45 Jahre
21-25 Jahre	>45 Jahre

6 Page_03a

1. Statistische Daten

"Unternehmen" im Kontext unserer Forschung bezieht sich auf das Rechtssubjekt, für das Sie arbeiten.
Beispiel: Falls Ihr Unternehmen Teil eines internationalen Konzerns ist, so bitten wir Sie sich bei der Beantwortung der Fragen auf Ihr lokales Unternehmen, anstatt auf die Konzernmutter zu beziehen.

Bitte geben Sie das Land an, in welchem Ihr Unternehmen angesiedelt ist:

Deutschland
Vereinigtes Königreich (Großbri
Frankreich
Italien
Schweiz
Österreich
Polen
Irland
Andere (bitte angeben):

Andere (bitte angeben):

Bitte geben Sie Ihre lokale Währung an:

EUR - EURO
CHF - Schweizer Franken
GBP - Pfund Sterling
USD - US Dollar
PLN - Polnischer Zloty
Andere (bitte angeben):

Andere (bitte angeben):

Wie hoch war Ihr Unternehmensumsatz in 2013 ?

(Skalierung: bitte geben Sie den Wert in 1.000 lokaler Währung ein - Beispiel: "1.000.000 EURO" sollte als "1000" eingeben werden - ohne Punkt oder Komma)

7 Page_03b

Plausibility checks:

Only production worker: (v_373) > In total: (v_372)

1. Statistische Daten

"Unternehmen" im Kontext unserer Forschung bezieht sich auf das Rechtssubjekt, für das Sie arbeiten.
Beispiel: Falls Ihr Unternehmen Teil eines internationalen Konzerns ist, so bitten wir Sie sich bei der Beantwortung der Fragen auf Ihr lokales Unternehmen, anstatt auf die Konzernmutter zu beziehen.

Wie viele Mitarbeiter beschäftigt Ihr Unternehmen aktuell?
(Bitte geben Sie Anzahl an vollzeit-äquivalenten Mitarbeitern (FTE) an. *Stichtag: 30. Juni 2014.*)

Gesamtzahl:

Davon Produktionsmitarbeiter:

Wie viele Produktionsstandorte hat Ihr Unternehmen?
(Bitte geben Sie die Anzahl an Produktionsstandorten zum Stichtag 30. Juni 2014 an)

In dem Land, in welchem Ihr Unternehmen ansässig ist:

In allen anderen Ländern:

8 Page_04_TransitionB1B2

Umfrage im Rahmen des Forschungsprojektes

„Business Performance Measurement in the Metal Finishing Industry"

Umfrageteil 2

Vielen Dank, dass Sie den ersten Umfrageteil abgeschlossen haben.

Der zweite Umfrageteil befasst sich mit den generellen Charakteristika Ihres Unternehmens. Diese Charakteristika werden benötigt um die Strategie, die Ausrichtung und das Verhalten Ihres Unternehmens verstehen zu können.

Diese Unternehmenscharakteristika sind die Basis, um darauf ein Business Performance Measurement Modell zu entwickeln.
Bitte beantworten Sie alle Fragen entsprechend Ihrer Einschätzung und Erfahrung - es gibt keine "richtigen" oder "falschen" Antworten.

"Unternehmen" im Kontext unserer Forschung bezieht sich auf das Rechtssubjekt, für das Sie arbeiten.
Beispiel: Falls Ihr Unternehmen Teil eines internationalen Konzerns ist, so bitten wir Sie sich bei der Beantwortung der Fragen auf Ihr lokales Unternehmen, anstatt auf die Konzernmutter zu beziehen.

9 Page_05

2. Charakteristika des Unternehmens

"Unternehmen" im Kontext unserer Forschung bezieht sich auf das Rechtssubjekt, für das Sie arbeiten.
Beispiel: Falls Ihr Unternehmen Teil eines internationalen Konzerns ist, so bitten wir Sie sich bei der Beantwortung der Fragen auf Ihr lokales Unternehmen, anstatt auf die Konzernmutter zu beziehen.

Bitte bewerten Sie die folgenden Aussagen bezüglich Ihrer Unternehmenscharakteristika:

	Stimme überhaupt nicht zu						Stimme voll und ganz zu
Wir sind auf sich verändernde Kundenwünsche und -bedürfnisse eingestellt und reagieren auf diese Veränderungen frühzeitig.	○	○	○	○	○	○	○
Wir beobachten die Entwicklung von formellen und informellen Kundenbeschwerden genau.	○	○	○	○	○	○	○
Wir bitten unsere Kunden um Ihr Feedback bezüglich Qualität und unserem Service.	○	○	○	○	○	○	○
Wir stimmen uns mit unseren Kunden ab, um Zuverlässigkeit, Antwortverhalten und andere Standards zu setzen.	○	○	○	○	○	○	○
Der zentrale Zweck unseres Geschäfts ist die Befriedigung von Kundenbedürfnissen.	○	○	○	○	○	○	○
Die Kundenorientierung spiegelt sich in unserer Geschäftsplanung wider.	○	○	○	○	○	○	○
Wir stellen Produkte her beziehungsweise bieten Dienstleistungen an, die Kundenerwartungen befriedigen und/oder übersteigen.	○	○	○	○	○	○	○
	Stimme						Stimme

	überhaupt nicht zu						voll und ganz zu
Wir sind immer bestrebt, unsere direkten persönlichen Kundenkontakte zu steigern.	○	○	○	○	○	○	○
Um die Anforderungen unserer Kunden festzustellen, gehen wir aktiv auf unsere Kunden zu.	○	○	○	○	○	○	○
Unsere Qualität basiert auf den Anforderungen der Kunden.	○	○	○	○	○	○	○
Den Kunden in die Ausgestaltung unserer Produkte/Dienstleistungen einzubinden, ist entscheidend für uns.	○	○	○	○	○	○	○

10 Page_06

2. Charakteristika des Unternehmens

"Unternehmen" im Kontext unserer Forschung bezieht sich auf das Rechtssubjekt, für das Sie arbeiten.
Beispiel: Falls Ihr Unternehmen Teil eines internationalen Konzerns ist, so bitten wir Sie sich bei der Beantwortung der Fragen auf Ihr lokales Unternehmen, anstatt auf die Konzernmutter zu beziehen.

Bitte bewerten Sie die folgenden Aussagen bezüglich Ihrer Unternehmenscharakteristika:

	Stimme überhaupt nicht zu						Stimme voll und ganz zu
Es ist nicht unsere Strategie, Produkte zum niedrigsten Preis anzubieten.	○	○	○	○	○	○	○
Unsere Strategie basiert auf Qualität anstatt auf dem Preis.	○	○	○	○	○	○	○
Innovation hat für uns eine größere Bedeutung als der Preis.	○	○	○	○	○	○	○
Wir legen mehr Wert auf den Kundenservice als auf den Preis.	○	○	○	○	○	○	○
Unsere Strategie zielt darauf ab, leistungsfähige Produkte zu liefern beziehungsweise effiziente Dienstleistungen anzubieten.	○	○	○	○	○	○	○
Wir legen Wert darauf, Produkte/Dienstleistungen schnell auf den Markt zu bringen.	○	○	○	○	○	○	○

Bitte ordnen Sie die folgenden Faktoren den Plätzen 1, 2 und 3 entsprechend ihrer Priorität in Ihrem Unternehmen zu:

(Bitte klicken Sie den Faktor mit der höchsten Priorität für Ihr Unternehmen zuerst an - „1" hat die höchste Priorität.)

	1	2	3
Kosten/Preis	○	○	○
Qualität	○	○	○
Liefertreue	○	○	○

11 Page_07

2. Charakteristika des Unternehmens

"Unternehmen" im Kontext unserer Forschung bezieht sich auf das Rechtssubjekt, für das Sie arbeiten.
Beispiel: Falls Ihr Unternehmen Teil eines internationalen Konzerns ist, so bitten wir Sie sich bei der Beantwortung der Fragen auf Ihr lokales Unternehmen, anstatt auf die Konzernmutter zu beziehen.

Bitte bewerten Sie die folgenden Aussagen bezüglich möglicher Unsicherheiten in Ihrem Unternehmen:

	Nie vorhersehbar						Immer vorhersehbar
Tatsächliche Konsumenten unserer Produkte/Dienstleistungen sind ...	○	○	○	○	○	○	○
Konkurrenten bei der Beschaffung von Rohmaterial sind ...	○	○	○	○	○	○	○
Konkurrenten um unsere Kunden sind ...	○	○	○	○	○	○	○
Die rahmengebende Gesetzgebung für unsere Industrie ist ...	○	○	○	○	○	○	○
Die politischen Ansichten und Einstellungen der Öffentlichkeit gegenüber unserer Industrie sind ...	○	○	○	○	○	○	○

	Stimme überhaupt nicht zu						Stimme voll und ganz zu
Unsere Industrie ist durch sich rasant ändernde Technologien charakterisiert.	○	○	○	○	○	○	○
Wenn wir uns bezüglich Technologieänderungen nicht auf dem Laufenden halten, wird es schwierig für uns, wettbewerbsfähig zu bleiben.	○	○	○	○	○	○	○

Der Anteil veralteter Prozesse ist in unserer Industrie hoch.							
Die Produktionstechnologien ändern sich oft und in ausreichendem Maß.							

	Stimme überhaupt nicht zu						Stimme voll und ganz zu
Unsere Produktionsprogrammplanung sieht einen hohen Anteil an Bedarfsschwankung vor.							
Der Bedarf der Kunden schwankt drastisch von Woche zu Woche.							
Unser interner Beschaffungsbedarf (bspw. für Betriebsmittel) schwankt drastisch von Woche zu Woche.							
Wir haben einen Lagerbestand der wichtigsten Materialien von mehreren Wochen, um auf den schwankenden Kundenbedarf reagieren zu können.							
Das Ausmaß und die Zusammensetzung des Bedarfs der Kunden sind schwierig vorauszusehen.							

12 Page_08

2. Charakteristika des Unternehmens

"Unternehmen" im Kontext unserer Forschung bezieht sich auf das Rechtssubjekt, für das Sie arbeiten.
Beispiel: Falls Ihr Unternehmen Teil eines internationalen Konzerns ist, so bitten wir Sie sich bei der Beantwortung der Fragen auf Ihr lokales Unternehmen, anstatt auf die Konzernmutter zu beziehen.

Bitte bewerten Sie die folgenden Aussagen bezüglich des Verhaltens Ihres Unternehmens:

	Stimme überhaupt nicht zu						Stimme voll und ganz zu
Unser Unternehmen nimmt Kundenbestellungen an, die über unsere bestehenden Produkte und Dienstleistungen hinausgehen.							
Wir vertreiben Produkte und bieten Dienstleistungen an, die komplett neu für unser Unternehmen sind.							
Häufig nutzen wir neue Möglichkeiten in neuen Absatzmärkten.							
Unser Unternehmen nutzt regelmäßig neue Vertriebskanäle.							

	Stimme überhaupt nicht zu						Stimme voll und ganz zu
Wir machen häufig kleine Anpassungen an unseren existierenden Produkten und Dienstleistungen.							
Wir verbessern die Effizienz unserer Produktion und/oder die Effizienz bei der Bereitstellung unserer Dienstleistungen.							
Wir reduzieren die Kosten durch Effizienzvorteile, die aus steigenden Produktionsmengen bestehender Produkte generiert werden.							
Unser Unternehmen erweitert den Service für bestehende Kunden.							

13 Page_09

2. Charakteristika des Unternehmens

"Unternehmen" im Kontext unserer Forschung bezieht sich auf das Rechtssubjekt, für das Sie arbeiten.
Beispiel: Falls Ihr Unternehmen Teil eines internationalen Konzerns ist, so bitten wir Sie sich bei der Beantwortung der Fragen auf Ihr lokales Unternehmen, anstatt auf die Konzernmutter zu beziehen.

Bitte bewerten Sie die Bedeutung der folgenden Flexibilitäts-Faktoren für Ihr Unternehmen:

	Extrem unbedeutend						Extrem bedeutend
Liefer-Flexibilität: Die Fähigkeit eines Unternehmens effektiv auf Änderungen bei geplanten Lieferterminen zu reagieren.							
Volumen-Flexibilität: Die Fähigkeit eines Unternehmens effektiv die Produktion zu steigern oder zu reduzieren als Reaktion auf den Kundenbedarf.							

Portfolio-Flexibilität: Die Fähigkeit eines Unternehmens effektiv eine Vielzahl verschiedener Produkte herzustellen oder eine Vielzahl verschiedener Dienstleistungen anzubieten.	○	○	○	○	○	○	○
Umstellungs-Flexibilität: Die Fähigkeit eines Unternehmens fortwährend effektiv mit Änderungen (Ergänzung, Wegfall) im Produkte- / Dienstleistungs-Portfolio umzugehen.	○	○	○	○	○	○	○
Modifizierungs-Flexibilität: Die Fähigkeit eines Unternehmens effektiv kleinere Anpassungen an aktuellen Produkten oder Dienstleistungen vorzunehmen, die aus Korrekturmaßnahmen oder sich ändernden Kundenanforderungen abgeleitet werden.	○	○	○	○	○	○	○

Bitte geben Sie die durchschnittliche Auftragsreichweite in Ihrem Unternehmen in Werktagen an:

○ 0 Werktage – wir arbeiten "just-in-time"

○ 1 bis 5 Werktage

○ 6 bis 10 Werktage

○ 11 bis 15 Werktage

○ 16 bis 20 Werktage

○ Mehr als 20 Werktage

14 Page_10_TransitionB2B3

Umfrage im Rahmen des Forschungsprojektes

„Business Performance Measurement in the Metal Finishing Industry"

Umfrageteil 3

Herzlichen Dank, dass Sie den zweiten Umfrageteil beantwortet haben.

Der dritte und letzte Umfrageteil befasst sich mit "Betrieblicher Leistungsmessung / Business Performance Measurement (BPM)" in Ihrem Unternehmen. *"Betriebliche Leistungsmessung - Business Performance Measurement"* kann im Kontext unserer Forschung als Prozess zur Überprüfung von Effektivität und Effizienz Ihrer Unternehmensaktivitäten* beschrieben werden.

"Kennzahlen" und / oder "Key Performance Indicators (KPIs)" können als Faktoren/Messgrößen in betrieblichen Leistungsmessungsprozessen eingesetzt werden:

- *"Kennzahlen"* im Kontext unserer Forschung sind sowohl finanzielle wie auch nicht-finanzielle Messgrößen wie bspw. Umsatz, Profit, Mitarbeiterzahl, Anzahl an Reklamationen.
- *"Key Performance Indicators"* sind mehr-dimensionale Messgrößen*, welche beispielsweise eine Kennzahl mit weiteren Dimensionen kombinieren (bspw. mit einer anderen Kennzahl oder anderen Faktoren wie der Zeit). Beispiele für KPIs: Umsatz pro Mitarbeiter, Arbeitsunfälle pro Monat.

Verschiedene Unternehmen nutzen Systeme zur betrieblichen Leistungsmessung ("Business Performance Measurement Systems - BPM Systems") in denen verschiedene Kennzahlen oder KPIs kombiniert und analysiert werden:

- Diese Systeme werden oft von sogenannten "generischen" BPM-System-Modellen adaptiert (bereits existierende, theoriegeleitete Modelle wie die Balanced Scorecard* oder das Performance-Prisma*). Diese generischen BPM-System-Modelle werden oftmals entsprechend der Unternehmenscharakteristika und -anforderungen angepasst, bevor sie als Unternehmens-BPM-System implementiert werden.
- Ein BPM-System kann natürlich auch direkt ohne Ableitung von einem generischen BPM-System-Modell entwickelt werden, um die Unternehmensanforderungen zu reflektieren.

Traditionelle Leistungsmessungssysteme sind zumeist auf finanzielle Daten und Kennzahlen fokussiert. Moderne betriebliche Leistungsmessungssysteme* decken neben den finanziellen auch nicht-finanzielle Messgrößen ab, um somit Wechselbeziehungen zwischen den Messgrößen zu analysieren und zu kontrollieren.

**Für weitere Informationen bezüglich BPM, KPI und BPM Systemen verweisen wir auf Gleich, R. (2011): Performance measurement. Konzepte, Fallstudien und Grundschema für die Praxis. 2nd Edition. München: Vahlen.*

In unserer Forschung zielen wir auf die Entwicklung eines BPM Systems für die Metallveredlungsindustrie ab.
In diesem Zusammenhang bitten wir Sie uns zu erläutern, wie Ihr Unternehmen *seine betriebliche Leistung misst.*

15 Page_11

Plausibility checks:

The responsibility for setting BPM rules/guidelines in our organization is NOT defined. (v_294) == quoted
AND (**Corporate Headquarters (v_137)** == quoted
OR **Executive Board (v_138)** == quoted
OR **Supervisory Board (v_139)** == quoted
OR **Owner (v_140)** == quoted
OR **Local Management (v_141)** == quoted
OR **Local Department Heads (v_142)** == quoted
OR **Other (please indicate): (v_143)** == quoted)

3. Betriebliche Leistungsmessung - Business Performance Measurement

"Business Performance Measurement" kann im Kontext unserer Forschung als Prozess zur Überprüfung von Effektivität und Effizienz Ihrer Unternehmensaktivitäten beschrieben werden.

Wer ist in Ihrem Unternehmen für die Erstellung von Leitlinien und Regeln bezüglich betrieblicher Leistungsmessung verantwortlich?

(Mehrfachnennungen erlaubt)

- Konzern-/Segment-Zentrale
- Vorstand
- Aufsichtsrat
- Eigentümer
- Lokale Geschäftsführung
- Lokale Abteilungsleiter
- Andere (bitte angeben):
- Die Verantwortlichkeit bezüglich der Erstellung von Leitlinien und Regeln zur betrieblichen Leistungsmessung ist in unserem Unternehmen NICHT definiert.

Bitte beurteilen Sie bezüglich Ihres Hauptproduktes oder Hauptdienstleistung die Position Ihres Unternehmens innerhalb der vergangenen zwei Jahre im Vergleich zu Ihren Wettbewerbern hinsichtlich der folgenden Dimensionen:

	Signifikant niedriger						Signifikant höher
Wachstum Marktanteil	○	○	○	○	○	○	○
Umsatzwachstum	○	○	○	○	○	○	○
Kapitalrentabilität - Return on Investment (ROI)	○	○	○	○	○	○	○
Wachstum der Kapitalrentabilität - Return on Investment (ROI)	○	○	○	○	○	○	○
Umsatzrentabilität - Return on Sales (ROS)	○	○	○	○	○	○	○

16 Page_12a

3. Betriebliche Leistungsmessung - Business Performance Measurement

"Kennzahlen" und / oder "Key Performance Indicators (KPIs)" können als Faktoren/Messgrößen in betrieblichen Leistungsmessungsprozessen eingesetzt werden:

- *"Kennzahlen"* im Kontext unserer Forschung sind sowohl finanzielle wie auch nicht-finanzielle Messgrößen wie bspw. Umsatz, Profit, Mitarbeiterzahl, Anzahl an Reklamationen.
- *"Key Performance Indicators"* sind mehr-dimensionale Messgrößen*, welche beispielsweise eine Kennzahl mit weiteren Dimensionen kombinieren (bspw. mit einer anderen Kennzahl oder anderen Faktoren wie der Zeit). Beispiele für KPIs: Umsatz pro Mitarbeiter, Arbeitsunfälle pro Monat.

Bitte beurteilen Sie die Bedeutung der folgenden Messgrößen in Ihrem Unternehmen:

	Extrem unbedeutend						Extrem bedeutend
Stückkosten	○	○	○	○	○	○	○
Kosten versus Budget	○	○	○	○	○	○	○
Qualität der angelieferten Teile vom Kunden	○	○	○	○	○	○	○
Prozessqualität	○	○	○	○	○	○	○
Reklamationen während der Gewährleistungszeit	○	○	○	○	○	○	○
Liefertreue zum Kunden	○	○	○	○	○	○	○
Lieferzeit	○	○	○	○	○	○	○
Produkt- / Maschinen-Rüstzeiten	○	○	○	○	○	○	○

Bitte beurteilen Sie das geschätzte Verhältnis von Fixkosten (bspw. Personalkosten) und variablen Kosten (bspw. Ausgaben für Material oder Betriebsmittel) zu den Gesamtkosten Ihres Unternehmens:

- Fixkosten 0% - Variable Kosten 100%
- Fixkosten 10% - Variable Kosten 90%
- Fixkosten 20% - Variable Kosten 80%
- Fixkosten 30% - Variable Kosten 70%
- Fixkosten 40% - Variable Kosten 60%
- Fixkosten 50% - Variable Kosten 50%
- Fixkosten 60% - Variable Kosten 40%
- Fixkosten 70% - Variable Kosten 30%
- Fixkosten 80% - Variable Kosten 20%
- Fixkosten 90% - Variable Kosten 10%
- Fixkosten 100% - Variable Kosten 0%

17 Page_12b

Plausibility checks:

Our organization is NOT using external business / market indicators for Business Performance Measurement. (v_334) == quoted
AND (**GDP - Gross Domestic Product (v_330)** == quoted
OR **PMI - Purchasing Managers Index (v_331)** == quoted
OR **Others (please indicate): (v_332)** == quoted)

3. Betriebliche Leistungsmessung - Business Performance Measurement

"Kennzahlen" und / oder "Key Performance Indicators (KPIs)" können als Faktoren/Messgrößen in betrieblichen Leistungsmessungsprozessen eingesetzt werden:

- *"Kennzahlen"* im Kontext unserer Forschung sind sowohl finanzielle wie auch nicht-finanzielle Messgrößen wie bspw. Umsatz, Profit, Mitarbeiterzahl, Anzahl an Reklamationen.
- *"Key Performance Indicators"* sind mehr-dimensionale Messgrößen*, welche beispielsweise eine Kennzahl mit weiteren Dimensionen kombinieren (bspw. mit einer anderen Kennzahl oder anderen Faktoren wie der Zeit). Beispiele für KPIs: Umsatz pro Mitarbeiter, Arbeitsunfälle pro Monat.

Bitte nennen Sie bis zu fünf der bedeutendsten Key Performance Indicators (KPI) und / oder Kennzahlen, welche Sie nutzen, um die Leistung Ihres Unternehmens zu beurteilen:

(Bitte nennen Sie die KPI / Kennzahl sowie eine kurze Erläuterung bezüglich der Bedeutung und des Informationsgehaltes für Ihr Unternehmen. In jedes Textfeld können maximal 255 Zeichen eingegeben werden.)

	KPI / Kennzahl	Erläuterung
KPI / Kennzahl Nr. 1		
KPI / Kennzahl Nr. 2		
KPI / Kennzahl Nr. 3		
KPI / Kennzahl Nr. 4		
KPI / Kennzahl Nr. 5		

Bitte benennen Sie die externen Wirtschafts-/Marktindikatoren, die für die betriebliche Leistungsmessung Ihres Unternehmens relevant sind:

(Mehrfachnennungen erlaubt)

- BIP - Bruttoinlandsprodukt
- EMI - Einkaufsmanager Index
- Andere (bitte angeben):
- Unser Unternehmen nutzt KEINE externen Wirtschafts-/Marktindikatoren zur betrieblichen Leistungsmessung.

18 Page_12c

Plausibility checks:

We are NOT using IT Systems / Software for measuring our organization's business performance. (v_311) == quoted
AND (**Standardized Office Software (for example Microsoft Excel or Access, OpenOffice) (v_295)** == quoted
OR **Enterprise Resource Planning (ERP) Software (v_296)** == quoted
OR **Reporting based on Data Warehouses (for example IBM Business Warehouse) (v_297)** == quoted
OR **Customer Relationship Management (CRM) Software (v_298)** == quoted
OR **Others (please indicate): (v_309)** == quoted)

3. Betriebliche Leistungsmessung - Business Performance Measurement

Bitte nennen Sie die EDV-/IT-Systeme, welche Sie zur Messung der betrieblichen Leistung Ihres Unternehmens nutzen:

(Mehrfachnennungen erlaubt)

- Standard Office-Software (bspw. Microsoft Excel oder Access, OpenOffice)
- Enterprise Resource Planning (ERP) Software
- Reporting basierend auf Data Warehouses (bspw. IBM Business Warehouse)
- Customer Relationship Management (CRM) Software
- Andere (bitte angeben):
- Wir nutzen KEINE EDV-/IT-Systeme zur Messung der betrieblichen Leistung unseres Unternehmens.

19 Page_13

3. Betriebliche Leistungsmessung - Business Performance Measurement

Verschiedene Unternehmen nutzen Systeme zur betrieblichen Leistungsmessung ("Business Performance Measurement Systems - BPM Systems") in denen verschiedene Kennzahlen oder KPIs kombiniert und analysiert werden:

- Diese Systeme werden oft von sogenannten "generischen" BPM-System-Modellen adaptiert (bereits existierende, theoriegeleitete Modelle wie die Balanced Scorecard oder das Performance-Prisma). Diese generischen BPM-System-Modelle werden oftmals entsprechend der Unternehmenscharakteristika und -anforderungen angepasst, bevor sie als Unternehmens-BPM-System implementiert werden.
- Ein BPM-System kann natürlich auch direkt ohne Ableitung von einem generischen BPM-System-Modell entwickelt werden, um die Unternehmensanforderungen zu reflektieren.

Nutzen Sie in Ihrem Unternehmen ein System zur betrieblichen Leistungsmessung (BPM-System)?

- JA
- NEIN

20 Filter Filter_BPM_EN_YES

v_113 BPM_ExistsSystem Are you using a BPM-System within your organization? - BPM_ExistsSystem (From page 19: Page_13) equal 1

20.1 Page_14.1_BPM_YES

3. Betriebliche Leistungsmessung - Business Performance Measurement

Verschiedene Unternehmen nutzen Systeme zur betrieblichen Leistungsmessung ("Business Performance Measurement Systems - BPM Systems") in denen verschiedene Kennzahlen oder KPIs kombiniert und analysiert werden:

- Diese Systeme werden oft von sogenannten "generischen" BPM-System-Modellen adaptiert (bereits existierende, theoriegeleitete Modelle wie die Balanced Scorecard oder das Performance-Prisma). Diese generischen BPM-System-Modelle werden oftmals entsprechend der Unternehmenscharakteristika und -anforderungen angepasst, bevor sie als Unternehmens-BPM-System implementiert werden.
- Ein BPM-System kann natürlich auch direkt ohne Ableitung von einem generischen BPM-System-Modell entwickelt werden, um die Unternehmensanforderungen zu reflektieren.

Bitte nennen Sie die Aufgaben, für die Ihr BPM-System eingesetzt wird:

(Mehrfachnennungen erlaubt)

- Aufbau Unternehmensstrategie ("Plan")
- Implementierung Unternehmensstrategie ("Do")
- Überprüfung Unternehmensstrategie ("Check")
- Anpassung Unternehmensstrategie ("Act" - im Falle von Strategieabweichungen)
- Persönliche Zielsetzung
- Persönlicher Feedback-Prozess (inklusive Anreiz, Bonus oder Malus)
- Informationsversorgung (bspw. Management Information System)
- Kommunikation (bspw. Strategie, Zielerreichung, werden über das BPM-System kommuniziert)
- Operative Steuerung des Unternehmens
- Andere (bitte angeben):

Auf welchem generischen System basiert Ihre betriebliche Leistungsmessung (BPM)?

- Balanced Scorecard
- Performance-Pyramide
- Performance-Prisma
- SCOR-Modell
- EFQM-Modell
- Andere (bitte angeben):
- Unser BPM-System basiert NICHT auf einem generischen BPM System Modell - es ist unsere eigene Entwicklung.

20.1.1 Filter Filter_BPM_GenSys_Yes

v_352 BPM_GenSys On which generic system is your BPM based? - BPM_GenSys (From page 20.1: Page_14.1_BPM_YES) equal 1
or v_352 BPM_GenSys On which generic system is your BPM based? - BPM_GenSys (From page 20.1: Page_14.1_BPM_YES) equal 2
or v_352 BPM_GenSys On which generic system is your BPM based? - BPM_GenSys (From page 20.1: Page_14.1_BPM_YES) equal 3
or v_352 BPM_GenSys On which generic system is your BPM based? - BPM_GenSys (From page 20.1: Page_14.1_BPM_YES) equal 4
or v_352 BPM_GenSys On which generic system is your BPM based? - BPM_GenSys (From page 20.1: Page_14.1_BPM_YES) equal 5
or v_352 BPM_GenSys On which generic system is your BPM based? - BPM_GenSys (From page 20.1: Page_14.1_BPM_YES) equal 6

20.1.1.1 Page_15.1_GenSys_YES

3. Betriebliche Leistungsmessung - Business Performance Measurement

Bitte beschreiben Sie die hauptsächlichen Änderungen, die Ihr Unternehmen bei der Umsetzung des generischen BPM-Systems in Ihr Unternehmens-BPM-System vorgenommen hat:

Sind Sie mit Ihrem BPM-System zufrieden?

- JA
- NEIN

20.1.1.1.1 Filter Filter_BPM_Satisfy_NO

v_114 BPM_Yes_Satisfaction Are you satisfied with your BPM-System? - BPM_Yes_Satisfaction (From page 20.1.1.1: Page_15.1_GenSys_YES) equal 2

20.1.1.1.1.1 Page_16_BPM_Satisfaction_NO

3. Betriebliche Leistungsmessung - Business Performance Measurement

Bezüglich Ihrer Antwort nicht mit Ihrem BPM-System zufrieden zu sein - was könnte verbessert werden?

20.1.2 Filter Filter_BPM_GenSys_No

v_352 BPM_GenSys On which generic system is your BPM based? - BPM_GenSys (From page 20.1: Page_14.1_BPM_YES) equal 7

20.1.2.1 Page_15.2_GenSys_No

3. Betriebliche Leistungsmessung - Business Performance Measurement

Bitte nennen Sie die hauptsächlichen Ausprägungen des BPM-Systems in Ihrem Unternehmen:
(Mehrfachnennungen erlaubt)

- Enthält Finanzkennzahlen
- Enthält Nicht-Finanzkennzahlen (bspw. Produktionskennzahlen)
- Die Geschäftsführung nutzt das BPM-System
- Die Abteilungsleiter nutzen das BPM-System
- Die Finanzabteilung nutzt das BPM-System
- Andere (bitte angeben):

Sind Sie mit Ihrem BPM-System zufrieden?

- JA
- NEIN

20.1.2.1.1 Filter Filter_BPM_Satisfy_NO

v_359 BPM_Yes_Satisfaction Are you satisfied with your BPM-System? - BPM_Yes_Satisfaction (From page 20.1.2.1: Page_15.2_GenSys_No) equal 2

20.1.2.1.1.1 Page_16_BPM_Satisfaction_NO

3. Betriebliche Leistungsmessung - Business Performance Measurement

Bezüglich Ihrer Antwort nicht mit Ihrem BPM-System zufrieden zu sein - was könnte verbessert werden?

21 Filter Filter_BPM_EN_NO

v_113 BPM_ExistsSystem Are you using a BPM-System within your organization? - BPM_ExistsSystem (From page 19: Page_13) equal 2

21.1 Page_14.2_BPM_NO

3. Betriebliche Leistungsmessung - Business Performance Measurement

Was sind die Gründe für das Fehlen eines BPM-Systems in Ihrem Unternehmen?
(Mehrfachnennungen erlaubt)

- Implementierung zu zeitaufwendig
- Implementierung zu kostspielig
- Laufender Betrieb zu zeitaufwendig
- Laufender Betrieb zu kostspielig
- Kosten unverhältnismäßig zu Nutzen
- Bereits eingeführtes System wird nicht mehr genutzt - was waren die Gründe?:
- Andere Gründe (bitte angeben):

Welche hauptsächlichen Ausprägungen sollte Ihrer Meinung nach ein potenzielles BPM-Systems für Ihr Unternehmen enthalten?
(Mehrfachnennungen erlaubt)

- Sollte Finanzkennzahlen enthalten
- Sollte Nicht-Finanzkennzahlen (bspw. Produktionskennzahlen) enthalten
- Die Geschäftsführung sollte das BPM-System nutzen können

- Die Abteilungsleiter sollten das BPM-System nutzen können
- Die Finanzabteilung sollte das BPM-System nutzen können
- Andere (bitte angeben):

Planen Sie in den kommenden sechs Monaten ein BPM-System einzuführen?

- JA
- NEIN

22 FINAL

Sie haben die Umfrage *Business Performance Measurement in the Metal Finishing Industry* fast abgeschlossen.
Auf dieser Seite haben Sie die Möglichkeit Anmerkungen, Hinweise oder Ergänzungen zu der Umfrage zu hinterlassen. Sie können ebenfalls Ihre E-Mail-Adresse hinterlassen, falls Sie einen Kurzbericht über die Umfrageergebnisse erhalten möchten.

Hier haben Sie die Möglichkeit Anmerkungen, Hinweise oder Ergänzungen zu der Umfrage zu hinterlassen:

Bitte geben Sie Ihre E-Mail-Adresse ein, falls Sie einen Kurzbericht über die Umfrageergebnisse erhalten möchten:

Bitte schließen die Umfrage mit einem Klick auf den "Weiter"-Knopf ab.

23 Endseite

Ihre Angaben und Daten werden anonymisiert und strikt vertraulich verarbeitet.

Wir versichern Ihnen, dass wir Ihre persönlichen Daten strikt getrennt von der anonymen Auswertung der Umfrage halten; Ihre Kontaktinformationen (E-Mail-Adresse falls angegeben) werden ausschließlich zur Zusendung des Kurzberichtes verwendet.

Für weitere Fragen kontaktieren Sie bitte **Herrn Mike Dietrich**: E-Mail: **mike.dietrich@students.ebs.edu** Telefon: [redacted]

Vielen Dank, dass Sie das Dissertations-Forschungsprojekt

Business Performance Measurement in the Metal Finishing Industry

unterstützt haben!

A.17 MFI Survey – Analyses

A.17.1: Turnover, Employees, and Production Sites of Responding Organizations

Country	Turnover 2013 [k loc.cur.]	Local Currency*	FX rate to EUR as of 31.12.2013	Turnover 2013 [kEUR]	Employees Total Number as of 30.06.2014	Turnover/ Employee [kEUR]	No. of Production Sites
Germany	24,000	EUR	1.00000	24,000	90	267	1
Germany	18,000	EUR	1.00000	18,000	100	180	1
Germany	15,000	EUR	1.00000	15,000	102	147	2
Germany	15,000	EUR	1.00000	15,000	103	146	2
Germany	6,200	EUR	1.00000	6,200	70	89	1
Germany	4,300	EUR	1.00000	4,300	37	116	1
Germany	2,600	EUR	1.00000	2,600	36	72	1
Germany	1,700	EUR	1.00000	1,700	20	85	1
Germany	1,600	EUR	1.00000	1,600	20	80	1
Germany	*No value*		*1.00000*	*No value*	50	---	1
Sub-Total Germany (Turn.: N=9 / Empl.: N=10)	*88,400*	*EUR*	*1.00000*	*88,400*	*628*	*131***	*12*
United Kingdom	100,000	GBP	1.19763	119,763	800	150	13
United Kingdom	40,000	GBP	1.19763	47,905	270	177	1
United Kingdom	18,000	GBP	1.19763	21,557	270	80	3
United Kingdom	5,700	GBP	1.19763	6,826	140	49	4
United Kingdom	5,265	GBP	1.19763	6,306	65	97	1
United Kingdom	4,943	GBP	1.19763	5,920	33	179	1
United Kingdom	3,000	GBP	1.19763	3,593	41	88	3
United Kingdom	2,000	GBP	1.19763	2,395	22	109	1
United Kingdom	1,800	GBP	1.19763	2,156	29	74	1
United Kingdom	1,300	GBP	1.19763	1,557	22	71	1
United Kingdom	1,100	GBP	1.19763	1,317	20	66	1
United Kingdom	1,050	GBP	1.19763	1,258	10	126	1
United Kingdom	850	GBP	1.19763	1,018	18	57	1
United Kingdom	350	GBP	1.19763	419	5	84	1
Sub-Total United Kingdom (N=14)	*185,358*	*GBP*	*1.19763*	*221,990*	*1,745*	*100*	*33*
Netherlands (N=1)	130,000	EUR	1.00000	130,000	980	133	1
Switzerland (N=1)	500	CHF	0.81573	408	3	136	1
TOTAL *(Turn.: N=25 / Empl.: N=26)*				**440,798**	**3,356**	**114****	**47**

* Source for foreign exchance (FX) rates - date set to December 31st, 2013: OANDA, 2015

** Excluding one German case with no value in turnover (N=25)

A.17.2a: Independent Samples – MFC Country – Means

Grouping Criterion: MFC Country		N	Mean	Std. Deviation	Std. Error Mean
We anticipate and respond to customers' evolving needs and wants.	Germany	10	6.70	.483	.153
	United Kingdom	14	6.00	.877	.234
We emphasize the evolution of formal and informal customer complaints.	Germany	10	6.10	0.738	.233
	United Kingdom	14	5.64	1.277	.341
We follow-up with customers for quality/service feedback.	Germany	10	6.20	0.789	.249
	United Kingdom	14	5.36	1.550	.414
We interact with customers to set reliability, responsiveness, and other standards.	Germany	10	5.70	1.567	.496
	United Kingdom	14	5.79	1.188	.318
Satisfying customer needs is the central purpose of our business.	Germany	10	6.40	0.699	.221
	United Kingdom	14	6.57	0.646	.173
Customer focus is reflected in our business planning.	Germany	10	6.40	0.699	.221
	United Kingdom	14	6.07	0.997	.267
We produce products and/or offer services that satisfy or exceed customer expectations.	Germany	10	6.60	0.516	.163
	United Kingdom	14	6.36	0.745	.199
We always try to increase our organization^s direct personal contacts with customers.	Germany	10	6.50	0.527	.167
	United Kingdom	14	5.79	0.802	.214
We are actively seeking customer inputs to determine their requirements.	Germany	10	6.20	1.033	.327
	United Kingdom	14	5.71	1.267	.339
We are using customer requirements as the basis for quality.	Germany	10	6.50	0.707	.224
	United Kingdom	14	6.14	0.864	.231
Involving customers in product/service design is crucial for our organization.	Germany	10	6.30	0.823	.260
	United Kingdom	14	5.14	1.657	.443
Our strategy cannot be described as concentrating solely on offering products with the lowest price.	Germany	10	6.30	1.889	.597
	United Kingdom	14	6.43	0.852	.228
Our strategy is based on quality performance rather than price.	Germany	10	6.90	0.316	.100
	United Kingdom	14	6.36	0.633	.169
We place greater emphasis on innovation than price.	Germany	10	6.10	0.876	.277
	United Kingdom	14	5.50	1.160	.310
We place greater emphasis on customer service than price.	Germany	10	5.80	1.229	.389
	United Kingdom	14	6.14	1.027	.275
Our strategy places importance on delivering products/services with high performance.	Germany	10	6.40	1.350	.427
	United Kingdom	14	6.29	0.994	.266
We emphasize launching new products/services quickly.	Germany	10	5.40	1.265	.400
	United Kingdom	14	4.71	1.541	.412
Actual users of our products/services are ...	Germany	10	4.50	1.509	.477
	United Kingdom	14	3.36	1.865	.498
Competitors for the supply of raw materials are ...	Germany	10	3.90	1.595	.504
	United Kingdom	14	4.57	1.284	.343
Competitors for our customers are ...	Germany	10	3.90	2.079	.657
	United Kingdom	14	3.64	1.216	.325
Government regulations controlling our industry are ...	Germany	10	3.00	1.155	.365
	United Kingdom	14	3.00	1.840	.492
The public^s political views and attitudes toward our industry are ...	Germany	10	2.70	1.059	.335
	United Kingdom	14	3.79	1.578	.422
Our industry is characterized by rapidly changing technology.	Germany	10	4.40	1.506	.476
	United Kingdom	14	3.36	1.499	.401
If we do not keep up with the changes in technology, it will be difficult for us to remain competitive.	Germany	10	5.90	1.287	.407
	United Kingdom	14	5.43	1.342	.359
The rate of process obsolescence is high in our industry.	Germany	10	4.00	1.333	.422
	United Kingdom	14	2.86	1.703	.455
The production technology changes frequently and sufficiently.	Germany	10	4.00	1.491	.471
	United Kingdom	14	2.64	1.082	.289
Our master production schedule has a high percentage of variation in demand.	Germany	10	5.20	1.874	.593
	United Kingdom	14	5.79	1.311	.350
Customer demands fluctuate drastically from week to week.	Germany	10	4.30	2.111	.667
	United Kingdom	14	5.79	1.929	.515
Our internal supply requirements (for example for consumables) vary drastically from week to week.	Germany	10	2.60	1.578	.499
	United Kingdom	14	3.21	1.369	.366
We keep weeks of inventory of critical materials to meet changing customer demands.	Germany	10	4.90	1.792	.567
	United Kingdom	14	4.71	2.268	.606

Grouping Criterion: MFC Country		N	Mean	Std. Deviation	Std. Error Mean
The volume and/or composition of customer demands are difficult to predict.	Germany	10	5.20	1.932	.611
	United Kingdom	14	5.57	1.453	.388
Our organization accepts demands that go beyond existing products and services.	Germany	10	5.00	1.944	.615
	United Kingdom	14	5.00	1.840	.492
We commercialize products and services that are completely new to our organization.	Germany	10	4.30	2.003	.633
	United Kingdom	14	4.07	1.639	.438
We frequently utilize new opportunities in new markets.	Germany	10	4.70	1.252	.396
	United Kingdom	14	4.79	0.893	.239
Our organization regularly uses new distribution channels.	Germany	10	2.50	1.179	.373
	United Kingdom	14	2.50	1.401	.374
We frequently make small adjustments to our existing products and services.	Germany	10	5.30	1.703	.539
	United Kingdom	14	4.79	2.190	.585
We improve the production efficiency for our products and/or the provision efficiency for our services.	Germany	10	6.00	0.943	.298
	United Kingdom	14	5.64	1.393	.372
We reduce costs by having efficiency effects channelled through the volume increase of existing products.	Germany	10	4.30	2.263	.716
	United Kingdom	14	4.50	1.506	.403
Our organization expands services for existing clients.	Germany	10	5.40	1.265	.400
	United Kingdom	14	5.36	0.842	.225
Delivery flexibility: The ability of an organization to effectively respond to changes in planned delivery dates.	Germany	10	6.40	0.699	.221
	United Kingdom	14	6.36	0.745	.199
Volume flexibility: The ability of an organization to effectively increase or decrease aggregate production in response to customers.	Germany	10	5.50	1.650	.522
	United Kingdom	14	6.14	0.864	.231
Mix flexibility: The ability of an organization to effectively produce a wide variety of different products or offer a wide variety of different services.	Germany	10	5.80	1.874	.593
	United Kingdom	14	6.29	0.726	.194
Changeover flexibility: The ability of an organization to effectively handle additions and subtractions to the product or service mix over time.	Germany	10	5.60	1.350	.427
	United Kingdom	14	5.86	1.406	.376
Modification flexibility: The ability of an organization to effectively implement minor changes in current products or services that result from corrective actions or changing customer requirements.	Germany	10	6.20	1.135	.359
	United Kingdom	14	6.00	1.038	.277
Market share growth	Germany	10	4.60	1.265	.400
	United Kingdom	14	4.79	0.893	.239
Sales growth	Germany	10	4.70	1.337	.423
	United Kingdom	14	4.93	0.730	.195
Return on investment (ROI)	Germany	10	4.70	1.059	.335
	United Kingdom	14	5.00	0.784	.210
Growth in ROI	Germany	10	4.70	1.252	.396
	United Kingdom	14	4.93	0.730	.195
Return on sales (ROS)	Germany	10	4.50	0.972	.307
	United Kingdom	14	5.07	0.730	.195
Unit product costs	Germany	10	3.40	2.271	.718
	United Kingdom	14	5.29	1.773	.474
Costs versus budget	Germany	10	5.40	1.174	.371
	United Kingdom	14	4.79	2.155	.576
Incoming customer parts quality	Germany	10	6.50	0.707	.224
	United Kingdom	14	5.93	1.328	.355
In process quality	Germany	10	6.90	0.316	.100
	United Kingdom	14	6.57	0.756	.202
Field failure under warranty	Germany	10	5.70	1.703	.539
	United Kingdom	14	4.93	2.056	.549
On-time delivery to customer	Germany	10	6.80	0.422	.133
	United Kingdom	14	6.43	0.514	.137
Delivery lead time	Germany	10	6.30	1.059	.335
	United Kingdom	14	6.36	0.842	.225
Product/machine change over times	Germany	10	4.50	2.550	.806
	United Kingdom	14	4.86	1.657	.443

A.17.2b: Independent Samples – MFC Country – T-Tests

Grouping Criterion: MFC Country		Levene's Test for Equality of Variances		t-test for Equality of Means						
		F	Sig.	t	df	Sig. (2-tailed)	Mean Difference	Std. Error Difference	95% Confidence Interval of the Difference	
									Lower	Upper
We anticipate and respond to customers' evolving needs and wants.	Equal variances assumed	3.481	.075	2.280	22	.033	.700	.307	.063	1.337
	Equal variances not assumed			2.502	20.933	.021	.700	.280	.118	1.282
We emphasize the evolution of formal and informal customer complaints.	Equal variances assumed	3.709	.067	1.013	22	.322	.457	.451	-.478	1.393
	Equal variances not assumed			1.105	21.276	.281	.457	.414	-.402	1.316
We follow-up with customers for quality/service feedback.	Equal variances assumed	3.505	.075	1.574	22	.130	.843	.536	-.268	1.954
	Equal variances not assumed			1.743	20.287	.096	.843	.483	-.165	1.850
We interact with customers to set reliability, responsiveness, and other standards.	Equal variances assumed	0.280	.602	-0.153	22	.880	-.086	.561	-1.250	1.079
	Equal variances not assumed			-0.146	16.039	.886	-.086	.589	-1.333	1.162
Satisfying customer needs is the central purpose of our business.	Equal variances assumed	0.156	.697	-0.619	22	.542	-.171	.277	-.745	0.402
	Equal variances not assumed			-0.611	18.551	.549	-.171	.281	-.760	0.417
Customer focus is reflected in our business planning.	Equal variances assumed	1.014	.325	0.894	22	.381	.329	.367	-.433	1.091
	Equal variances not assumed			0.949	22.000	.353	.329	.346	-.390	1.047
We produce products and/or offer services that satisfy or exceed customer expectations.	Equal variances assumed	2.233	.149	0.887	22	.385	.243	.274	-.325	0.810
	Equal variances not assumed			0.943	21.996	.356	.243	.257	-.291	0.777
We always try to increase our organization's direct personal contacts with customers.	Equal variances assumed	0.313	.581	2.456	22	.022	.714	.291	.111	1.317
	Equal variances not assumed			2.631	21.906	.015	.714	.271	.151	1.277
We are actively seeking customer inputs to determine their requirements.	Equal variances assumed	0.837	.370	0.997	22	.330	.486	.487	-.525	1.496
	Equal variances not assumed			1.033	21.526	.313	.486	.470	-.491	1.462
We are using customer requirements as the basis for quality.	Equal variances assumed	0.762	.392	1.073	22	.295	.357	.333	-.333	1.047
	Equal variances not assumed			1.111	21.505	.279	.357	.322	-.311	1.025
Involving customers in product/service design is crucial for our organization.	Equal variances assumed	7.031	.015	2.027	22	.055	1.157	.571	-.027	2.341
	Equal variances not assumed			2.252	20.072	.036	1.157	.514	.086	2.229
Our strategy cannot be described as concentrating solely on offering products with the lowest price.	Equal variances assumed	1.216	.282	-0.226	22	.823	-.129	.569	-1.308	1.051
	Equal variances not assumed			-0.201	11.634	.844	-.129	.639	-1.526	1.269
Our strategy is based on quality performance rather than price.	Equal variances assumed	11.475	.003	2.487	22	.021	.543	.218	.090	0.996
	Equal variances not assumed			2.761	20.118	.012	.543	.197	.133	0.953
We place greater emphasis on innovation than price.	Equal variances assumed	0.781	.386	1.376	22	.183	.600	.436	-.304	1.504
	Equal variances not assumed			1.443	21.892	.163	.600	.416	-.262	1.462
We place greater emphasis on customer service than price.	Equal variances assumed	0.007	.933	-0.743	22	.465	-.343	.461	-1.300	0.614
	Equal variances not assumed			-0.720	17.245	.481	-.343	.476	-1.346	0.660
Our strategy places importance on delivering products/services with high performance.	Equal variances assumed	0.247	.624	0.239	22	.813	.114	.477	-.876	1.104
	Equal variances not assumed			0.227	15.698	.823	.114	.503	-.953	1.182
We emphasize launching new products/services quickly.	Equal variances assumed	1.288	.269	1.155	22	.261	.686	.594	-.546	1.917
	Equal variances not assumed			1.194	21.481	.245	.686	.574	-.506	1.878

Grouping Criterion: MFC Country		Levene's Test for Equality of Variances		t-test for Equality of Means						
		F	Sig.	t	df	Sig. (2-tailed)	Mean Difference	Std. Error Difference	95% Confidence Interval of the Difference Lower	Upper
Actual users of our products/services are ...	Equal variances assumed	2.809	.108	1.597	22	.125	1.143	.716	-.341	2.627
	Equal variances not assumed			1.656	21.572	.112	1.143	.690	-.290	2.576
Competitors for the supply of raw materials are ...	Equal variances assumed	0.051	.823	-1.142	22	.266	-.671	.588	-1.890	0.547
	Equal variances not assumed			-1.101	16.770	.287	-.671	.610	-1.960	0.617
Competitors for our customers are ...	Equal variances assumed	4.660	.042	0.382	22	.706	.257	.673	-1.138	1.653
	Equal variances not assumed			0.351	13.381	.731	.257	.733	-1.323	1.837
Government regulations controlling our industry are ...	Equal variances assumed	2.411	.135	0.000	22	1.000	.000	.661	-1.370	1.370
	Equal variances not assumed			0.000	21.741	1.000	.000	.612	-1.271	1.271
The public's political views and attitudes toward our industry are ...	Equal variances assumed	0.577	.456	-1.888	22	.072	-1.086	.575	-2.279	0.107
	Equal variances not assumed			-2.016	21.956	.056	-1.086	.539	-2.203	0.031
Our industry is characterized by rapidly changing technology.	Equal variances assumed	0.044	.836	1.677	22	.108	1.043	.622	-.247	2.332
	Equal variances not assumed			1.676	19.493	.110	1.043	.622	-.257	2.343
If we do not keep up with the changes in technology, it will be difficult for us to remain competitive.	Equal variances assumed	0.009	.926	0.863	22	.398	.471	.547	-.662	1.605
	Equal variances not assumed			0.869	20.047	.395	.471	.542	-.660	1.603
The rate of process obsolescence is high in our industry.	Equal variances assumed	1.340	.259	1.767	22	.091	1.143	.647	-.199	2.485
	Equal variances not assumed			1.842	21.750	.079	1.143	.620	-.145	2.431
The production technology changes frequently and sufficiently.	Equal variances assumed	1.330	.261	2.591	22	.017	1.357	.524	.271	2.443
	Equal variances not assumed			2.454	15.524	.026	1.357	.553	.182	2.532
Our master production schedule has a high percentage of variation in demand.	Equal variances assumed	0.935	.344	-0.903	22	.376	-.586	.648	-1.930	0.759
	Equal variances not assumed			-0.851	15.118	.408	-.586	.688	-2.052	0.881
Customer demands fluctuate drastically from week to week.	Equal variances assumed	1.099	.306	-1.790	22	.087	-1.486	.830	-3.207	0.236
	Equal variances not assumed			-1.762	18.404	.095	-1.486	.843	-3.255	0.283
Our internal supply requirements (for example for consumables) vary drastically from week to week.	Equal variances assumed	0.396	.536	-1.018	22	.320	-.614	.604	-1.866	0.638
	Equal variances not assumed			-0.993	17.732	.334	-.614	.619	-1.915	0.687
We keep weeks of inventory of critical materials to meet changing customer demands.	Equal variances assumed	1.102	.305	0.215	22	.832	.186	.864	-1.606	1.977
	Equal variances not assumed			0.224	21.705	.825	.186	.830	-1.536	1.908
The volume and/or composition of customer demands are difficult to predict.	Equal variances assumed	0.092	.765	-0.539	22	.596	-.371	.690	-1.802	1.059
	Equal variances not assumed			-0.513	15.935	.615	-.371	.724	-1.907	1.164
Our organization accepts demands that go beyond existing products and services.	Equal variances assumed	0.433	.517	0.000	22	1.000	.000	.780	-1.617	1.617
	Equal variances not assumed			0.000	18.858	1.000	.000	.787	-1.648	1.648
We commercialize products and services that are completely new to our organization.	Equal variances assumed	0.560	.462	0.307	22	.762	.229	.744	-1.314	1.771
	Equal variances not assumed			0.297	16.981	.770	.229	.770	-1.396	1.853
We frequently utilize new opportunities in new markets.	Equal variances assumed	1.013	.325	-0.196	22	.846	-.086	.437	-.991	0.820
	Equal variances not assumed			-0.185	15.326	.855	-.086	.462	-1.069	0.897

Grouping Criterion: MFC Country		Levene's Test for Equality of Variances		t-test for Equality of Means						
		F	Sig.	t	df	Sig. (2-tailed)	Mean Difference	Std. Error Difference	95% Confidence Interval of the Difference Lower	Upper
Our organization regularly uses new distribution channels.	Equal variances assumed	0.801	.380	0.000	22	1.000	.000	.544	-1.129	1.129
	Equal variances not assumed			0.000	21.306	1.000	.000	.528	-1.097	1.097
We frequently make small adjustments to our existing products and services.	Equal variances assumed	1.064	.313	0.619	22	.542	.514	.830	-1.207	2.236
	Equal variances not assumed			0.647	21.781	.525	.514	.795	-1.136	2.165
We improve the production efficiency for our products and/or the provision efficiency for our services.	Equal variances assumed	2.106	.161	0.702	22	.490	.357	.509	-.698	1.412
	Equal variances not assumed			0.749	21.970	.462	.357	.477	-.632	1.346
We reduce costs by having efficiency effects channelled through the volume increase of existing products.	Equal variances assumed	3.200	.087	-0.261	22	.797	-.200	.768	-1.792	1.392
	Equal variances not assumed			-0.244	14.586	.811	-.200	.821	-1.955	1.555
Our organization expands services for existing clients.	Equal variances assumed	1.750	.200	0.100	22	.921	.043	.429	-.847	0.932
	Equal variances not assumed			0.093	14.586	.927	.043	.459	-.938	1.023
Delivery flexibility: The ability of an organization to effectively respond to changes in planned delivery dates.	Equal variances assumed	0.106	.748	0.142	22	.888	.043	.301	-.581	0.667
	Equal variances not assumed			0.144	20.281	.887	.043	.298	-.577	0.663
Volume flexibility: The ability of an organization to effectively increase or decrease aggregate production in response to customers.	Equal variances assumed	4.203	.052	-1.245	22	.226	-.643	.516	-1.714	0.428
	Equal variances not assumed			-1.127	12.542	.281	-.643	.571	-1.880	0.594
Mix flexibility: The ability of an organization to effectively produce a wide variety of different products or offer a wide variety of different services.	Equal variances assumed	8.548	.008	-0.887	22	.385	-.486	.547	-1.621	0.650
	Equal variances not assumed			-0.779	10.948	.453	-.486	.624	-1.859	0.887
Changeover flexibility: The ability of an organization to effectively handle additions and subtractions to the product or service mix over time.	Equal variances assumed	0.013	.911	-0.449	22	.658	-.257	.573	-1.445	0.931
	Equal variances not assumed			-0.452	20.030	.656	-.257	.569	-1.443	0.929
Modification flexibility: The ability of an organization to effectively implement minor changes in current products or services that result from corrective actions or changing customer requirements.	Equal variances assumed	0.838	.370	0.448	22	.659	.200	.447	-.726	1.126
	Equal variances not assumed			0.441	18.409	.664	.200	.454	-.752	1.152
Market share growth	Equal variances assumed	1.628	.215	-0.423	22	.677	-.186	.439	-1.097	0.725
	Equal variances not assumed			-0.399	15.209	.696	-.186	.466	-1.177	0.806
Sales growth	Equal variances assumed	5.955	.023	-0.540	22	.595	-.229	.424	-1.107	0.650
	Equal variances not assumed			-0.491	12.836	.632	-.229	.466	-1.236	0.779
Return on investment (ROI)	Equal variances assumed	2.552	.124	-0.799	22	.433	-.300	.376	-1.079	0.479
	Equal variances not assumed			-0.759	15.757	.459	-.300	.395	-1.139	0.539
Growth in ROI	Equal variances assumed	5.310	.031	-0.565	22	.578	-.229	.405	-1.068	0.611
	Equal variances not assumed			-0.518	13.359	.613	-.229	.441	-1.179	0.722
Return on sales (ROS)	Equal variances assumed	1.831	.190	-1.648	22	.114	-.571	.347	-1.291	0.148
	Equal variances not assumed			-1.570	15.926	.136	-.571	.364	-1.343	0.201
Unit product costs	Equal variances assumed	1.246	.276	-2.287	22	.032	-1.886	.825	-3.596	-0.176
	Equal variances not assumed			-2.192	16.392	.043	-1.886	.860	-3.706	-0.066

Grouping Criterion: MFC Country		Levene's Test for Equality of Variances		t-test for Equality of Means						
		F	Sig.	t	df	Sig. (2-tailed)	Mean Difference	Std. Error Difference	95% Confidence Interval of the Difference	
									Lower	Upper
Costs versus budget	Equal variances assumed	3.358	.080	0.816	22	.423	.614	.753	-.947	2.176
	Equal variances not assumed			0.897	20.848	.380	.614	.685	-.811	2.040
Incoming customer parts quality	Equal variances assumed	4.349	.049	1.236	22	.229	.571	.462	-.387	1.530
	Equal variances not assumed			1.362	20.665	.188	.571	.420	-.302	1.445
In process quality	Equal variances assumed	8.698	.007	1.290	22	.211	.329	.255	-.200	0.857
	Equal variances not assumed			1.458	18.543	.162	.329	.225	-.144	0.801
Field failure under warranty	Equal variances assumed	1.403	.249	0.971	22	.342	.771	.795	-.876	2.419
	Equal variances not assumed			1.003	21.420	.327	.771	.769	-.826	2.369
On-time delivery to customer	Equal variances assumed	5.728	.026	1.876	22	.074	.371	.198	-.039	0.782
	Equal variances not assumed			1.941	21.481	.065	.371	.191	-.026	0.769
Delivery lead time	Equal variances assumed	0.771	.390	-0.147	22	.884	-.057	.388	-.862	0.747
	Equal variances not assumed			-0.142	16.612	.889	-.057	.404	-.910	0.796
Product/machine change over times	Equal variances assumed	4.623	.043	-0.417	22	.681	-.357	.857	-2.134	1.420
	Equal variances not assumed			-0.388	14.349	.704	-.357	.920	-2.326	1.611

Items with significance levels of being below 5% show significant statistical differences between the two independent samples.

A.17.3a: Independent Samples – MFC Size Class – Means

Grouping Criterion: MFC Size Class		N	Mean	Std. Deviation	Std. Error Mean
We anticipate and respond to customers' evolving needs and wants.	Larger MFCs >49 Emps.	12	6.58	.669	.193
	Smaller MFCs <50 Emps.	14	6.07	.829	.221
We emphasize the evolution of formal and informal customer complaints.	Larger MFCs >49 Emps.	12	6.17	0.718	.207
	Smaller MFCs <50 Emps.	14	5.57	1.222	.327
We follow-up with customers for quality/service feedback.	Larger MFCs >49 Emps.	12	6.33	0.778	.225
	Smaller MFCs <50 Emps.	14	5.21	1.424	.381
We interact with customers to set reliability, responsiveness, and other standards.	Larger MFCs >49 Emps.	12	5.83	1.403	.405
	Smaller MFCs <50 Emps.	14	5.71	1.204	.322
Satisfying customer needs is the central purpose of our business.	Larger MFCs >49 Emps.	12	6.42	0.669	.193
	Smaller MFCs <50 Emps.	14	6.50	0.650	.174
Customer focus is reflected in our business planning.	Larger MFCs >49 Emps.	12	6.50	0.674	.195
	Smaller MFCs <50 Emps.	14	5.93	0.917	.245
We produce products and/or offer services that satisfy or exceed customer expectations.	Larger MFCs >49 Emps.	12	6.50	0.674	.195
	Smaller MFCs <50 Emps.	14	6.36	0.633	.169
We always try to increase our organization^s direct personal contacts with customers.	Larger MFCs >49 Emps.	12	6.33	0.651	.188
	Smaller MFCs <50 Emps.	14	5.79	0.802	.214
We are actively seeking customer inputs to determine their requirements.	Larger MFCs >49 Emps.	12	6.17	1.030	.297
	Smaller MFCs <50 Emps.	14	5.71	1.204	.322
We are using customer requirements as the basis for quality.	Larger MFCs >49 Emps.	12	6.42	0.669	.193
	Smaller MFCs <50 Emps.	14	6.14	0.864	.231
Involving customers in product/service design is crucial for our organization.	Larger MFCs >49 Emps.	12	6.33	0.651	.188
	Smaller MFCs <50 Emps.	14	5.07	1.639	.438
Our strategy cannot be described as concentrating solely on offering products with the lowest price.	Larger MFCs >49 Emps.	12	6.08	1.730	.499
	Smaller MFCs <50 Emps.	14	6.57	0.852	.228
Our strategy is based on quality performance rather than price.	Larger MFCs >49 Emps.	12	6.50	0.674	.195
	Smaller MFCs <50 Emps.	14	6.57	0.646	.173
We place greater emphasis on innovation than price.	Larger MFCs >49 Emps.	12	5.83	1.030	.297
	Smaller MFCs <50 Emps.	14	5.64	1.216	.325
We place greater emphasis on customer service than price.	Larger MFCs >49 Emps.	12	6.00	1.206	.348
	Smaller MFCs <50 Emps.	14	5.93	0.997	.267
Our strategy places importance on delivering products/services with high performance.	Larger MFCs >49 Emps.	12	6.08	1.311	.379
	Smaller MFCs <50 Emps.	14	6.50	0.941	.251
We emphasize launching new products/services quickly.	Larger MFCs >49 Emps.	12	5.25	1.215	.351
	Smaller MFCs <50 Emps.	14	4.79	1.578	.422
Actual users of our products/services are ...	Larger MFCs >49 Emps.	12	3.75	1.603	.463
	Smaller MFCs <50 Emps.	14	3.64	1.985	.530
Competitors for the supply of raw materials are ...	Larger MFCs >49 Emps.	12	4.33	1.557	.449
	Smaller MFCs <50 Emps.	14	4.29	1.383	.370
Competitors for our customers are ...	Larger MFCs >49 Emps.	12	4.00	1.651	.477
	Smaller MFCs <50 Emps.	14	3.57	1.453	.388
Government regulations controlling our industry are ...	Larger MFCs >49 Emps.	12	3.33	1.723	.497
	Smaller MFCs <50 Emps.	14	2.93	1.439	.385
The public^s political views and attitudes toward our industry are ...	Larger MFCs >49 Emps.	12	3.17	1.337	.386
	Smaller MFCs <50 Emps.	14	3.79	1.672	.447
Our industry is characterized by rapidly changing technology.	Larger MFCs >49 Emps.	12	4.58	1.311	.379
	Smaller MFCs <50 Emps.	14	3.43	1.742	.465
If we do not keep up with the changes in technology, it will be difficult for us to remain competitive.	Larger MFCs >49 Emps.	12	5.83	1.193	.345
	Smaller MFCs <50 Emps.	14	5.50	1.401	.374
The rate of process obsolescence is high in our industry.	Larger MFCs >49 Emps.	12	3.83	1.115	.322
	Smaller MFCs <50 Emps.	14	3.00	1.840	.492
The production technology changes frequently and sufficiently.	Larger MFCs >49 Emps.	12	3.75	1.215	.351
	Smaller MFCs <50 Emps.	14	2.79	1.369	.366
Our master production schedule has a high percentage of variation in demand.	Larger MFCs >49 Emps.	12	5.58	1.564	.452
	Smaller MFCs <50 Emps.	14	5.14	1.748	.467
Customer demands fluctuate drastically from week to week.	Larger MFCs >49 Emps.	12	4.67	2.188	.632
	Smaller MFCs <50 Emps.	14	5.43	1.950	.521
Our internal supply requirements (for example for consumables) vary drastically from week to week.	Larger MFCs >49 Emps.	12	3.00	1.477	.426
	Smaller MFCs <50 Emps.	14	3.14	1.512	.404
We keep weeks of inventory of critical materials to meet changing customer demands.	Larger MFCs >49 Emps.	12	5.67	1.073	.310
	Smaller MFCs <50 Emps.	14	3.79	2.392	.639

Grouping Criterion: MFC Size Class		N	Mean	Std. Deviation	Std. Error Mean
The volume and/or composition of customer demands are difficult to predict.	Larger MFCs >49 Emps.	12	5.25	1.288	.372
	Smaller MFCs <50 Emps.	14	5.50	1.829	.489
Our organization accepts demands that go beyond existing products and services.	Larger MFCs >49 Emps.	12	4.83	1.946	.562
	Smaller MFCs <50 Emps.	14	5.21	1.672	.447
We commercialize products and services that are completely new to our organization.	Larger MFCs >49 Emps.	12	3.92	1.929	.557
	Smaller MFCs <50 Emps.	14	4.57	1.555	.416
We frequently utilize new opportunities in new markets.	Larger MFCs >49 Emps.	12	5.08	1.165	.336
	Smaller MFCs <50 Emps.	14	4.57	0.852	.228
Our organization regularly uses new distribution channels.	Larger MFCs >49 Emps.	12	2.58	1.165	.336
	Smaller MFCs <50 Emps.	14	2.71	1.541	.412
We frequently make small adjustments to our existing products and services.	Larger MFCs >49 Emps.	12	5.33	1.923	.555
	Smaller MFCs <50 Emps.	14	4.86	1.956	.523
We improve the production efficiency for our products and/or the provision efficiency for our services.	Larger MFCs >49 Emps.	12	6.17	0.835	.241
	Smaller MFCs <50 Emps.	14	5.57	1.399	.374
We reduce costs by having efficiency effects channelled through the volume increase of existing products.	Larger MFCs >49 Emps.	12	4.83	2.167	.626
	Smaller MFCs <50 Emps.	14	4.21	1.477	.395
Our organization expands services for existing clients.	Larger MFCs >49 Emps.	12	5.83	0.835	.241
	Smaller MFCs <50 Emps.	14	5.07	1.072	.286
Delivery flexibility: The ability of an organization to effectively respond to changes in planned delivery dates.	Larger MFCs >49 Emps.	12	6.42	0.669	.193
	Smaller MFCs <50 Emps.	14	6.21	0.802	.214
Volume flexibility: The ability of an organization to effectively increase or decrease aggregate production in response to customers.	Larger MFCs >49 Emps.	12	6.00	1.128	.326
	Smaller MFCs <50 Emps.	14	5.57	1.505	.402
Mix flexibility: The ability of an organization to effectively produce a wide variety of different products or offer a wide variety of different services.	Larger MFCs >49 Emps.	12	6.08	1.240	.358
	Smaller MFCs <50 Emps.	14	5.86	1.460	.390
Changeover flexibility: The ability of an organization to effectively handle additions and subtractions to the product or service mix over time.	Larger MFCs >49 Emps.	12	5.67	1.155	.333
	Smaller MFCs <50 Emps.	14	5.71	1.490	.398
Modification flexibility: The ability of an organization to effectively implement minor changes in current products or services that result from corrective actions or changing customer requirements.	Larger MFCs >49 Emps.	12	5.83	1.030	.297
	Smaller MFCs <50 Emps.	14	6.21	1.051	.281
Market share growth	Larger MFCs >49 Emps.	12	5.25	0.965	.279
	Smaller MFCs <50 Emps.	14	4.36	0.929	.248
Sales growth	Larger MFCs >49 Emps.	12	5.17	0.937	.271
	Smaller MFCs <50 Emps.	14	4.64	1.008	.269
Return on investment (ROI)	Larger MFCs >49 Emps.	12	5.17	0.835	.241
	Smaller MFCs <50 Emps.	14	4.71	0.914	.244
Growth in ROI	Larger MFCs >49 Emps.	12	5.25	1.055	.305
	Smaller MFCs <50 Emps.	14	4.64	0.842	.225
Return on sales (ROS)	Larger MFCs >49 Emps.	12	5.08	0.900	.260
	Smaller MFCs <50 Emps.	14	4.79	0.893	.239
Unit product costs	Larger MFCs >49 Emps.	12	4.33	2.348	.678
	Smaller MFCs <50 Emps.	14	4.50	2.175	.581
Costs versus budget	Larger MFCs >49 Emps.	12	5.50	1.679	.485
	Smaller MFCs <50 Emps.	14	4.50	1.912	.511
Incoming customer parts quality	Larger MFCs >49 Emps.	12	6.00	1.044	.302
	Smaller MFCs <50 Emps.	14	6.21	1.311	.350
In process quality	Larger MFCs >49 Emps.	12	6.67	0.651	.188
	Smaller MFCs <50 Emps.	14	6.64	0.745	.199
Field failure under warranty	Larger MFCs >49 Emps.	12	5.83	1.337	.386
	Smaller MFCs <50 Emps.	14	4.93	2.200	.588
On-time delivery to customer	Larger MFCs >49 Emps.	12	6.83	0.389	.112
	Smaller MFCs <50 Emps.	14	6.36	0.497	.133
Delivery lead time	Larger MFCs >49 Emps.	12	6.50	0.798	.230
	Smaller MFCs <50 Emps.	14	6.07	0.997	.267
Product/machine change over times	Larger MFCs >49 Emps.	12	4.83	2.250	.649
	Smaller MFCs <50 Emps.	14	4.79	1.805	.482

A.17.3b: Independent Samples – MFC Size Class – T-Tests

Grouping Criterion: MFC Size Class		Levene's Test for Equality of Variances		t-test for Equality of Means						
		F	Sig.	t	df	Sig. (2-tailed)	Mean Difference	Std. Error Difference	95% Confidence Interval of the Difference Lower	Upper
We anticipate and respond to customers' evolving needs and wants.	Equal variances assumed	0.452	.508	1.713	24	.100	.512	.299	-.105	1.129
	Equal variances not assumed			1.742	23.931	.094	.512	.294	-.095	1.118
We emphasize the evolution of formal and informal customer complaints.	Equal variances assumed	3.504	.073	1.480	24	.152	.595	.402	-.235	1.425
	Equal variances not assumed			1.539	21.457	.139	.595	.387	-.208	1.399
We follow-up with customers for quality/service feedback.	Equal variances assumed	2.187	.152	2.425	24	.023	1.119	.461	.167	2.071
	Equal variances not assumed			2.532	20.677	.020	1.119	.442	.199	2.039
We interact with customers to set reliability, responsiveness, and other standards.	Equal variances assumed	0.156	.696	0.233	24	.818	.119	.511	-.936	1.174
	Equal variances not assumed			0.230	21.890	.820	.119	.517	-.954	1.192
Satisfying customer needs is the central purpose of our business.	Equal variances assumed	0.012	.912	-0.322	24	.751	-.083	.259	-.618	0.452
	Equal variances not assumed			-0.321	23.179	.751	-.083	.260	-.620	0.454
Customer focus is reflected in our business planning.	Equal variances assumed	0.229	.636	1.783	24	.087	.571	.320	-.090	1.233
	Equal variances not assumed			1.826	23.515	.081	.571	.313	-.075	1.218
We produce products and/or offer services that satisfy or exceed customer expectations.	Equal variances assumed	0.086	.772	0.557	24	.583	.143	.257	-.387	0.673
	Equal variances not assumed			0.554	22.865	.585	.143	.258	-.391	0.677
We always try to increase our organization's direct personal contacts with customers.	Equal variances assumed	0.046	.832	1.890	24	.071	.548	.290	-.050	1.146
	Equal variances not assumed			1.921	23.947	.067	.548	.285	-.041	1.136
We are actively seeking customer inputs to determine their requirements.	Equal variances assumed	0.175	.679	1.020	24	.318	.452	.444	-.463	1.368
	Equal variances not assumed			1.032	24.000	.312	.452	.438	-.452	1.357
We are using customer requirements as the basis for quality.	Equal variances assumed	1.188	.287	0.891	24	.382	.274	.307	-.360	0.908
	Equal variances not assumed			0.910	23.786	.372	.274	.301	-.348	0.895
Involving customers in product/service design is crucial for our organization.	Equal variances assumed	9.860	.004	2.497	24	.020	1.262	.505	.219	2.305
	Equal variances not assumed			2.647	17.528	.017	1.262	.477	.258	2.265
Our strategy cannot be described as concentrating solely on offering products with the lowest price.	Equal variances assumed	1.387	.250	-0.934	24	.360	-.488	.523	-1.567	0.590
	Equal variances not assumed			-0.889	15.480	.387	-.488	.549	-1.655	0.678
Our strategy is based on quality performance rather than price.	Equal variances assumed	0.077	.783	-0.275	24	.785	-.071	.259	-.607	0.464
	Equal variances not assumed			-0.275	23.051	.786	-.071	.260	-.610	0.467
We place greater emphasis on innovation than price.	Equal variances assumed	0.255	.618	0.427	24	.673	.190	.446	-.731	1.111
	Equal variances not assumed			0.433	23.999	.669	.190	.440	-.718	1.099
We place greater emphasis on customer service than price.	Equal variances assumed	0.010	.921	0.165	24	.870	.071	.432	-.820	0.963
	Equal variances not assumed			0.163	21.440	.872	.071	.438	-.839	0.982
Our strategy places importance on delivering products/services with high performance.	Equal variances assumed	2.048	.165	-0.941	24	.356	-.417	.443	-1.331	0.497
	Equal variances not assumed			-0.917	19.612	.370	-.417	.454	-1.366	0.532
We emphasize launching new products/services quickly.	Equal variances assumed	2.692	.114	0.829	24	.415	.464	.560	-.691	1.620
	Equal variances not assumed			0.846	23.768	.406	.464	.549	-.668	1.597

Grouping Criterion: MFC Size Class		Levene's Test for Equality of Variances		t-test for Equality of Means						
		F	Sig.	t	df	Sig. (2-tailed)	Mean Difference	Std. Error Difference	95% Confidence Interval of the Difference	
									Lower	Upper
Actual users of our products/services are ...	Equal variances assumed	0.783	.385	0.150	24	.882	.107	.716	-1.370	1.585
	Equal variances not assumed			0.152	23.933	.880	.107	.704	-1.346	1.560
Competitors for the supply of raw materials are ...	Equal variances assumed	0.003	.958	0.083	24	.935	.048	.576	-1.142	1.237
	Equal variances not assumed			0.082	22.283	.936	.048	.582	-1.158	1.254
Competitors for our customers are ...	Equal variances assumed	0.010	.919	0.704	24	.488	.429	.609	-.827	1.685
	Equal variances not assumed			0.697	22.175	.493	.429	.615	-.846	1.703
Government regulations controlling our industry are ...	Equal variances assumed	0.449	.509	0.653	24	.520	.405	.620	-.875	1.684
	Equal variances not assumed			0.644	21.563	.527	.405	.629	-.901	1.710
The public^s political views and attitudes toward our industry are ...	Equal variances assumed	0.109	.745	-1.030	24	.313	-.619	.601	-1.860	0.621
	Equal variances not assumed			-1.048	23.907	.305	-.619	.591	-1.838	0.600
Our industry is characterized by rapidly changing technology.	Equal variances assumed	0.571	.457	1.883	24	.072	1.155	.613	-.111	2.421
	Equal variances not assumed			1.925	23.655	.066	1.155	.600	-.084	2.394
If we do not keep up with the changes in technology, it will be difficult for us to remain competitive.	Equal variances assumed	0.687	.415	0.647	24	.524	.333	.515	-.730	1.397
	Equal variances not assumed			0.655	24.000	.519	.333	.509	-.717	1.383
The rate of process obsolescence is high in our industry.	Equal variances assumed	3.196	.086	1.367	24	.184	.833	.610	-.425	2.092
	Equal variances not assumed			1.418	21.795	.170	.833	.588	-.386	2.053
The production technology changes frequently and sufficiently.	Equal variances assumed	0.073	.789	1.884	24	.072	.964	.512	-.092	2.020
	Equal variances not assumed			1.902	23.958	.069	.964	.507	-.082	2.011
Our master production schedule has a high percentage of variation in demand.	Equal variances assumed	0.068	.797	0.672	24	.508	.440	.655	-.912	1.793
	Equal variances not assumed			0.678	23.941	.504	.440	.650	-.901	1.782
Customer demands fluctuate drastically from week to week.	Equal variances assumed	0.435	.516	-0.939	24	.357	-.762	.811	-2.437	0.913
	Equal variances not assumed			-0.930	22.321	.362	-.762	.819	-2.459	0.935
Our internal supply requirements (for example for consumables) vary drastically from week to week.	Equal variances assumed	0.223	.641	-0.243	24	.810	-.143	.589	-1.358	1.072
	Equal variances not assumed			-0.243	23.554	.810	-.143	.587	-1.356	1.071
We keep weeks of inventory of critical materials to meet changing customer demands.	Equal variances assumed	14.371	.001	2.511	24	.019	1.881	.749	.335	3.427
	Equal variances not assumed			2.648	18.611	.016	1.881	.710	.392	3.370
The volume and/or composition of customer demands are difficult to predict.	Equal variances assumed	1.345	.258	-0.396	24	.695	-.250	.631	-1.552	1.052
	Equal variances not assumed			-0.407	23.211	.688	-.250	.614	-1.520	1.020
Our organization accepts demands that go beyond existing products and services.	Equal variances assumed	0.637	.433	-0.537	24	.596	-.381	.709	-1.845	1.083
	Equal variances not assumed			-0.531	21.905	.601	-.381	.718	-1.870	1.108
We commercialize products and services that are completely new to our organization.	Equal variances assumed	1.199	.284	-0.959	24	.347	-.655	.683	-2.064	0.755
	Equal variances not assumed			-0.942	21.123	.357	-.655	.695	-2.099	0.790
We frequently utilize new opportunities in new markets.	Equal variances assumed	0.037	.848	1.292	24	.209	.512	.396	-.306	1.330
	Equal variances not assumed			1.261	19.865	.222	.512	.406	-.335	1.359

Grouping Criterion: MFC Size Class		Levene's Test for Equality of Variances		t-test for Equality of Means						
		F	Sig.	t	df	Sig. (2-tailed)	Mean Difference	Std. Error Difference	95% Confidence Interval of the Difference Lower	Upper
Our organization regularly uses new distribution channels.	Equal variances assumed	3.347	.080	-0.241	24	.812	-.131	.543	-1.252	0.990
	Equal variances not assumed			-0.246	23.675	.808	-.131	.532	-1.229	0.967
We frequently make small adjustments to our existing products and services.	Equal variances assumed	0.000	.998	0.624	24	.539	.476	.763	-1.099	2.052
	Equal variances not assumed			0.625	23.513	.538	.476	.762	-1.099	2.051
We improve the production efficiency for our products and/or the provision efficiency for our services.	Equal variances assumed	1.808	.191	1.288	24	.210	.595	.462	-.358	1.549
	Equal variances not assumed			1.338	21.636	.195	.595	.445	-.328	1.518
We reduce costs by having efficiency effects channelled through the volume increase of existing products.	Equal variances assumed	2.036	.166	0.862	24	.397	.619	.718	-.864	2.102
	Equal variances not assumed			0.837	18.959	.413	.619	.740	-.929	2.168
Our organization expands services for existing clients.	Equal variances assumed	0.251	.621	1.996	24	.057	.762	.382	-.026	1.550
	Equal variances not assumed			2.036	23.816	.053	.762	.374	-.011	1.535
Delivery flexibility: The ability of an organization to effectively respond to changes in planned delivery dates.	Equal variances assumed	0.444	.512	0.692	24	.496	.202	.293	-.401	0.806
	Equal variances not assumed			0.702	23.989	.490	.202	.288	-.393	0.798
Volume flexibility: The ability of an organization to effectively increase or decrease aggregate production in response to customers.	Equal variances assumed	0.743	.397	0.810	24	.426	.429	.529	-.664	1.521
	Equal variances not assumed			0.828	23.631	.416	.429	.517	-.640	1.497
Mix flexibility: The ability of an organization to effectively produce a wide variety of different products or offer a wide variety of different services.	Equal variances assumed	0.173	.682	0.422	24	.677	.226	.536	-.881	1.333
	Equal variances not assumed			0.427	24.000	.673	.226	.530	-.867	1.319
Changeover flexibility: The ability of an organization to effectively handle additions and subtractions to the product or service mix over time.	Equal variances assumed	0.934	.343	-0.090	24	.929	-.048	.530	-1.141	1.046
	Equal variances not assumed			-0.092	23.795	.928	-.048	.519	-1.120	1.025
Modification flexibility: The ability of an organization to effectively implement minor changes in current products or services that result from corrective actions or changing customer requirements.	Equal variances assumed	0.104	.750	-0.930	24	.362	-.381	.410	-1.226	0.465
	Equal variances not assumed			-0.931	23.535	.361	-.381	.409	-1.226	0.464
Market share growth	Equal variances assumed	0.027	.871	2.400	24	.025	.893	.372	.125	1.661
	Equal variances not assumed			2.393	23.084	.025	.893	.373	.121	1.665
Sales growth	Equal variances assumed	0.298	.590	1.364	24	.185	.524	.384	-.269	1.317
	Equal variances not assumed			1.372	23.816	.183	.524	.382	-.265	1.312
Return on investment (ROI)	Equal variances assumed	0.120	.732	1.309	24	.203	.452	.346	-.261	1.166
	Equal variances not assumed			1.318	23.882	.200	.452	.343	-.256	1.161
Growth in ROI	Equal variances assumed	1.659	.210	1.632	24	.116	.607	.372	-.161	1.375
	Equal variances not assumed			1.603	20.990	.124	.607	.379	-.180	1.395
Return on sales (ROS)	Equal variances assumed	0.104	.750	0.844	24	.407	.298	.353	-.430	1.025
	Equal variances not assumed			0.844	23.330	.407	.298	.353	-.432	1.027
Unit product costs	Equal variances assumed	0.490	.491	-0.188	24	.853	-.167	.888	-1.999	1.665
	Equal variances not assumed			-0.187	22.726	.854	-.167	.893	-2.015	1.682

Grouping Criterion: MFC Size Class		Levene's Test for Equality of Variances		t-test for Equality of Means						
		F	Sig.	t	df	Sig. (2-tailed)	Mean Difference	Std. Error Difference	95% Confidence Interval of the Difference Lower	Upper
Costs versus budget	Equal variances assumed	0.921	.347	1.406	24	.173	1.000	.711	-.468	2.468
	Equal variances not assumed			1.420	23.977	.168	1.000	.704	-.453	2.453
Incoming customer parts quality	Equal variances assumed	0.413	.526	-0.455	24	.653	-.214	.471	-1.186	0.757
	Equal variances not assumed			-0.463	23.895	.647	-.214	.462	-1.169	0.740
In process quality	Equal variances assumed	0.130	.722	0.086	24	.932	.024	.277	-.547	0.595
	Equal variances not assumed			0.087	23.983	.931	.024	.274	-.541	0.589
Field failure under warranty	Equal variances assumed	7.287	.013	1.240	24	.227	.905	.730	-.601	2.411
	Equal variances not assumed			1.286	21.828	.212	.905	.703	-.555	2.364
On-time delivery to customer	Equal variances assumed	5.087	.033	2.684	24	.013	.476	.177	.110	0.842
	Equal variances not assumed			2.736	23.835	.012	.476	.174	.117	0.836
Delivery lead time	Equal variances assumed	0.000	.988	1.196	24	.244	.429	.358	-.311	1.168
	Equal variances not assumed			1.217	23.908	.236	.429	.352	-.299	1.156
Product/machine change over times	Equal variances assumed	1.269	.271	0.060	24	.953	.048	.795	-1.593	1.689
	Equal variances not assumed			0.059	21.063	.954	.048	.809	-1.634	1.730

Items with significance levels of being below 5% show significant statistical differences between the two independent samples.

A.17.4: Organizational Ambidexterity – Item Means

	A: Exploratory Innovation					*B: Exploitative Innovation*					Variance Means
Respondent	Item 1	Item 2	Item 3	Item 4	Means	Item 1	Item 2	Item 3	Item 4	Means	*B* vs. *A*
1	5	5	5	5	5.00	7	7	5	6	6.25	1.25
2	2	2	6	1	2.75	7	7	4	5	5.75	3.00
3	7	7	6	4	6.00	7	5	1	6	4.75	-1.25
4	6	5	5	5	5.25	6	6	4	5	5.25	0.00
5	5	6	6	4	5.25	6	7	7	7	6.75	1.50
6	6	6	5	4	5.25	6	6	3	4	4.75	-0.50
7	3	6	4	2	3.75	6	7	6	5	6.00	2.25
8	6	4	4	4	4.50	5	5	5	6	5.25	0.75
9	5	3	3	2	3.25	3	2	2	5	3.00	-0.25
10	7	6	5	2	5.00	7	7	3	6	5.75	0.75
11	7	5	6	2	5.00	7	7	4	7	6.25	1.25
12	4	5	4	2	3.75	6	6	6	5	5.75	2.00
13	5	6	4	1	4.00	3	5	4	4	4.00	0.00
14	7	6	6	4	5.75	5	5	2	4	4.00	-1.75
15	2	5	6	2	3.75	7	6	5	6	6.00	2.25
16	1	1	5	4	2.75	1	7	7	5	5.00	2.25
17	5	4	5	4	4.50	5	6	6	6	5.75	1.25
18	5	3	4	1	3.25	2	5	5	5	4.25	1.00
19	7	5	5	1	4.50	7	5	4	6	5.50	1.00
20	4	2	5	2	3.25	2	5	6	5	4.50	1.25
21	6	5	4	1	4.00	4	4	3	3	3.50	-0.50
22	6	1	5	3	3.75	6	6	7	7	6.50	2.75
23	2	2	2	2	2.00	4	6	6	5	5.25	3.25
24	6	5	5	2	4.50	4	7	7	7	6.25	1.75
25	7	4	6	4	5.25	7	7	1	5	5.00	-0.25
26	5	2	4	1	3.00	2	6	4	6	4.50	1.50
Means	5.04	4.27	4.81	2.65	4.19	5.08	5.85	4.50	5.42	5.21	1.02

A.18 MFI Survey – Scale Reliability

Scale	Original English Scales Cronbach's Alpha (as Reported by the Authors)	BPM in the MFI - Survey					
		English Scales (Sample Size N=15)			German Scales (Sample Size N=11)		
		Cronbach's Alpha	Cronbach's Alpha Based on Standardized Items	N of Items	Cronbach's Alpha	Cronbach's Alpha Based on Standardized Items	N of Items
"Customer Focus"	0.86	0.904	0.919	7	*0.341*	*0.596*	7
"Closer to Customers"	0,78-0,90	0.833	0.859	4	0.719	0.746	4
"Competitive Priorities"	0.83	0.818	0.859	6	*0.244*	*0.263*	6
"Environmental Uncertainty"	Not reported	*0.310*	*0.194*	5	*0.152*	*0.081*	5
"Technology Uncertainty"	0.83	*0.555*	*0.576*	4	0.733	0.728	4
"Demand Uncertainty"	0.84	*0.556*	*0.588*	5	0.766	0.771	5
"Exploratory Innovation"	0.86	*0.543*	*0.516*	4	0.695	0.734	4
"Exploitative Innovation"	0.80	*0.571*	*0.666*	4	*0.325*	*0.483*	4
"Flexibility"	0.90	0.805	0.828	5	0.848	0.867	5
"Profit"	0.93	0.907	0.909	3	0.963	0.967	3
"Growth"	Not reported	0.802	0.812	2	0.955	0.956	2

Cronbach's coefficient alpha (Cronbach, 1951) represents a common measure to estimate the reliability and internal consistency of multi-item scales (Cortina, 1993). In social science literature an alpha value being greater than 0.70 is regarded as acceptance level for the reliability of the applied scales (Cortina, 1993, p. 101).

I calculated the Cronbach's alpha for each English and German multi-item scale in SPSS to test the scale reliability. As I adopted these scales from existing English publications, I anticipated receiving a comparable alpha value as reported by the authors of these source publications. The calculations however revealed significant differences between the three types of scales as shown in the table above. The alpha values for the five scales in the English language, which I adopted for uncertainty and organizational ambidexterity were all below 0.70 (marked grey). The reliability criterion for these scales has therefore not been met; however, three out of the five scales in the German language (using the back-translation method for the translation) met the criterion.

Charter (2008, p. 249) suggests one answer to explain this curious situation, as the minimum sample size for the calculation of Cronbach's alpha and related coefficients should not be less than N=30. Charter's criterion for the calculation of the Cronbach's alpha for the English and German scales applied in the MFI survey is not met as the sample size is N=15 for the English and N=11 for the German scales. Therefore I neither approve nor reject the reliability of the applied scales within the MFI survey based on the calculated Cronbach's alpha values. However, I included the calculation of the alphas as presented in the table above in my dissertation thesis to document the consideration of this reliability measure. This circumstance limits the assertion of the applied scales' internal consistencies to the reported Cronbach's alphas by the authors of the publications from which I adopted the scales.